New Energy Special Vehicle, Tractor and Agricultural Machinery

New Energy Special Vehicle, Tractor and Agricultural Machinery

Editors

Fachao Jiang
Yongyu Li
Weiwei Kong

MDPI

Basel • Beijing • Wuhan • Barcelona • Belgrade • Novi Sad • Cluj • Manchester

Editors

Fachao Jiang
China Agricultural University
Beijing
China

Yongyu Li
China Agricultural University
Beijing
China

Weiwei Kong
China Agricultural University
Beijing
China

Editorial Office
MDPI AG
Grosspeteranlage 5
4052 Basel, Switzerland

This is a reprint of articles from the Special Issue published online in the open access journal *World Electric Vehicle Journal* (ISSN 2032-6653) (available at: https://www.mdpi.com/journal/wevj/special_issues/BH9Z4V9E36).

For citation purposes, cite each article independently as indicated on the article page online and as indicated below:

Lastname, A.A.; Lastname, B.B. Article Title. *Journal Name* **Year**, *Volume Number*, Page Range.

ISBN 978-3-7258-1697-2 (Hbk)
ISBN 978-3-7258-1698-9 (PDF)
doi.org/10.3390/books978-3-7258-1698-9

Contents

World Electric Vehicle Journal

MDPI

Article

Energy-Saving Control of Hybrid Tractors Based on Instantaneous Optimization

Junjiang Zhang [1,2,3], Ganghui Feng [1], Liyou Xu [1,3,*], Xianghai Yan [1,2,3], Wei Wang [4] and Mengnan Liu [2,4]

1 College of Vehicle and Traffic Engineering, Henan University of Science and Technology, Luoyang 471003, China
2 State Key Laboratory of Power System of Tractor, Luoyang 471039, China
3 Henan Province Collaborative Innovation Center for Advanced Manufacturing of Mechanical Equipment, Luoyang 471003, China
4 YTO Group Corporation, Luoyang 471004, China
* Correspondence: xlyou@haust.edu.cn; Tel.: +86-136-6387-3262

Abstract: In this study, an energy-saving control strategy based on instantaneous optimization is proposed to improve the energy efficiency of hybrid tractors. Using a parallel diesel–electric hybrid tractor as the research object, the topological and working characteristics were analyzed, and a coupled dynamic model of rotary tillage and tractor plow was constructed. Aiming to minimize the equivalent fuel consumption of the entire machine, the motor and diesel engine torques were taken as the control variables, and the state of charge of the power battery was taken as the state variable. Subsequently, an energy-saving control strategy based on instantaneous optimization is proposed. Finally, a simulation experiment was carried out using MATLAB to verify the effectiveness of the energy-saving control strategy based on instantaneous optimization. Compared with the energy-saving control strategy based on power-following, the results show that energy-saving control strategy based on instantaneous optimization can reasonably control the operating state of the diesel engine and motor. Therefore, the diesel engine and motor work in the high-efficiency area, and effectively reduce the equivalent fuel consumption of the tractor during field operation. Under rotary tillage and plowing conditions, equivalent fuel consumption is reduced by 4.70% and 6.31%, respectively.

Keywords: hybrid tractor; instantaneous optimization; energy-saving control; equivalent fuel consumption

Citation: Zhang, J.; Feng, G.; Xu, L.; Yan, X.; Wang, W.; Liu, M. Energy-Saving Control of Hybrid Tractors Based on Instantaneous Optimization. *World Electr. Veh. J.* **2023**, *14*, 27. https://doi.org/10.3390/wevj14020027

Academic Editors: Fachao Jiang, Yongyu Li and Weiwei Kong

Received: 3 December 2022
Revised: 13 January 2023
Accepted: 14 January 2023
Published: 19 January 2023

1. Introduction

Recently, with the introduction of various agricultural machinery subsidy policies, there has been an increase in the use of tractors as the main power machinery for agricultural production operations. The national "14th Five-Year Plan and 2035 Long-term Goals" proposed goals regarding a carbon emission peak in 2030 and carbon neutrality in 2060. At this stage, agricultural machinery mainly uses diesel engines as the main power source, which consume a significant amount of fuel during operation and result in greenhouse gas emissions [1]. Under the double pressure of the global energy crisis and environmental pollution, it is particularly important to design and develop energy-saving and environmentally friendly agricultural machinery vehicles [2–4]. With the improvements in hybrid vehicle technology, hybrid tractors have been developed [5]. Hybrid tractors are energy-saving and environmentally friendly agricultural machinery vehicles that have the advantages of both traditionally fueled tractors and electric tractors [6–8]. The energy conversion of a series hybrid tractor is repeated during operation, which results in low energy utilization efficiency. Meanwhile, parallel hybrid tractors can be powered directly by diesel engines or electric motors. It has no secondary energy conversion and realizes high energy utilization efficiency. Therefore, based on the topology of a parallel diesel-electric hybrid tractor, this study develops an energy-saving control strategy to improve the energy utilization efficiency of the entire machinery.

Using the energy-saving control strategy as the core control strategy of a hybrid tractor directly affects the performance of a hybrid tractor. Currently, energy-saving control strategies can be mainly divided into two categories: rule-based control strategies and optimization-based control strategies [9,10].

Rule-based control strategies are inexpensive to develop and simple to implement, and they are widely used in various types of hybrid vehicles. Luo et al. [11] proposed a fuzzy reasoning energy management strategy for a series of diesel-electric hybrid tractors. The engine demand power is determined according to the self-set fuzzy reasoning rule table. The corresponding results demonstrate that the fuel economy is improved by 20.92% compared with the power following control strategy. Xu et al. [12] proposed a predefined energy management strategy for extended-range electric tractors; they indicated that the fuel consumption was reduced by 34.22% in the continuous transition operation mode. Based on series hybrid tractors, Fang et al. [13] proposed a fuzzy control energy management strategy that differentiated between electric vehicles and electric tractors. Their results showed that the battery of the state of charge (SOC) curve shows the slowest change when the fuzzy control energy management strategy is adopted. However, rule-based control strategies are certain and according to the designer's experience, they do not exhibit good adaptability to working conditions [14].

Optimization-based control strategies are solved by minimizing or maximizing a cost function, which is generally a measure of the control target. Lee et al. [15] established a power shunt ratio strategy based on a drivetrain simulation model. The power allocation strategy of a hybrid electric tractor was optimized using a deterministic dynamic programming algorithm. Simulation results show that the proposed control strategy can reduce the fuel consumption of a hybrid electric tractor. Spano et al. [16] proposed a multi-objective particle swarm optimization algorithm to determine the optimal power system layout of parallel P2 hybrid electric vehicles, aiming to maximize fuel economy and minimize production costs. The results showed that the control algorithm can improve the fuel economy of HEV and reduce HEV production costs. Qian et al. [17] proposed a calculation method based on the fuzzy PID torque recognition coefficient K. Subsequently, they used the particle swarm-ant colony combination optimization algorithm to optimize the key control parameters in the control strategy. Their results indicated that the fuel consumption and emissions are reduced by ensuring the dynamics of the entire vehicle. However, the solution of deterministic dynamic programming algorithms requires extracting control rules, which is computationally intensive and time-consuming. The particle swarm optimization algorithm has strong global search ability and is a simple algorithm, but it has poor local search ability. Meanwhile, the control strategy of particle swarm-ant colony combination optimization algorithm is more complicated. Recently, energy-saving control strategies based on instantaneous optimization have become a research hotspot in the vehicle control field, owing to their advantages of fast calculation speed, good control effect, smaller calculation amount than dynamic programming algorithms, and simple control algorithms.

In this study, a diesel-electric parallel hybrid tractor is considered as the research object. Meanwhile, an energy-saving control strategy based on instantaneous optimization of the required torque is proposed [18–23]. By optimizing the torque of both the diesel engine and motor, the required torque of the entire machine can be distributed in real time, which reduces the equivalent fuel consumption of the entire machine while ensuring power stability. First, the topology and performance parameters of the hybrid tractor are introduced; then, the main components are simulated and modeled [24,25]. Based on the entire machine model, an energy-saving control strategy based on instantaneous optimization was designed. The simulation results were analyzed and compared with the power-following energy-saving control strategy. Finally, the conclusions of this study are presented.

2. Tractor Topology and Main Parameters

2.1. Subsection Model of Hybrid Tractor Drivetraino

Figure 1 shows the topology of a parallel diesel–electric hybrid tractor, which has two drive systems, namely a diesel engine and an electric motor. The diesel engine is the main power source, and the electric motor is the auxiliary power source [26]. The engine and motor output torques are transmitted to the transmission input shaft through the torque coupler, followed by the transmission output power. It is used as the power input of the central transmission device and power take-off (PTO) shaft. The vehicle controller and power battery, low-voltage battery, diesel engine, clutch, drive motor, transmission, AC/DC module, and DC/DC module are connected through the CAN bus. According to the total power demand of the entire machine and SOC value of the power battery, the torques of the diesel engine and motor are dynamically distributed according to the control strategy and algorithm of the entire machine. Accordingly, the tractor can achieve the best power and economic performance.

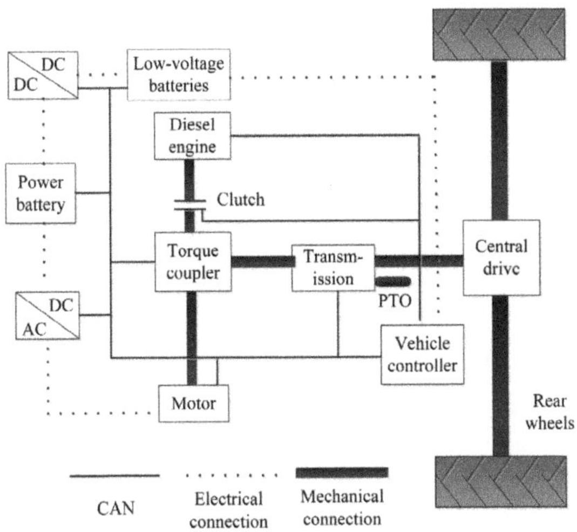

Figure 1. Topology of a diesel–electric parallel hybrid tractor.

2.2. Subsection Model of Hybrid Tractor Drivetraino

In this study, the energy-saving control strategy of a 220 hp hybrid tractors is investigated. The main components of the hybrid tractor were selected according to its working conditions [27]. The specific parameters are listed in Table 1.

Table 1. Topology of a diesel–electric parallel hybrid tractor.

Name	Parameter	Value (Unit)
Diesel engine	Rated power	162 (kW)
	Rated speed	2500 (rpm)
	Maximum torque	800 (Nm)
Motor	Rated power	30 (kW)
	Rated speed	3000 (rpm)
	Rated torque	96 (Nm)
Power battery	Energy capacity	70 (Ah)
	Rated voltage	360 (V)
	SOC	0.90–0.25

2.3. Determination of Theoretical Speed and Transmission Ratio

The design of the transmission gear ratio is based on the working characteristics of the tractor, speed range of various working conditions of the current tractor, and the output characteristics of the hybrid tractor coupling system. Subsequently, the main reducer and gearbox transmission ratio, and its theoretical speeds are matched and calculated [28,29].

When rotary tillage is performed, the traveling speed of the tractor is 4–5 km/h, and the theoretical value of the PTO speed is 540 rpm. From this, the speed ratio of the main reducer is determined as i_0, which is taken as 19.10. Seven forward gears are designed, including three transport gears, two working gears, and two amble gears [30,31]. The corresponding transmission ratio increases sequentially, while the theoretical speed decreases sequentially. The specific parameters used are listed in Table 2.

Table 2. Hybrid tractor transmission ratios and theoretical speeds.

Forward Gear	Transport III	Transport II	Transport I	Working II	Working I	Amble II	Amble I
Ratio	0.864	1.377	2.307	3.296	4.963	7.405	11.208
Theoretical speed (km/h)	39.980	28.668	17.164	11.977	7.654	5.231	3.522

3. Hybrid Tractor Model Building

Based on the topology of the hybrid tractor, its main components were modelled, including the transmission system, rotary tillage unit dynamic, plowing unit dynamic, tire, motor, diesel engine, power battery model. Subsequently, the simulation model of the machine was built.

3.1. Model of Hybrid Tractor Drivetraino

The required torque of the hybrid tractors provided by the motor and diesel engine. The required torque of the entire machine is obtained at the input end of the torque coupler, which can be expressed as follows:

$$T_{req} = (T_m \cdot \eta_m + T_e \cdot \eta_e) \tag{1}$$

where T_m and T_e are the motor and diesel engine torques, respectively. η_m and η_e represent the working efficiencies of the motor and diesel engine, respectively. T_{req} is the torque required for the torque coupler input.

According to the tractor working speed and parameters of each component used to calculate the power source speed, this study considers the diesel engine speed as the power source speed, which can be expressed as follows:

$$n_{tire} = \frac{v}{0.377 \cdot r} \tag{2}$$

$$n_e = n_{tire} \cdot i_t \cdot i_0 \tag{3}$$

where i_t and i_0 are the transmission and main reducer speed ratios, respectively. n_{tire} and n_e are the drive and diesel engine speeds, respectively. v is the speed of the hybrid tractor during operation. r is the driving wheel radius of the hybrid tractor.

3.2. Dyanmic Model of Rotary Tillage Unit

The power balance relationship characterizing the working time group of the hybrid tractor traction rotary cultivator is formulated as follows:

$$P_{req}(T_{req}, n_{req}) = \left(\frac{P_{drive}}{\eta_{zj}\eta_b} + \frac{P_r}{\eta_b} \right) / \eta_o \tag{4}$$

$$P_{drive} = \frac{\left(mgf\cos\alpha + mg\sin\alpha + m\delta\dot{v} + \frac{C_d A v^2}{21.15}\right)v}{3600} \tag{5}$$

$$P_r = \frac{P_c + P_{th} + P_a + P_h}{\eta_r} \tag{6}$$

where P_{drive} and P_r are the tractor travel power and rotary cultivator power consumptions, respectively. η_r, η_{zj}, η_b, and η_o represent the rotary tillage unit mechanical transmission efficiency, main reducer transmission efficiency, transmission efficiency, and torque-coupler efficiency, respectively. P_c, P_{th}, P_a, and P_h are the cutting power consumption, throwing earth power consumption, rotary cultivator forward power consumption and power required to overcome the soil horizontal reaction forces, respectively. m is the tractor mass. f is the rolling resistance coefficient. δ is the mass conversion of the factor. α is the tilt of the ground. C_d and A are the tractor drag coefficient and windward area, respectively.

When the tractor is operating at a low speed, the influence of air resistance and acceleration resistance on the tractor can be ignored [28].

The hybrid tractor is equipped with a double-acting clutch, which can realize independent control of the PTO power [29]. The relationship formula of the rotary tillage operation timing group is as follows:

$$P_r = P_{PTO} = \frac{n_{PTO} T_{PTO}}{9550} \tag{7}$$

$$v = v_r \tag{8}$$

where P_{PTO}, T_{PTO}, and n_{PTO} are the power, torque, and speed of PTO, respectively. v_r is the forward speed of the rotary cultivator.

3.3. *Dyanmic Model of Plowing Unit*

When the tractor is working, its driving force must overcome the rolling resistance and other driving resistances before it can be operated. The balance between the driving force F_{TN} and various resistances when the tractor is operating is formulated as follows:

$$F_{TN} = F_g + F_f + F_p + F_{Af} + F_i \tag{9}$$

where F_g, F_f, F_p, F_{Af}, and F_i are the tillage, rolling, slope, air, and acceleration resistances, respectively. F_{TN} is the driving force.

When the tractor is operating at a low speed, the influence of air and acceleration resistances on the tractor can be ignored.

Under normal circumstances, the tractor drive force F_{TN} is primarily determined by the tillage resistance F_g when the supporting agricultural tools are working. The calculation formula is stated as follows:

$$F_g = Z \cdot b_l \cdot h_k \cdot k \tag{10}$$

where Z denotes the number of plowshares. b_l and h_k represent the individual plow width and depth, respectively. k is the soil specific resistance coefficient.

The power demand at the input end of the torque coupler when the hybrid tractor pulls the plow unit can be expressed as follows:

$$P_{req}(T_{req}, n_{req}) = \frac{F_{TN} v}{\eta_{zj}\eta_b\eta_o} \tag{11}$$

3.4. Tire Model

The Duggof model belongs to the theoretical model and is suitable for the studying vehicle dynamics control algorithms. Therefore, the Duggof tire model is used to calculate the driving force of the driving wheel [32], which can be expressed as follows:

$$F_q = \begin{cases} F_Z\left[\varphi - \varphi^2 \frac{F_Z(1-\zeta)}{4c\zeta}\right], & \frac{c\zeta}{1-\zeta} \geq \frac{\varphi F_Z}{2} \\ \frac{c\zeta}{1-\zeta}, & \frac{c\zeta}{1-\zeta} \leq \frac{\varphi F_Z}{2} \end{cases} \tag{12}$$

where F_q and F_z are the driving and loading forces of the drive wheel, respectively. φ and ζ represent the slip rate of the corresponding drive wheel and adhesion factor of the corresponding drive wheel, respectively. c is the horizontal distance of the hitch traction point from the center of the rear wheel.

3.5. Motor Model

A permanent magnet synchronous motor with superior performance is selected as the electric drive system of the tractor, which can operate not only in the forward direction, but also in the reverse direction. Moreover, it has the working characteristics of low-speed constant torque and high-speed constant power. The formula characterizing the relationship between the power, speed, and torque is stated as follows:

$$P_m = \frac{n_m \cdot T_m}{9550} \tag{13}$$

where n_m is the speed of the motor.

The motor model is established via a numerical model method, and the relationship between motor system efficiency, torque, and speed are obtained via the spline interpolation method based on experimental motor efficiency experimental data. The relationship is determined and unique, which is suitable for the studied control strategies. The numerical model of motor efficiency is shown in Figure 2.

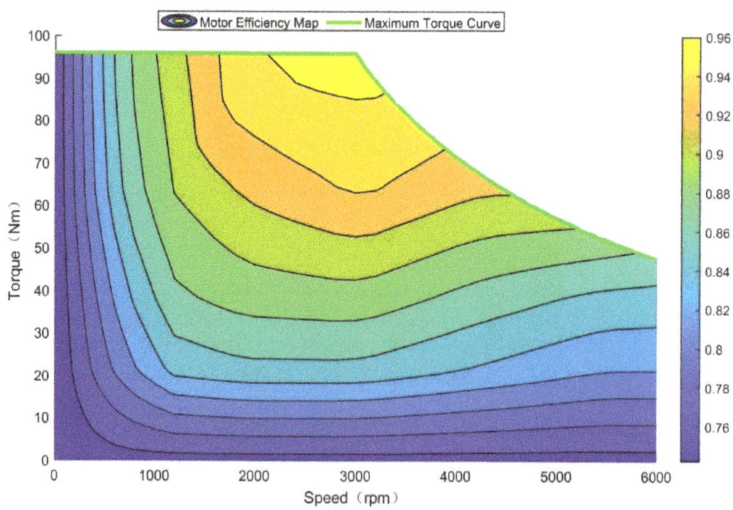

Figure 2. Motor model MAP diagram.

3.6. Diesel Engine Model

The diesel engine of the corresponding power is selected according to the use conditions of the hybrid tractor considering only the input and output parameters of the engine. The formula characterizing their relationship is stated as follows:

$$P_e = \frac{n_e \cdot T_e}{9550} \tag{14}$$

where n_e denotes the speed of the diesel engine.

Diesel engine modeling methods are mainly divided into two types: theoretical modeling and numerical modeling methods. The theoretical modeling method is based on the structural parameters of the engine, using thermodynamic theory, combustion theory, fluid mechanics, and heat transfer theory to establish a mathematical model of the engine working process. The numerical modeling method tests the load characteristics and speed characteristic curves of the engine by building an experimental bench of the engine and then constructing a numerical model by interpolation fitting. This study mainly uses the engine output characteristics to study the drive system of the whole machine, and only considers the relationship between the input and output parameters of the engine. Therefore, the method of measured modeling is adopted. On the basis of the engine steady-state test data, the number table or formula is used to fit to obtain an accurate and simple engine numerical model [33], as shown in Figure 3.

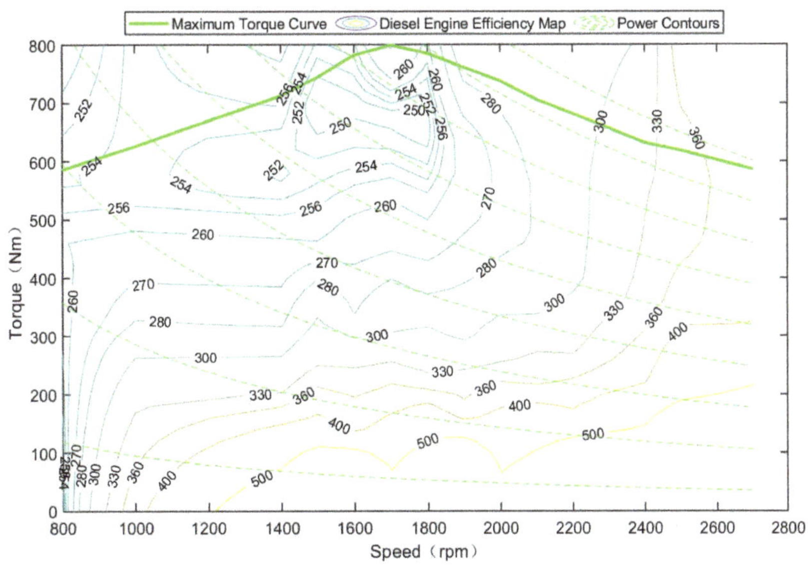

Figure 3. Diesel engine model MAP diagram.

3.7. Power Battery Model

The power battery model describes the external characteristics of the power battery during operation, and most of the equivalent circuit models are currently used. Because the model has good applicability to various working states of power batteries, the equation of state of the model can be derived.

Therefore, the equivalent internal resistance model in the equivalent circuit model is used here. The power battery is equivalent to an ideal voltage source and a circuit model with a resistor connected in series. The mathematical equations are simple and easy to calculate and model, as shown in Figure 4.

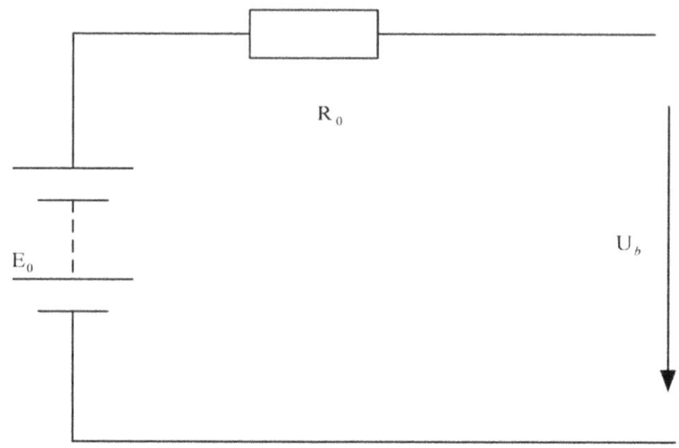

Figure 4. Power battery internal resistance model.

According to Ohm's law, the voltage characteristic equation of a power battery is stated as follows:

$$U_b = E_0 - I_b R_0 \tag{15}$$

where U_b and E_0 are the power battery output voltage and terminal voltage, respectively. I_b and R_0 represent the output current and internal resistance of the power battery, respectively.

Ignoring the influence of the internal resistance and discharge factors of the power battery on the electromotive force E_0 and setting it as a constant, the output power equation of the power battery can be stated as follows:

$$P_{b\max} = \frac{U_b I_b}{1000} = \frac{(E_0 - I_b R_0) I_b}{1000} \tag{16}$$

The charge and discharge power $P_{bat}(t)$ of the battery is positive when discharged and negative when charging. This parameter is determined as follows:

$$P_{bat} = \frac{P_m}{\eta_{bat}} \tag{17}$$

where η_{bat} is the battery charging and discharging efficiency.

The ampere-hour integral method is used to calculate the change in the SOC value of the power battery, which is formulated as follows:

$$SOC(t) = SOC_0 - \frac{\int_0^t I_b(t)dt}{Q_b} \tag{18}$$

where Q_b denotes the rated power battery capacity. SOC_0 represents the initial the state of charge value.

3.8. Power Battery Model

Based on the characteristics of the hybrid tractor transmission system, the simulation model of the entire machine is built using MATLAB. A simplified diagram of the entire machine model is shown in Figure 5. The simulation model includes the dynamic model of the unit (rotary tillage and plowing), motor model, diesel engine model, transmission system model, battery model, and tire model. F_{TN} and v are the resistance and travel speed of the tractor, respectively, which are determined based on the operating conditions of the tractor and are the output parameters of the dynamic model of the unit. Drivetrain model output torque and speed (T_{req} and n_{req}) at the input end of the torque coupler. The motor and diesel engine models receive the instructions (T_{mreq}, n_{mreq}, T_{ereq}, and n_{ereq}).

$$\begin{cases} T_{mreq} = T_m \cdot \eta_m \\ T_{ereq} = T_e \cdot \eta_e \end{cases} \tag{19}$$

where T_{mreq} and T_{ereq} are the torques required for the motor and diesel engine, respectively, at the input of the torque coupler.

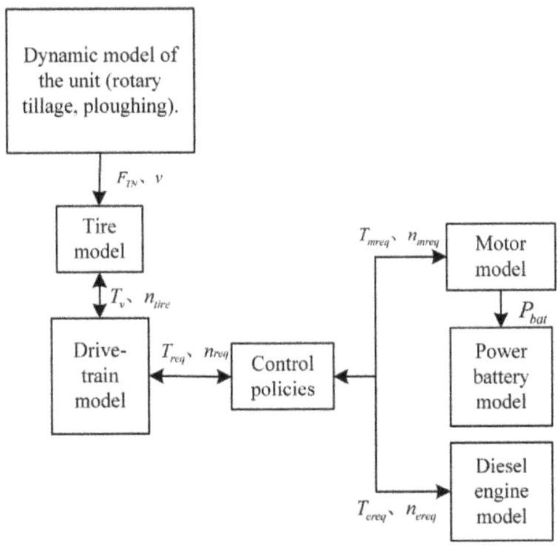

Figure 5. Schematic diagram of the whole machine simulation model.

Finally, the diesel engine and motor work according to the command and output the corresponding torque and speed. Through the drivetrain model, the power is transmitted to the tire model to ensure the normal operation of the whole machine.

4. Energy Saving Control Strategy Design

Based on the entire machine model, two energy-saving control strategies (based on instantaneous optimization and power follower) were designed. The design process of an energy-saving control strategy based on instantaneous optimization is introduced in detail, while that of a comparative control strategy (power-following) is briefly introduced.

4.1. Energy-Saving Control Strategy Based on Instantaneous Optimization

First, according to the topological structure and main component model of hybrid electric tractor, an energy-saving control optimization model of the entire machine is designed. Then, the optimization solution is carried out according to the instantaneous optimization algorithm. Finally, the solution flow is discussed in detail.

4.1.1. Optimization Model of Energy-Saving Control Strategy

A hybrid tractor has two energy sources, electric power and fuel. To unify the energy, the equivalent fuel consumption is used for the evaluation. The goal of energy management is to minimize the equivalent fuel consumption by optimizing and rationalizing the operating state between the diesel engine and motor. The equivalent fuel consumption in the operation process of the hybrid electric tractor, namely, the objective function, can be expressed as follows:

$$Q_c(t) = \int_0^t f\left(Q_f(T_e, n_e) + \frac{j_m P_m(T_m, n_m)}{j_e \eta_{bat} \eta_m}\right) dt \tag{20}$$

$$Q_f(t) = \frac{f_e \cdot P_e(T_e, n_e)}{1000 \cdot 3600 \cdot 0.84} \tag{21}$$

where $Q_c(t)$ and $Q_f(t)$ are the equivalent fuel consumption and instantaneous fuel consumption, respectively. t_f represents the terminal moment. j_e and j_m are the prices per liter of oil and per kWh of electricity, respectively. f_e is engine fuel consumption at that moment.

According to the calculation of the battery SOC value using Equation (18), the system state equation can be obtained as follows:

$$S\dot{O}C(t) = -\frac{I_b(t)}{Q_b} = -\frac{U_b(t) - \sqrt{U_b^2 - 4P_b(t)R_0(t)}}{2R_0(t)Q_b} \tag{22}$$

The control variables of the system are $T_e(t)$ of the diesel engine and $T_m(t)$ of the motor torque. The relationship between them and the required torque is introduced based on Equation (1).

Because the working capacity of each component is limited by realistic conditions, the system must satisfy the following constraints:

$$\begin{cases} T_{m\min}(n_{m,t}) \leq T_m(t) \leq T_{m\max}(n_{m,t}) \\ T_{e\min}(n_{e,t}) \leq T_e(t) \leq T_{e\max}(n_{e,t}) \\ SOC_{\min} \leq SOC(t) \leq SOC_{\max} \end{cases} \tag{23}$$

where $T_{m\min}$ and $T_{m\max}$ are the minimum and maximum torques of the motor, respectively. $T_{e\min}$ and $T_{e\max}$ are the minimum and maximum torque of the diesel engine, respectively. $SOC_{\min}$ and $SOC_{\max}$ represent the minimum and maximum values allowed by the SOC value of the power battery, respectively.

Equation (22) constitutes the permissive reach of the control variables.

4.1.2. Establish the Optimal Torque Distribution Table

To optimize the energy-saving control, an instantaneous optimization control strategy is adopted to solve this problem. The SOC value of the power battery is taken as the state variable. Moreover, the diesel engine torque T_e and motor torque T_m are taken as the control variables to address the optimal torque distribution table. The specific process is illustrated in Figure 6.

1. According to the typical working conditions of a tractor, a set of operating parameters (speed ratio i_t, required torque T_{req}, and power source speed n_e) within a short period of time are used as the system input parameters;
2. In the value range, step sizes ΔT_{req} and Δn_e are used to discretize the required torque and power source speed, respectively;

$$\begin{cases} k = 0 : \Delta T_{req}(t) : T_{req}(t) \\ j = 0 : \Delta n_e(t) : n_e(t) \end{cases} \tag{24}$$

3. According to the speed n_e, determine the maximum torques $T_{e\max}$ and $T_{m\max}$ that the diesel engine and motor can achieve at this speed;
4. Take the SOC state value of the power battery as the state variable, the torques of motor and diesel engine as the control variable, and minimum equivalent fuel consumption as the objective function Q_c, to determines the optimal instantaneous torque of the diesel engine and motor;
5. Record the torque of the diesel engine and motor corresponding to the required instantaneous torque and speed until the end of t_f at the final moment, summarize the data at all moments, and form the optimal torque distribution table.

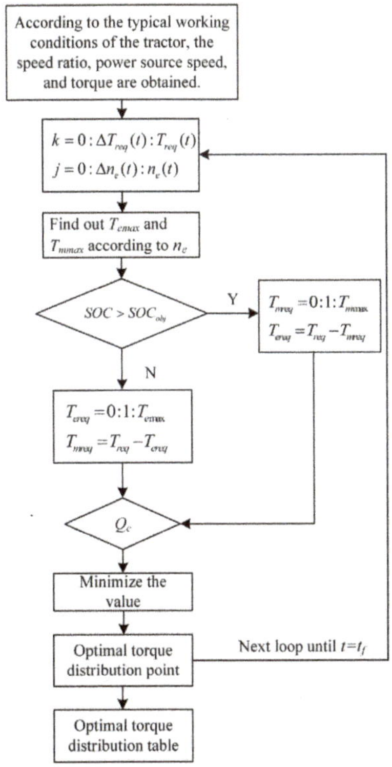

Figure 6. The flowchart of optimal torque distribution table solves.

According to the solution process presented in Figure 6, the torque distribution MAP of the diesel engine and motor can be obtained under at given speed and torque setpoint of the input of the torque coupler. This is illustrated in Figure 7.

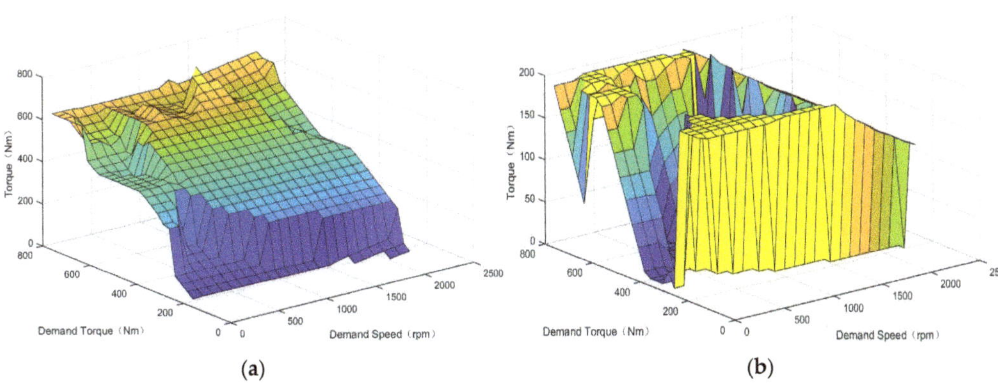

Figure 7. Torque distribution under hybrid drive. (**a**) Torque of the diesel engines. (**b**) Torque of the motor.

4.1.3. Instantaneous Optimization Control

According to the instantaneous optimization control strategy, the optimal torque distribution of the diesel engine and motor during tractor operation is solved, as shown in Figure 8.

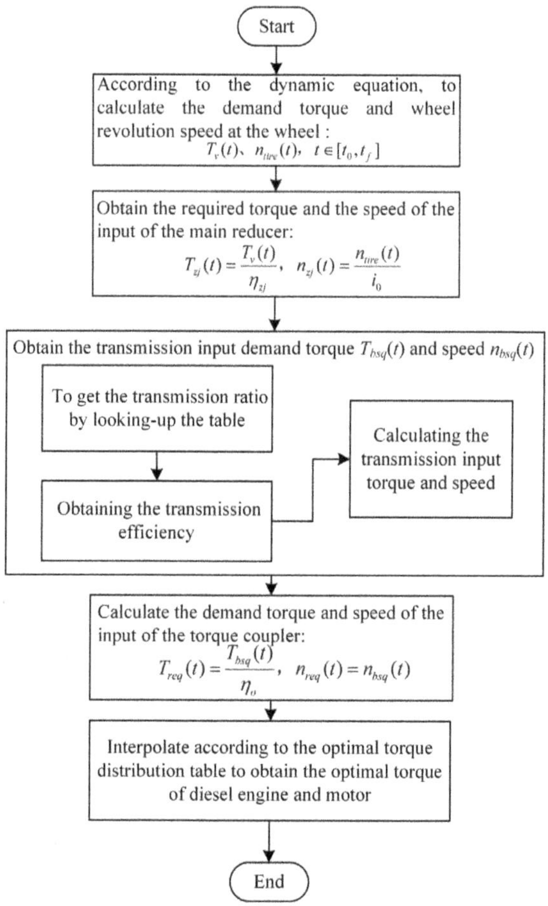

Figure 8. Torque distribution flowchart based on instantaneous optimization.

The specific solving steps are shown as follows:

1. According to the dynamic equation, calculate the required torque $T_v(t)$ and wheel speed $n_{tire}(t)$ at the wheel of the entire machine.

$$n_{tire}(t) = \frac{v(t)}{0.377 \cdot r} \tag{25}$$

$$T_v(t) = \frac{9550 \cdot P_v(t)}{n_{tire}(t)} \tag{26}$$

where $P_v(t)$ is the required power at the wheels of the entire machine.

2. Obtain the required torque $T_{zj}(t)$ and speed $n_{zj}(t)$ for the input of the main reducer.

$$\begin{cases} n_{zj}(t) = \frac{n_v(t)}{i_0} \\ T_{zj}(t) = \frac{T_v(t) \cdot i_0}{\eta_{zj}} \end{cases} \tag{27}$$

3. Obtain the required transmission input torque $T_{bsq}(t)$ and speed $n_{bsq}(t)$. First, according to the torque required by the main reducer and the tractor speed, the transmission ratio is calculated by looking up the table. Then, the transmission efficiency is obtained by looking up the transmission ratio and torque table. Finally, the transmission input torque and speed are calculated using the transmission speed ratio and efficiency.

$$\begin{cases} n_{bsq}(t) = \frac{n_{zj}(t)}{i_t} \\ T_{bsq}(t) = \frac{T_{zj}(t)}{i_t \cdot \eta_b} \end{cases} \tag{28}$$

4. Calculate the required torque $T_{req}(t)$ and speed $n_{req}(t)$ of the torque coupler input.

$$\begin{cases} n_{req}(t) = n_{bsq}(t) \\ T_{req}(t) = \frac{T_{bsq}(t)}{\eta_o} \end{cases} \tag{29}$$

5. The required torque $T_{req}(t)$ and speed $n_{req}(t)$ at the input end of the torque coupler obtained in the previous step are interpolated according to the optimal torque distribution table. Obtain the optimal torque distribution of the diesel engine and motor during tractor operation.

4.2. Energy-Saving Control Strategy Based on Power Following

4.2.1. Control Principle Based on Power Following

As a comparative control strategy, the power following energy-saving control strategy, which is also a rule-based control strategy, is considered. Taking the ratio of the rated power between the diesel engine and motor as the distribution ratio, the power demand of the entire machine is allocated according to the fixed proportion to fully leverage the working capacity of the diesel engine and motor.

4.2.2. Solving Process of Power Following Energy-Saving Control

According to the operating conditions of the hybrid electric tractor and dynamic model analysis conducted on the entire machine, the power required of the entire machine P_{req} and diesel engine speed n_e can be obtained. Accordingly, the required torque of the entire machine T_{req} can be obtained.

According to the rated power of the diesel engine and motor, the torque distribution proportion coefficient K can be determined, as follows:

$$\mathrm{K} = \frac{P_{erated}}{P_{erated} + P_{mrated}} \tag{30}$$

where P_{erated} and P_{mrated} are the rated powers of the diesel engine and motor, respectively.

According to the torque distribution proportion coefficient K, the respective working torques of the diesel engine and motor can be obtained using the following equations:

$$\begin{cases} T_{ereq} = \mathrm{K} \cdot T_{req} \\ T_{mreq} = (1 - \mathrm{K}) \cdot T_{req} \end{cases} \tag{31}$$

The constraint conditions are as follows:

$$\begin{cases} T_{m\min}(n_{m,t}) \leq T_m(t) \leq T_{m\max}(n_{m,t}) \\ T_{e\min}(n_{e,t}) \leq T_e(t) \leq T_{e\max}(n_{e,t}) \\ SOC_{\min} \leq SOC(t) \leq SOC_{\max} \end{cases} \tag{32}$$

5. Result Analysis

Two types of energy-saving control strategies were simulated and tested for the two typical working conditions of rotary tillage and plowing. Accordingly, the simulation results were compared and analyzed. The superiority of the energy-saving control strategy based on instantaneous optimization was verified.

5.1. Analysis of Results Obtained Rotary Tillage Condition

During the rotary tillage operation of the tractor, the PTO works independently and is unaffected by the driving conditions of the tractor. The torque and speed characteristics are presented in Figure 9. The driving speed of the tractor rotary tillage operation is shown in Figure 10.

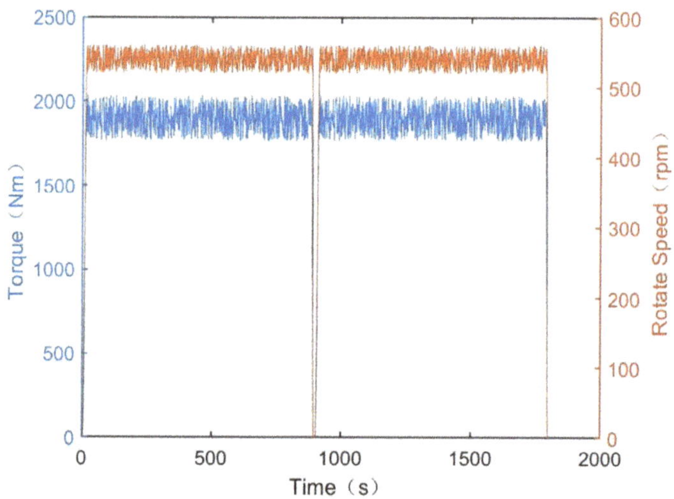

Figure 9. Torque and tachograph of the PTO.

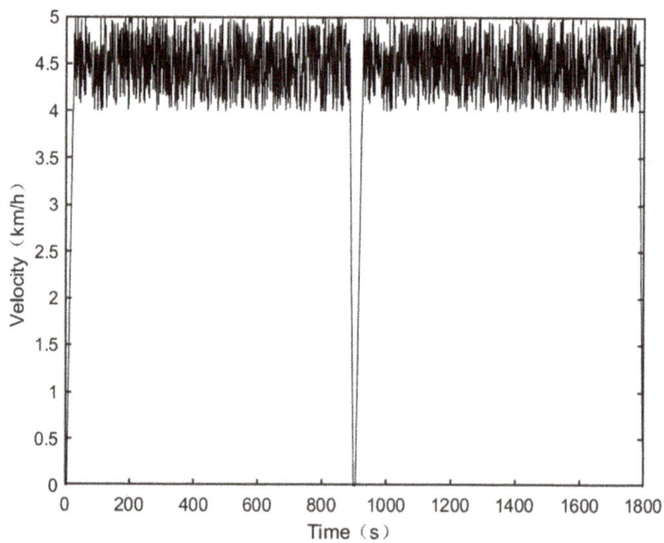

Figure 10. Speed of rotary tillage operations.

With respect to the tractor rotary tillage operations in the two control strategies, the motor power is shown in Figure 11; diesel power is shown in Figure 12, and battery SOC value variations are shown in Figure 13.

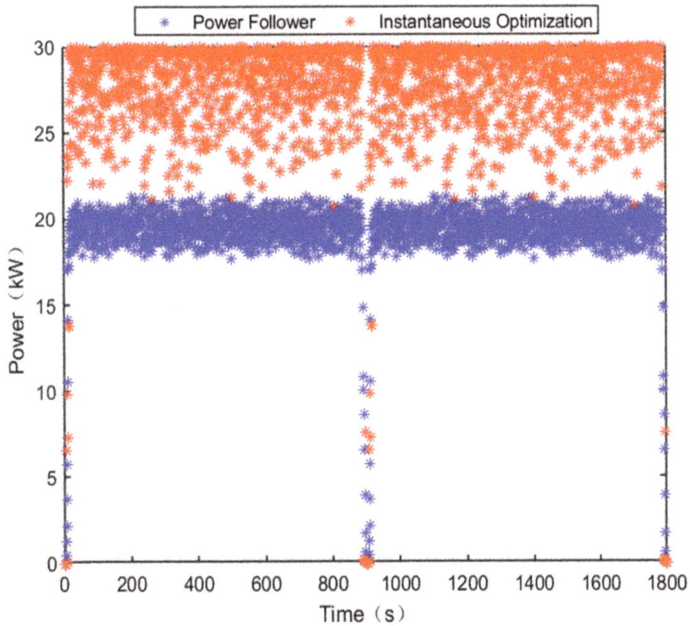

Figure 11. Working power of motor.

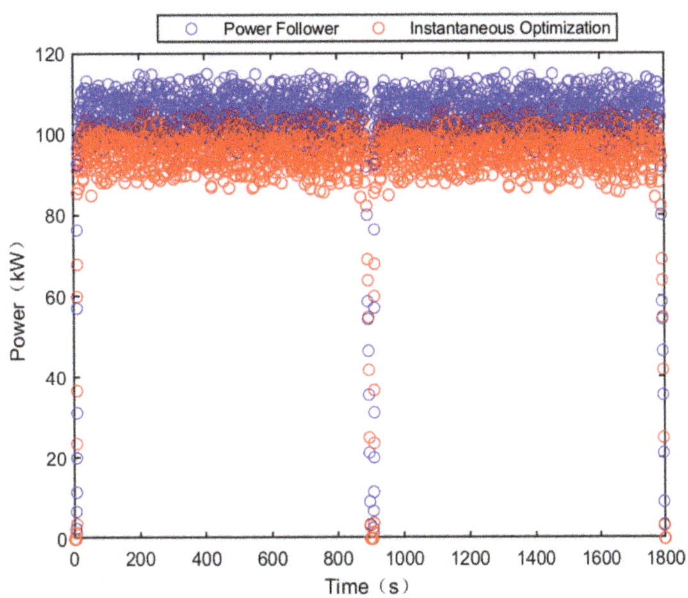

Figure 12. Working power of diesel engine.

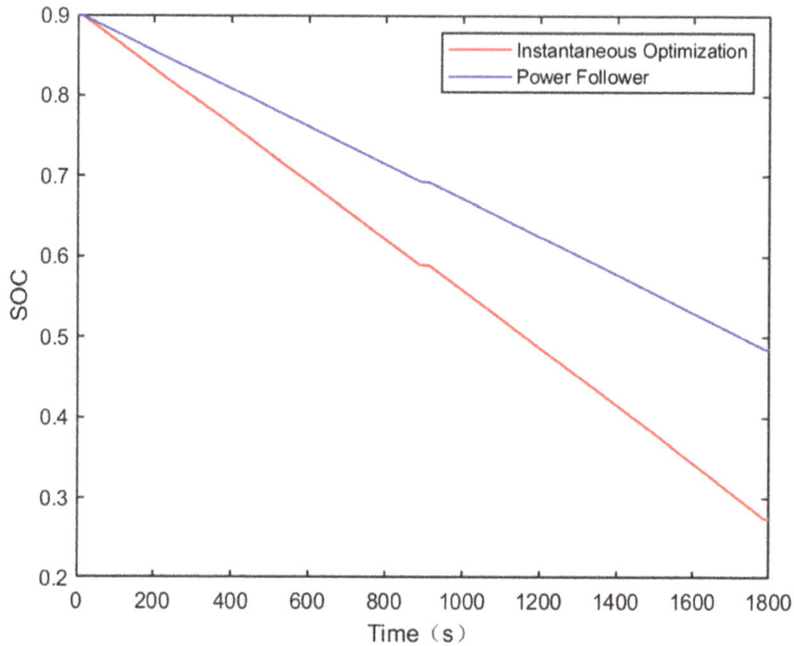

Figure 13. SOC state value changes.

From the simulation results, it can be seen that the load of the hybrid tractor is large, reaching 140 kW under rotary tilling conditions. Utilizing the energy-saving control strategy based on instantaneous optimization, the working power of the motor reaches more than 80% of the load. The working power of the diesel engine is mostly concentrated at 95 kW. The initial SOC value is 0.90, while the final value is 0.27. Using the power following energy-saving control strategy, the working power of the motor is relatively small and concentrated at approximately 20 kW. The maximum power of the diesel engine is approximately 115 kW, while the minimum power is approximately 100 kW. The initial SOC value is 0.90, while the final value is 0.48.

The MAP of the two control strategies for the diesel engine and motor under rotary tilling operation are presented in Figures 14 and 15, respectively.

The control strategy based on instantaneous optimization can dynamically adjust the operating regions of the motor and diesel engine according to the change of torque demand; subsequently, the diesel engine works near the optimal power curve with good dynamics and small power. The torque distribution of the energy-saving control strategy based on power following is based on the given ratio, where the adjustable range is small. The working area of the diesel engine is large, whereas that of the motor is small.

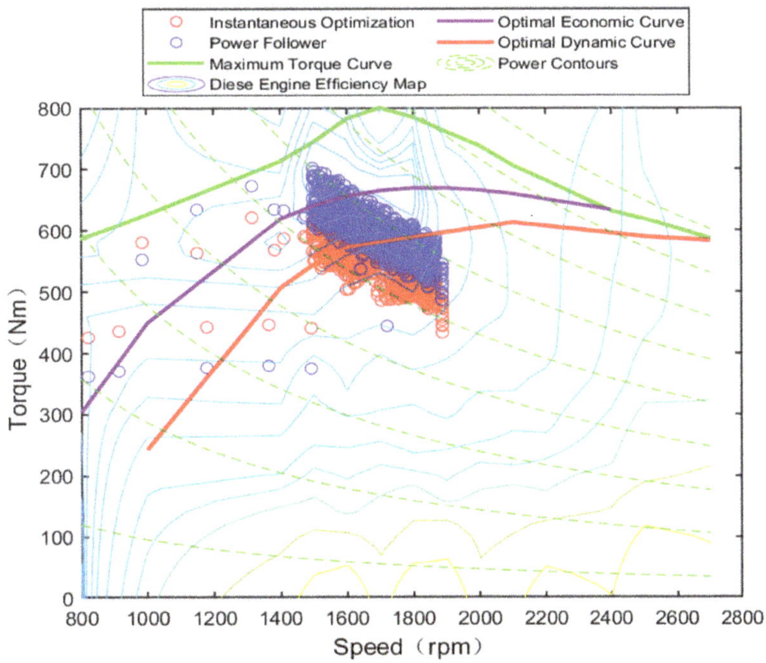

Figure 14. Diesel engine operating points.

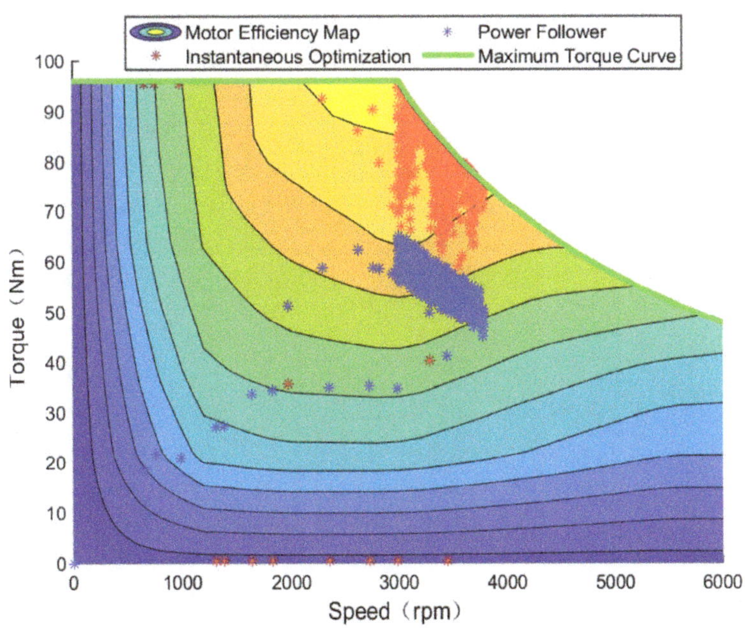

Figure 15. Motor operating points.

5.2. Analysis of Results Under Plow Condition

The tractor plowing speed is shown in Figure 16, and the plowing resistance is shown in Figure 17.

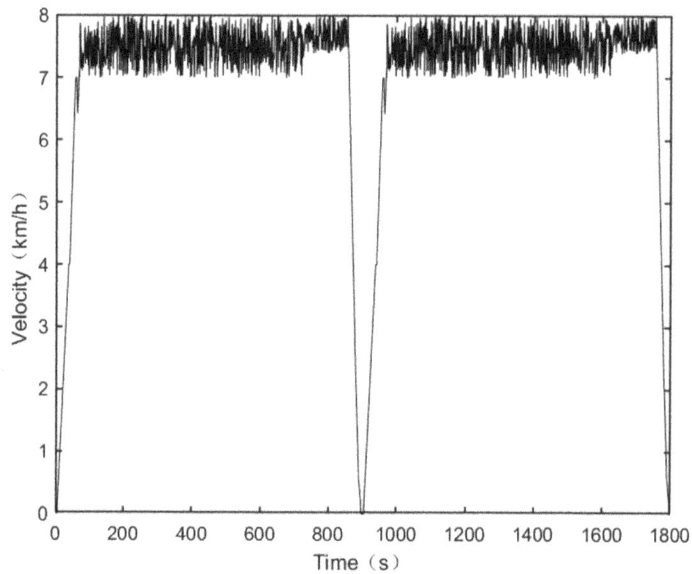

Figure 16. Plowing speed.

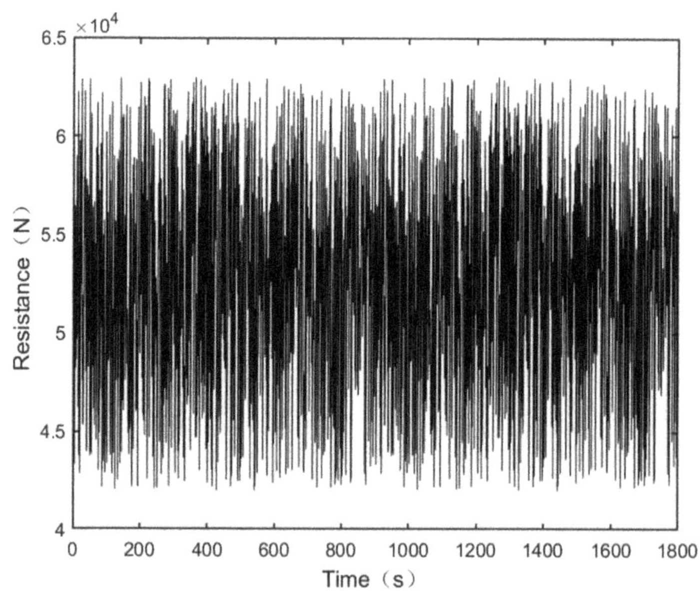

Figure 17. Resistance of plowing operation.

For tractor plowing operations conducted using the two control strategies, the motor power is shown in Figure 18, diesel power is shown in Figure 19, and battery SOC value variations are shown in Figure 20.

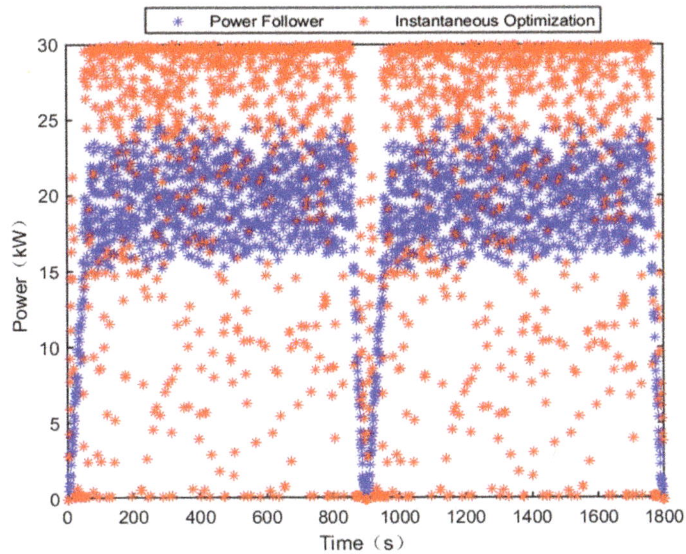

Figure 18. Working power of motor.

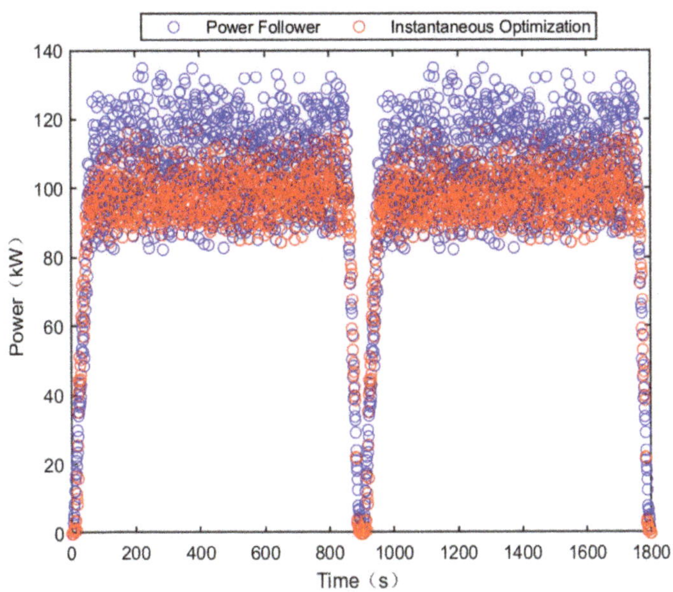

Figure 19. Working power of diesel engine.

It can be seen from the simulation results that under plowing conditions, based on the energy-saving control strategy of instantaneous optimization, the working power of the motor is dispersed while that of the diesel engine is concentrated. The diesel engine power is mostly between 85 and 105 kW. The initial SOC value is 0.90, while the final value is 0.42. Based on the power following the energy-saving control strategy, the working power of the motor was concentrated between 17 and 24 kW. The power of the diesel engine is large over a significant range between 85 and 130 kW. The initial value of SOC is 0.90, while the final value is 0.50.

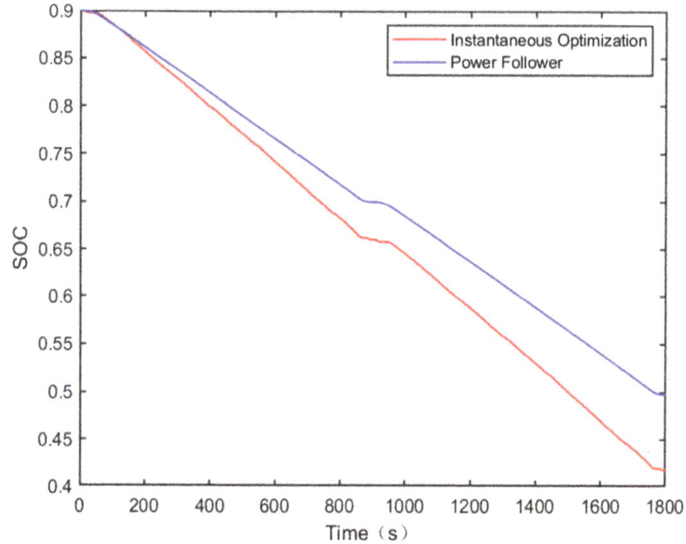

Figure 20. SOC state value changes.

The MAP of the diesel engine and motor for the two control strategies under plowing conditions are shown in Figures 21 and 22, respectively.

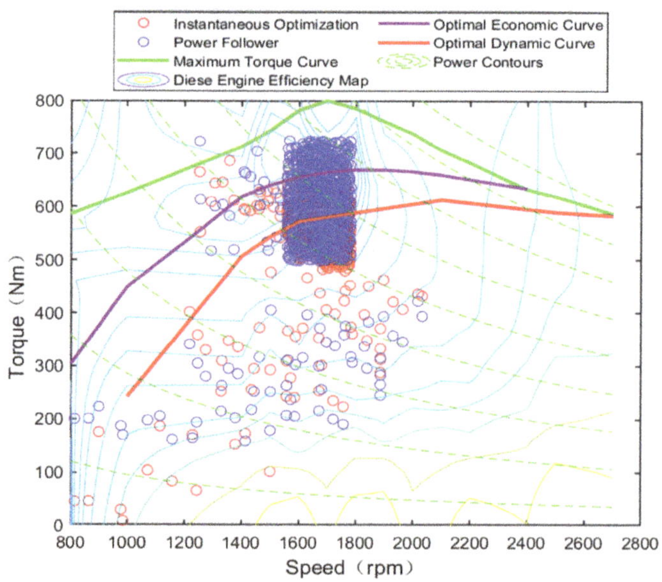

Figure 21. Diesel engine operating points.

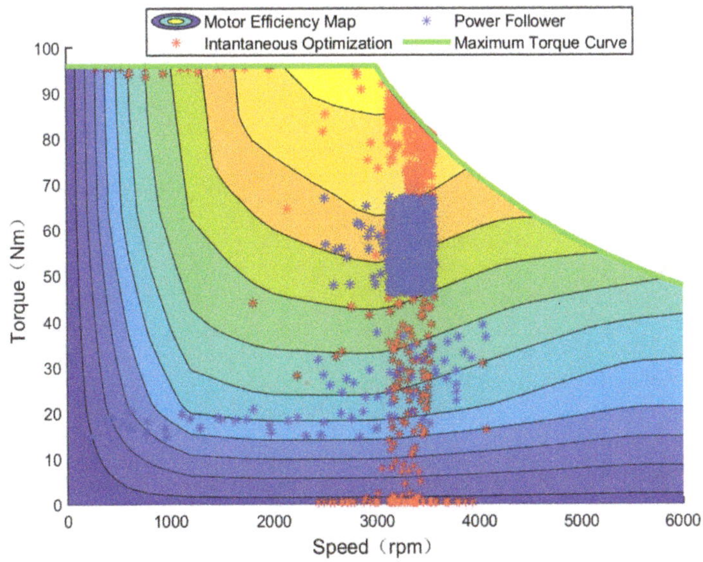

Figure 22. Motor operating points.

Based on the instantaneous optimization of the energy-saving control strategy, the overall working area of the diesel engine is low, near the optimal power curve. The fuel consumption is low with good dynamic performance. The efficiency within the overall operating region of the motor is also high, which improves the efficiency of the energy utilization. Based on the power following energy-saving control strategy, the working torque of the diesel engine at constant speed is large, the working power is large, and fuel consumption is increased. The overall operating region of the motor has a low efficiency and poor energy utilization.

6. Conclusions

This study describes an energy-saving control strategy for a hybrid tractor based on an instantaneous optimization algorithm. The objective is to minimize the equivalent fuel consumption of the entire machine. The motor torque and diesel engine torque are the control variables, while the state of charge of the power battery is the state variable. Finally, the instantaneous optimal distribution of torque is obtained.

Considering 220 hp tractors as the research object, the simulation models of the main components were built based on the topology of a parallel diesel-electric hybrid tractor. Finally, a control simulation model was built using MATLAB. To solve the problem of low energy utilization efficiency of hybrid tractors, an energy-saving control strategy based on instantaneous optimization is proposed and compared with the energy-saving control strategy based on power following. The results demonstrate that the proposed energy-saving control strategy based on instantaneous optimization effectively improves the energy efficiency of the tractor during plowing and rotary tillage and reduces equivalent fuel consumption. Under rotary tillage condition, the equivalent fuel consumption of an energy-saving control strategy based on power following is 14.46 L, whereas the equivalent fuel consumption of an energy-saving control strategy based on instantaneous optimization is 13.78 L. Accordingly, the equivalent fuel consumption decreases by 4.70%. Under plowing condition, the equivalent fuel consumption is 14.10 L based on the power following energy saving control strategy and 13.21 L based on the instantaneous optimization energy saving control strategy. Accordingly, the equivalent fuel consumption decreases by 6.31%.

The instantaneous optimal control strategy was designed for a 220 hp hybrid tractor, and acceptable results were obtained. In the future, we will design and test an instantaneous optimization control strategy framework that can meet a variety of different types of hybrid tractors. In addition, researchers can improve the energy-saving control effect of hybrid tractors by constructing models of the diesel engine and motor with more accurate dynamic simulation.

Author Contributions: Conceptualization, J.Z. and G.F.; methodology, L.X.; software, G.F.; validation, J.Z., G.F. and L.X.; formal analysis, M.L.; investigation, X.Y.; resources, L.X.; data curation, W.W.; writing—original draft preparation, G.F.; writing—review and editing, J.Z.; visualization, X.Y.; supervision, M.L.; funding acquisition, L.X. All authors have read and agreed to the published version of the manuscript.

Funding: This research was funded by National Key Research and Development Program (Grant No. 2022YFD2001203), Key Scientific and Technological Research Projects in Henan Province (Grant No. 222102240088), and the State Key Laboratory Open Project (Grant No. SKT202200).

Data Availability Statement: Not applicable.

Conflicts of Interest: Wei Wang and Mengnan Liu are employees of YTO Group Corporation, Luoyang 471004, China. The paper reflects the views of the scientists, and not the company.

References

1. Liu, Y.Q. New energy development and key technology research and development direction of agricultural machinery. *Agric. Mach. Using Maint.* **2021**, *4*, 33–34.
2. Xie, B.; Wu, Z.B.; Mao, E.N. Development and prospect of key technologies on agricultural tractor. *Trans. Chin. Soc. Agric. Mach.* **2018**, *49*, 1–17.
3. Mariano, G.; Emmi, L.; Benavides, C.; Garcia, I.; Gonzalez-de-Santos, P. Reducing air pollution with hybrid-powered robotic tractors for precision agriculture. *Biosyst. Eng.* **2016**, *143*, 79–94.
4. Till, G.; Speth, D.; Krail, M.; Wietschel, M.; Oberle, S. Pathways to Carbon-Free Transport in Germany until 2050. *World Electr. Veh. J.* **2022**, *13*, 136.
5. Francesco, M.; Aurelio, S. Analysis of a Parallel Hybrid Electric Tractor for Agricultural Applications. *Energies.* **2020**, *13*, 3055.
6. Yu, P.L.; Li, M.; Wang, Y.; Chen, Z. Fuel Cell Hybrid Electric Vehicles: A Review of Topologies and Energy Management Strategies. *World Electr. Veh. J.* **2022**, *13*, 172. [CrossRef]
7. Deng, X.T.; Zhu, S.H.; Qian, Z.X.; Zhang, Y. Research and development on power coupling device of hybrid electric tractor. *Trans. Chin. Soc. Agric. Eng.* **2012**, *28*, 29–34.
8. Li, H.; Song, Z.H.; Xie, B. Plowing Performance Simulation and Analysis for Hybrid Electric Tractor. *Appl. Mech. Mater.* **2013**, *2555*, 505–511. [CrossRef]
9. He, H.G.; Meng, X.F. A Review on Energy Management Technology of Hybrid Electric Vehicles. *Trans. Beijing Inst. Technol.* **2022**, *42*, 773–783.
10. Zhang, F.Q.; Hu, X.S.; Xu, K.H.; Tang, X.L.; Cui, Y.H. Status and Prospects for Model Predictive Energy Management in Hybrid Electric vehicles. *J. Mech. Eng.* **2019**, *55*, 86–108. [CrossRef]
11. Luo, G.J. Research on Energy Management Strategy of Series Diesel-Electric Hybrid Tractor. Master's Thesis, Department of Agricultural Engineering, Naju, China, 2019.
12. Xu, L.Y.; Zhang, J.J.; Liu, M.N.; Zhou, Z.; Liu, C. Control algorithm and energy management strategy for extended range electric tractors. *Int. J. Agric. Biol. Eng.* **2017**, *10*, 35–44.
13. Fang, S.P.; Zhou, Z.L.; Xu, L.Y. Energy management strategy for tandem hybrid tractors. *J. Henan Univ. Sci. Technol. Nat. Sci.* **2015**, *36*, 61–66.
14. Zhang, Z.; Zhang, T.; Hong, J.; Zhang, H.; Yang, J. Energy Management Optimization of Master–Slave Hybrid Electric Vehicle under Rule-Based Control Strategy. *Energy Technol.* **2022**, *10*, 2200630. [CrossRef]
15. Lee, H.S.; Kim, J.S.; Park, Y.I.; Cha, S.W. Rule-based power distribution in the power train of a parallel hybrid tractor for fuel savings. *Int. J. Precis. Eng. Manuf.-Green Technol.* **2016**, *3*, 231–237. [CrossRef]
16. Matteo, S.; Anselma, P.G.; Misul, D.A.; Belingardi, G. Exploitation of a Particle Swarm Optimization Algorithm for Designing a Lightweight Parallel Hybrid Electric Vehicle. *Appl. Sci.* **2021**, *11*, 6833.
17. Qian, L.J.; Qiu, L.H. Optimal control of a hybrid electric vehicle based on fuzzy-PID torque identification. *China Mech. Eng.* **2015**, *26*, 1752–1759.
18. Shi, D.P.; Chu, L.; Guo, J.; Tian, G.; Feng, Y.; Li, Z. Energy Control Strategy of HEB Based on the Instantaneous Optimization Algorithm. *IEEE Access* **2017**, *5*, 19876–19888. [CrossRef]
19. Liu, M.N.; Zhao, S.; Han, B.; Lei, S.; Xu, L. Multi-Objective Optimization and Test of a Tractor Drive Motor. *World Electr. Veh. J.* **2022**, *13*, 43. [CrossRef]

20. Wang, Y.; Zeng, X.H.; Song, D.F.; Yang, N.N. Optimal rule design methodology for energy management strategy of a power-split hybrid electric bus. *Energy.* **2019**, *185*, 1086–1099. [CrossRef]
21. Adeleke, O.P.; Li, Y.; Chen, Q.; Zhou, W.; Xu, X.; Cui, X. Torque Distribution Based on Dynamic Programming Algorithm for Four In-Wheel Motor Drive Electric Vehicle Considering Energy Efficiency Optimization. *World Electr. Veh. J.* **2022**, *13*, 181. [CrossRef]
22. Sun, D.Y.; Lin, X.Y.; Qin, D.T.; Deng, T. Power-balancing instantaneous optimization energy management for a novel series-parallel hybrid electric bus. *Chin. J. Mech. Eng.* **2012**, *25*, 1161–1170. [CrossRef]
23. Gökce, K.; Ozdemir, A. An instantaneous optimization strategy based on efficiency maps for internal combustion engine battery hybrid vehicles. *Energy Convers. Manag.* **2014**, *81*, 255–269. [CrossRef]
24. Minh, V.T.; Moezzi, R.; Cyrus, J.; Hlava, J.; Petru, M. Parallel Hybrid Electric Vehicle Modelling and Model Predictive Control. *Appl. Sci.* **2021**, *11*, 10668.
25. Zhao, G.F. Modeling and control strategy of hybrid electric vehicle system. *Sichuan Aerosp. Voc. Coll.* **2022**, *12329*, 1232902.
26. Choi, S.C.; Song, B.S.; Kim, Y.J. Torque Assist Strategy for Hybrid Agricultural Tractor with Consideration of Field Operations. *Trans. Korean Soc. Mech. Eng.—A* **2014**, *38*, 593–600. [CrossRef]
27. Xu, L.Y.; Liu, M.N.; Zhou, Z.L. Design of drive system for series hybrid electric tractor. *Trans. Chin. Soc. Agric. Eng.* **2014**, *30*, 11–18.
28. Li, J.F.; Wu, X.; Zhang, X.; Song, Z.; Li, W. Design of distributed hybrid electric tractor based on axiomatic design and Extenics. *Adv. Eng. Inform.* **2022**, *54*, 101765. [CrossRef]
29. Xie, B.; Hao, L.; Zheng-He, S.; En-Rong, M. Powertrain System Design of Medium-sized Hybrid Electric Tractor. *Inf. Technol. J.* **2013**, *12*, 7228–7233.
30. Zhou, Z.L.; Ni, Q.; Xu, L.Y. Parallel hybrid tractor drive train parameter design and performance analysis. *J. Henan Univ. Sci. Technol.* **2016**, *37*, 9–15.
31. Liu, M.N.; Xu, L.Y.; Zhou, Z.L.; Liu, W.G. Establishment of Extended Range Electric Tractor and Its Rotary Cultivator's Simulation Platforms. *China Mech. Eng.* **2016**, *27*, 413–419.
32. Xu, L.Y.; Zhang, J.J.; Liu, M.N. Torque distribution strategy for extended-range four-wheel drive electric tractor. *J. Henan Univ. Sci. Technol.* **2017**, *38*, 80–85.
33. Li, X.H.; Luo, F.Q.; Tang, D. Drawing engine universal performance characteristics map using ploynomial interpolation. *Trans. CSAE* **2004**, *5*, 138–141.

World Electric Vehicle Journal

MDPI

Article

Research on Global Optimal Energy Management Strategy of Agricultural Hybrid Tractor Equipped with CVT

Junjiang Zhang [1,2,3,4], Ganghui Feng [1], Mengnan Liu [2], Xianghai Yan [1,2,3], Liyou Xu [1,2,3,*] and Chengyan Shang [1]

1 College of Vehicle and Traffic Engineering, Henan University of Science and Technology, Luoyang 471003, China; 9906179@haust.edu.cn (J.Z.); 210321030258@stu.haust.edu.cn (G.F.); 9905167@haust.edu.cn (X.Y.)
2 State Key Laboratory of Intelligent Agricultural Power Equipment, Luoyang 471039, China; 210321121601@stu.haust.edu.cn
3 Henan Province Collaborative Innovation Center for Advanced Manufacturing of Mechanical Equipment, Luoyang 471003, China
4 YTO Group Corporation, Luoyang 471004, China
* Correspondence: xlyou@haust.edu.cn; Tel.: +86-136-6387-3262

Abstract: This paper presents a proposed global optimal energy management strategy based on dynamic programming to enhance the energy consumption efficiency of an agricultural hybrid tractor that is equipped with a continuously variable transmission (CVT). Firstly, using a diesel-electric parallel agricultural hybrid tractor as the research object, a tractor-rotary tillage coupling dynamics model is constructed. Secondly, with the torque and speed of the motor, the torque and speed of the diesel engine, and the CVT speed ratio as the control variables, the state of charge (SOC) of the power battery as the state variable, and the goal of minimizing the total energy consumption of the whole machine, a global optimal energy management model based on dynamic programming is established. Finally, the field operation measured data is injected into the MATLAB simulation model, and experiments are carried out to verify the effectiveness of the energy management strategy. The results show that compared with the power-following energy management strategy, the proposed energy management strategy can make the diesel engine and electric motor work in the optimal area, and effectively reduce the total cost of energy consumption of the tractor during field operations. Under the condition of rotary tillage, the total cost of energy consumption is decreased by 16.89%.

Keywords: agricultural hybrid tractor; dynamic programming; energy management; total cost of energy consumption

Citation: Zhang, J.; Feng, G.; Liu, M.; Yan, X.; Xu, L.; Shang, C. Research on Global Optimal Energy Management Strategy of Agricultural Hybrid Tractor Equipped with CVT. *World Electr. Veh. J.* **2023**, *14*, 127. https://doi.org/10.3390/wevj14050127

Academic Editors: Fachao Jiang, Yongyu Li and Weiwei Kong

Received: 12 April 2023
Revised: 13 May 2023
Accepted: 15 May 2023
Published: 17 May 2023

1. Introduction

In the global agricultural machinery industry, tractors are the largest category. However, traditional internal combustion engine tractors consume a large amount of fossil fuels and have poor emissions, resulting in energy shortages and environmental pollution [1–3]. With the advocacy of protecting the environment and reducing energy consumption around the world, it is of great significance to study new energy-saving tractors [4–6]. In the last decade, strict regulations were progressively applied to the Non-Road Mobile Machineries (NRMM). The field of NRMM is now more than ever considering the adoption of electric systems to reduce the amount of pollutant emissions and to improve energy efficiency per unit of work [7]. Pure electric agricultural tractors have a short operating mileage and are not suitable for long-term heavy-duty traction operations, whereas agricultural hybrid tractors can solve this problem by configuring two or more power sources under appropriate control strategies [8].

As the core control strategy of hybrid tractor, energy management strategy has a direct impact on the power, economy, comfort, and emission of the whole machine, so it has become the focus and difficulty of research in the field of hybrid electric vehicles [9].

Currently, there are primarily two classifications of energy management approaches: control strategies based on rules and those based on optimization [10–12]. The development cost of the rule-based control strategy is low, simple, and intuitive, and has strong practicability. It is extensively utilized in a diverse range of hybrid automobiles. For four-wheel drive hybrid electric vehicles, Ma et al. [13] proposed a rule-based logic threshold energy management strategy model to control the torque distribution of the hybrid system and realize the reasonable selection of the working mode. The results show that compared with the comprehensive fuel consumption of the prototype traditional vehicle, the designed control strategy can reduce the fuel consumption. Lv et al. [14] proposed an energy management strategy based on rule-based fuzzy control for plug-in hybrid electric vehicles. The results show that the proposed strategy significantly reduces fuel consumption. Luo et al. [15] proposed a control strategy based on the optimal use of electric energy to solve the poor fuel economy problem of hybrid vehicles on longer driving distances. The results show that this strategy can significantly improve the vehicle's fuel economy over a longer driving range. However, deterministic rule-based control strategies are determined based on the experience of the designer. It has high reliability but poor adaptability to working conditions. For the fuzzy control strategy, the utilization of a basic fuzzy information processing approach in the control strategy may result in compromised control precision and decreased dynamic quality of the system [16].

Control strategies that utilize optimization require minimizing or maximizing a cost function, which typically serves as a gauge for the control objective. Geng et al. [17] proposed a multi-objective energy management strategy based on particle swarm optimization (PSO) aiming to reduce vehicle energy consumption and control battery power at the same time. The findings indicate that the suggested approach is capable of accomplishing two objectives simultaneously, namely decreasing car energy usage and managing battery energy. Zhang et al. [18] proposed an energy management strategy based on instantaneous torque optimization for hybrid tractors. The findings indicate that the suggested approach effectively manages the operational status of the electric motor and diesel engine, leading to lower overall fuel usage when compared to the power-following energy management technique. Chen et al. [19] proposed a power distribution energy management strategy for plug-in hybrid electric vehicles. A series of quadratic equations are used to estimate the fuel rate of the vehicle, and the battery current is used as input, and the pontryagin minimum principle (PMP) is introduced to solve the problem. The results show that the proposed algorithm is able to reduce fuel consumption compared to charge depletion (CD) and charge maintenance (CS) modes. However, the instantaneous optimization method can only ensure the optimal fuel economy at each moment, but cannot guarantee the final global optimal. The PMP control strategy can only achieve an approximate global optimum.

In order to comply with the new energy trend and improve the energy utilization rate of agricultural machinery, a hybrid tractor used in agriculture is taken as the research object, which combines diesel and electric power sources. Then a global optimal energy management strategy based on dynamic programming (DP) is proposed in this study [20–24]. By optimizing the torque distribution of the diesel engine and the electric motor, the total energy consumption cost of the whole machine can be reduced while ensuring the power performance. The remainder of this study is organized as follows. In Section 2, presents a hybrid tractor topology and its main performance parameters. In Section 3, simulation modeling of major components of the hybrid tractor is described. A global optimal energy management strategy for hybrid tractors is designed in Section 4. Section 5 delineates the description of simulation verification and discussion. Finally, the conclusions drawn from the study are summarized in Section 6.

2. Topology and Principal Parameters of Hybrid Tractors

Taking a diesel-electric parallel agricultural hybrid tractor as the research object, the establishment of the hybrid tractor topology and the performance parameters of its main components are stated.

2.1. Hybrid Tractor Topology

Figure 1 shows the topology of a hybrid tractor equipped with continuously variable transmission (CVT), which has two power sources, namely a diesel engine and an electric motor. The two can work together or independently. The output torque of the diesel engine and the motor are transmitted to the CVT input shaft and the power take off (PTO) power output shaft through the torque coupling, which are served as the source of power for the central transmission and the PTO, respectively.

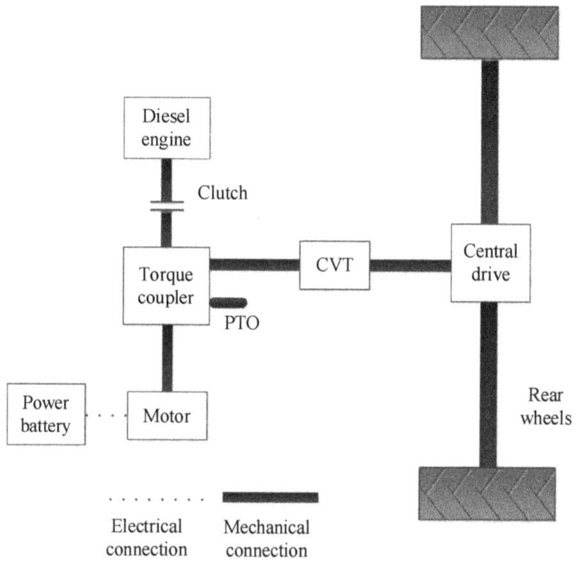

Figure 1. Topology of a hybrid tractor equipped with CVT.

The vehicle controller and diesel engine, power battery, clutch, electric motor, and CVT are connected through the controller area network (CAN) bus. According to the overall power demand of the whole system and the corresponding state of charge (SOC) index of the power battery, the torque output of the diesel engine and the motor is dynamically allocated to the whole machine through the control strategy. Consequently, the hybrid tractor has the capacity to attain the most optimal combination of power and economic efficiency.

2.2. Main Performance Parameters of Hybrid Tractor

The present investigation employs a diesel-electric parallel hybrid tractor as the research object and aims to investigate its energy management strategy [25–27]. The parameters characterizing primary components of the hybrid tractor are shown in Table 1.

Table 1. Parameters characterizing primary components of hybrid tractors.

Component	Parameter	Value (Unit)
Diesel engine	Rated power	60 (kW)
	Rated speed	2200 (rpm)
	Maximum torque	280 (Nm) (1700 rpm)
Motor	Maximum power	40 (kW)
	Rated speed	2800 (rpm)
	Maximum torque	140 (Nm)

Table 1. *Cont.*

Component	Parameter	Value (Unit)
Power battery	Rated capacity Rated voltage SOC	70 (Ah) 320 (V) 0.25–0.90
Central drive	Speed ratio	19.27
CVT	Speed ratio	0.864–9.728

3. Model Construction of Hybrid Tractor

Taking into account the topology of the hybrid tractor, an analysis was carried out to model its principal components, which encompass the transmission system, the diesel engine, the motor, the dynamics of the rotary tillage unit, the CVT, and the power battery model. Finally, the simulation model of the hybrid tractor was developed.

3.1. Hybrid Tractor Transmission System Model

The motor and diesel engine are responsible for supplying the necessary energy to operate the hybrid tractors. The total power of the tractor is derived from the input of the torque coupler, represented by the following expression:

$$P_{req}(T_{req}, n_{req}) = P_{mreq} + P_{ereq} \tag{1}$$

$$Pmreq = P_m(T_m, n_m)\eta_m \tag{2}$$

$$Pereq = P_e(T_e, n_e)\eta_e \tag{3}$$

where P_{ereq} and P_{mreq} are the diesel engine and motor required power, respectively. n_{req} and T_{req} denote the required speed and torque at the input end of the torque coupler, respectively. η_e and η_m are the working efficiencies of the diesel engine and motor, respectively. P_e and P_m denote the power required for the diesel engine and motor, respectively. T_e, T_m, n_e and n_m denote the torques and speeds of the diesel engine and motor, respectively.

Based on the tractor topology depicted in Figure 1, analyze the power transmission process. The relationship between the wheel and the diesel engine speeds is expressed as follows:

$$n_e = n_{req} = n_{tire} i_{cvt} i_0 \tag{4}$$

$$n_{tire} = \frac{v}{0.377r} \tag{5}$$

where i_0 and i_{cvt} denote the main reducer and CVT speed ratios, respectively. n_e and n_{tire} represent the diesel engine and driving wheel speeds, respectively. r denotes the driving wheel radius of the hybrid tractor. v represents the speed of the hybrid tractor during working.

3.2. Dynamic Model of Rotary Tillage Unit

The power balance relationship of the unit when the hybrid tractor is towing the rotary tillage unit is similar to that in the literature [18]. Based on the previous research results of the research group, combined with the hybrid tractor topology in this paper. In particular, CVT and PTO shafts are connected with torque couplings [28–30]. The power balance equation for the rotary tillage process of the hybrid tractor can be expressed as follows:

$$P_{req}(T_{req}, n_{req}) = \left(\frac{P_{drive}}{\eta_{zy}\eta_{cvt}} + P_r\right) / \eta_o \tag{6}$$

where η_{zy}, η_{cvt}, and η_o represent the central drive efficiency, CVT efficiency, and torque-coupler efficiency, respectively. P_r and P_{drive} denote the rotary cultivator power and tractor travel power consumptions, respectively.

3.3. Motor Model

Electric motors can convert between electrical energy and mechanical energy. The relationship between the conversion efficiency, torque and speed is expressed as follows:

$$\eta_m = f_m(n_m, T_m) \tag{7}$$

The numerical model method is employed to establish the motor model, whereby the spline interpolation method is utilized to determine the correlation between the efficiency, torque, and speed of the motor system. This approach is underpinned by experimental data pertaining to motor efficiency. This method has a fast calculation speed and can meet the rapidity requirements of simulation. Therefore, it is suitable for control strategy research. The numerical model representing the efficiency of the motor is shown in Figure 2.

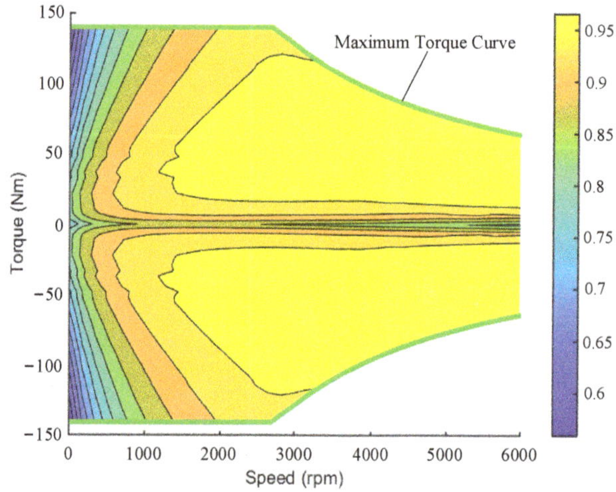

Figure 2. MAP diagram of motor model.

3.4. Diesel Engine Model

The modeling of diesel engines can be classified into two principal categories: mathematical modeling method and experimental modeling method [31]. The mathematical modeling method needs to analyze the transient characteristics of the diesel engine, and test the detailed parameters of each process. Also, the process is complex and the calculation speed is slow. The experimental modeling method is based on the measured data of the diesel engine, and constructs the MAP diagram by formulating the mapping relationship of the diesel engine torque and fuel consumption rate corresponding to the speed or throttle opening. In the simulation process, only the output and input parameters of the diesel engine are concerned, and its transient characteristics are ignored. Therefore, the experimental modeling method is used here to obtain the numerical model of the fuel consumption rate of the diesel engine, as shown in Figure 3.

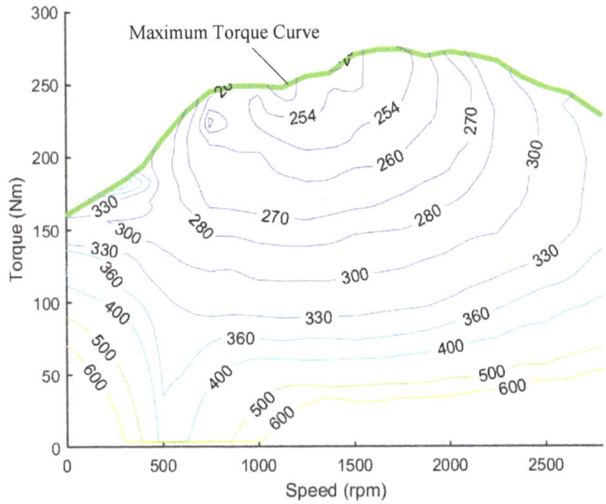

Figure 3. MAP diagram of diesel engine model.

3.5. CVT Model

The CVT has a continuously variable speed ratio, and can adjust the speed ratio in real time to adjust the working point of the motor and the diesel engine. Hence, it is evident that both the motor and the diesel engine can operate in the realm of optimal efficiency. This paper only considers the transmission efficiency of CVT, ignores its dynamic response characteristics, and does not consider the complex hydraulic actuators in CVT. A CVT numerical model was established based on bench test data [32]. CVT transmission efficiency is related to its input torque and speed ratio, which can be expressed as:

$$\eta_{cvt} = f_{cvt}(T_{cvt\_in}, i_{cvt}) \tag{8}$$

where T_{cvt_in} is the input torque of the CVT.

The relationship surface of CVT efficiency, speed, and torque are obtained by interpolation fitting method. That is the numerical model of CVT efficiency, as shown in Figure 4.

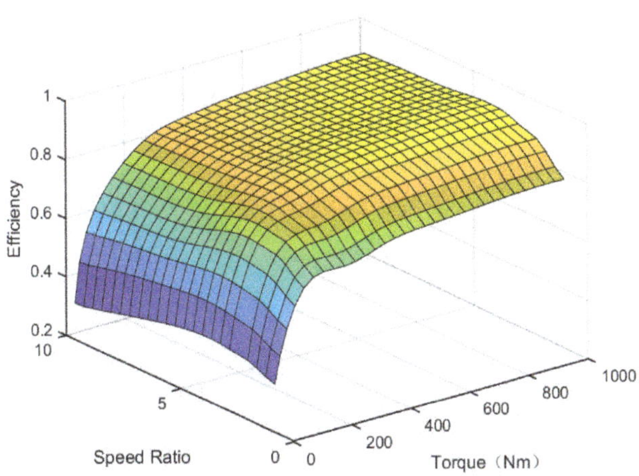

Figure 4. CVT efficiency model.

The calculation formula of CVT output torque and output speed is as follows:

$$T_{cvt\_out} = i_{cvt} T_{cvt\_in} \eta_{cvt} \tag{9}$$

$$n_{cvt\_out} = \frac{n_{cvt\_in}}{i_{cvt}} \tag{10}$$

where n_{cvt_in} and n_{cvt_out} are the CVT driving pulley and driven pulley speeds, respectively. T_{cvt_out} is the CVT output torque.

3.6. Power Battery Model

There are two prevalent battery models in common use, namely the resistance-capacitance model and the internal resistance model [33]. The internal resistance model regards the battery pack as an equivalent circuit in series with an ideal voltage source and an internal resistance, which belongs to the first-order model. The resistance-capacitance model regards the battery paçk as a circuit composed of two capacitors and three resistors, which belongs to the second-order model. In comparison, the internal resistance model is simple to model and has good applicability to various working conditions of the power battery. Therefore, the internal resistance model is adopted here. The power battery is equivalent to a circuit model of an ideal voltage source and a resistor connected in series, the mathematical equation is simple, and it is convenient for calculation and modeling.

The relationship curve between the electromotive force of the power battery, the internal resistance of charge or discharge and the SOC of the power battery is shown in Figure 5.

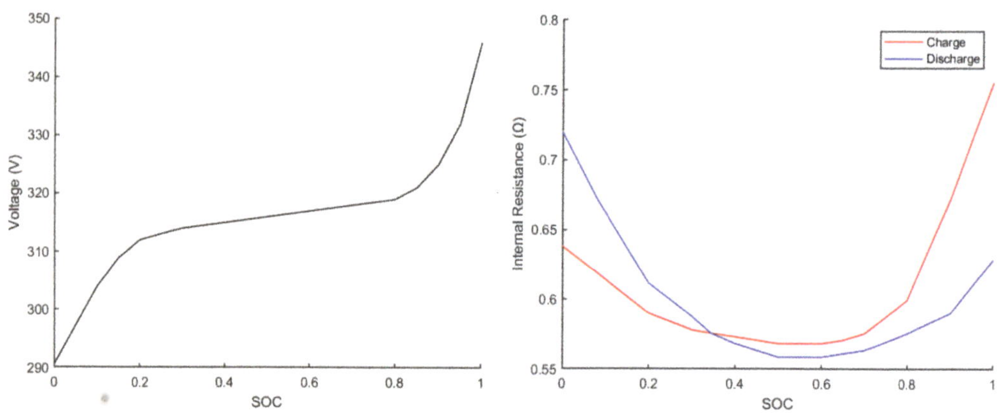

(a) **(b)**

Figure 5. (**a**) The relationship curve between electromotive force and SOC; (**b**) The relationship curve between charge or discharge internal resistance and SOC.

The equation for calculating the required power of the power battery is as follows:

$$P_{bat} = \begin{cases} \frac{P_m}{\eta_{bat}}, P_m > 0 \\ P_m \eta_{bat}, P_m < 0 \end{cases} \tag{11}$$

where η_{bat} denotes the battery charging or discharging efficiency. P_m greater than 0 is discharging, and P_m less than 0 is charging. P_{bat} represents the required power of the power battery.

The formulas for calculating the SOC and current of the power battery are as follows:

$$I(t) = \frac{U(t) - \sqrt{[U(t)]^2 - 4R(t)P_m(t)}}{2R(t)} \tag{12}$$

$$SOC(t+1) = SOC(t) - \frac{I(t)\Delta t}{Q_b} \tag{13}$$

where $I(t)$, $U(t)$, and $R(t)$ are the output current, output voltage, and internal resistance of the power battery, respectively. Q_b denotes the rated power battery capacity. $SOC(t)$ and $SOC(t + 1)$ are the SOC at the current and next moment, respectively. Δt is the time step.

3.7. Machine Simulation Model

The simulation model of the entire hybrid tractor has been constructed using MATLAB, taking into account the characteristics of its transmission system, as depicted in Figure 6. The machine simulation model includes the transmission system model, dynamic model of rotary tiller, diesel engine model, motor model, and power battery model. The controller collects signals based on the working conditions of the tractor. The required power and speed (P_{req} and n_{req}) of the whole machine are obtained through calculation and processing according to Equations (1), (4) and (6). Inside the controller, the required power of the whole machine is allocated according to the established control strategy (including the proposed and compared strategies). Then output the corresponding motor required power (P_{mreq}) and diesel engine required power (P_{ereq}) as the input of the motor model and diesel engine model.

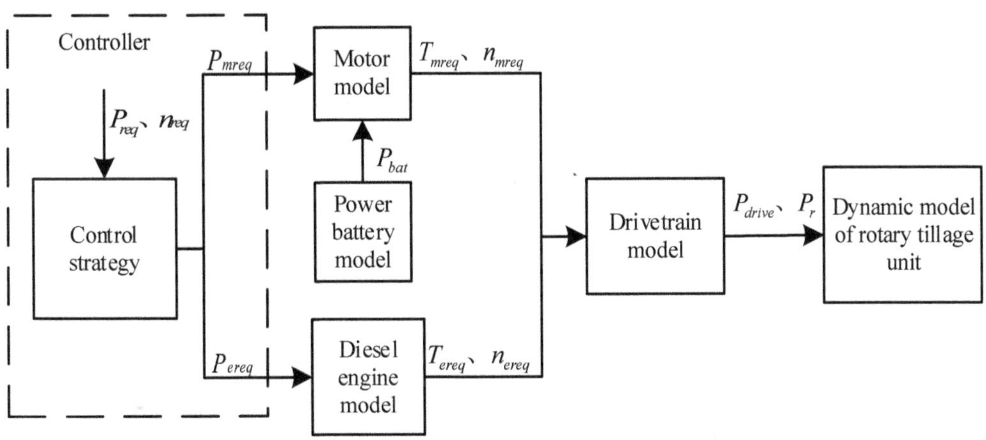

Figure 6. Schematic diagram of the whole machine simulation model.

The motor model and diesel engine model work in accordance with prescribed instructions, output the corresponding required torque and speed (T_{mreq}, n_{mreq}, T_{ereq}, and n_{ereq}) and transmit the power to the rotary tiller dynamic model (P_{drive} and P_r) through the transmission system model. At the same time, the power battery model performs energy transfer according to the required power of the motor model (P_{bat}).

4. Energy Management Strategy Design

This section details the design process of a globally optimized energy management strategy based on dynamic programming. Firstly, an energy management optimization model is built, and then the principle of dynamic programming algorithm is expounded. Finally, the dynamic programming algorithm is combined with the energy management optimization model, and the solution process of the dynamic programming control strategy is described. In addition, the control principle of the power-following energy management strategy is briefly described.

4.1. Global Optimal Energy Management Strategy Based on Dynamic Programming

4.1.1. Energy Management Optimization Model

Hybrid tractors derive their power from two distinct sources, namely electricity and diesel fuel. In order to measure the two kinds of energy uniformly, the energy consumption economy function is defined as the sum of electric energy cost and diesel engine fuel cost. The goal of energy management is to minimize the total cost of energy consumption by optimally allocating the operating states between the diesel engine and the electric motor. The total cost of energy consumption during hybrid tractor operation can be defined as:

$$Q_c(t) = \int_0^{t_f} \left(j_e Q_f(T_e, n_e) + \frac{j_m P_{bat}}{\eta_{bat}} \right) dt \tag{14}$$

where $Q_c(t)$ is the total cost of energy consumption. $Q_f(t)$ denotes the instantaneous fuel consumption. t_f denotes the end moment. j_m and j_e represent the prices per kWh of electricity and liter of oil, respectively.

According to the Equation (13), the power battery SOC changes as follows:

$$S\dot{O}C(t) = -\frac{I(t)}{Q_b} = -\frac{U(t) - \sqrt{[U(t)]^2 - 4P_{bat}(t)R(t)}}{2R(t)Q_b} \tag{15}$$

4.1.2. Dynamic Programming Algorithm

Dynamic programming is a multi-step optimization algorithm, which divides the multi-stage decision-making process into stages. Then properly selects state variables and decision variables to define the optimal objective function. Thus, the problem can be transformed into a family of sub-problems of the same type, and finally solved one by one. The solution begins with the boundary conditions. The process proceeds in reverse order, and the optimization is recursively searched segment by segment. When solving each sub-problem, the optimal result of the previous sub-problems must be used. The optimal solution to the last subproblem is the optimal solution to the entire problem. Using reverse solution, according to the given tractor operating conditions, it is divided into m parts.

When establishing the dynamic programming algorithm, set the sampling time as 1 s. Computation starts at m stage and goes forward. When calculating each stage, with the optimal control as the goal, the optimal control variable $u(k)$ is calculated by global search under the given component parameters. To ensure the lowest fuel consumption of the diesel engine and maintain the power SOC within the range of the set value. Accordingly, the optimal optimization result is calculated. The decision variables, namely the control variables, include motor torque, diesel engine torque, motor speed, diesel engine speed, and CVT gear ratio. The power battery SOC is defined as the state variable, and the total energy consumption cost is set as the optimal objective function. Based on this, a DP energy management strategy is established.

The state variables are as follows:

$$x(k) = [SOC(k)]^{\mathrm{T}} \tag{16}$$

At the same time, the state variables need to be discretized, as follows:

$$SOC(k) \in \{SOC_1, SOC_2, \cdots, SOC_m\} \tag{17}$$

where m is the discrete state space dimension, that is, the number of discrete points.

The control variables are as follows:

$$u(k) = [T_m(k), T_e(k), n_m(k), n_e(k), i_{cvt}(k)]^{\mathrm{T}} \tag{18}$$

In the calculation process, the control variables are discretized as follows:

$$u_m(k) \in \{u_{m1}, u_{m2}, \cdots, u_{mj}\} \tag{19}$$

where j denotes the number of discrete points.

The state equation of the system is as follows:

$$x(k+1) = f(x(k), u(k)) \tag{20}$$

According to Equations (14) and (16), the optimal objective function of dynamic programming is constructed as follows:

$$J = \min \sum_{k=0}^{m-1} [Q(SOC(k), P_e(k))] \tag{21}$$

where J is the minimum cumulative energy consumption total cost in the operation cycle. Q is the total cost of energy consumption at a certain stage.

Constraints can be expressed as:

$$\begin{cases} T_{m\min}(n_m(k), SOC(k)) \le T_m(k) \le T_{m\max}(n_m(k), SOC(k)) \\ T_{e\min}(n_e(k)) \le T_m(k) \le T_{e\max}(n_e(k)) \\ n_{m\min} \le n_m(k) \le n_{m\max} \\ n_{e\min} \le n_m(k) \le n_{e\max} \\ i_{cvt\min} \le i_{cvt}(k) \le i_{cvt\max} \\ SOC_{\min} \le SOC(k) \le SOC_{\max} \end{cases} \tag{22}$$

where $T_{m\ \min}$, $T_{m\ \max}$, $T_{e\ \min}$, and $T_{e\ \max}$ represent the minimum and maximum torques of the electric motor and diesel engine at the current moment, respectively. $n_{m\ \min}$, $n_{m\ \max}$, $n_{e\ \min}$, and $n_{e\ \max}$ represent the minimum and maximum speeds of the electric motor and diesel engine at the current moment, respectively. $i_{cvt\ \min}$ and $i_{cvt\ \max}$ denote the minimum and maximum speed ratios of the CVT, respectively. $SOC_{\min}$ and $SOC_{\max}$ are the least and greatest values permitted by the SOC, respectively.

4.1.3. Dynamic Programming Control Strategy Solution Process

The control strategy based on dynamic programming refers to the use of cyclic iteration method to reasonably optimize the working torque of diesel engine and electric motor, so as to minimize the total cost of energy consumption of the whole machine. The solution process is depicted in Figure 7.

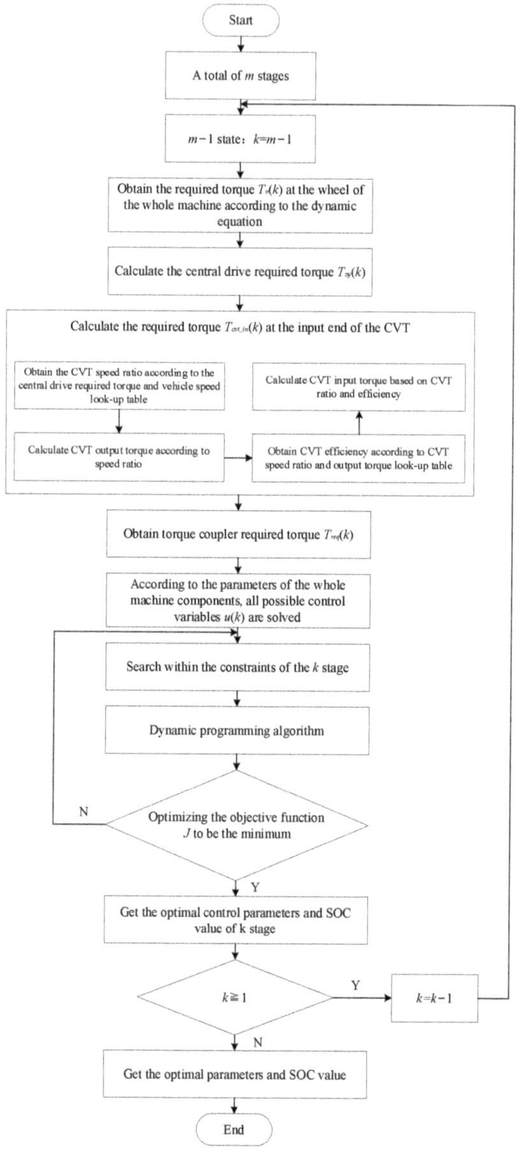

Figure 7. The solution process of dynamic programming algorithm control strategy.

1. Based on the operative circumstances, the dynamic programming algorithm is set to 1800 stages (m = 1800). According to the known vehicle speed and vehicle parameters in each stage, the required torque $T_v(k)$ of the tractor at the wheels of each stage is obtained through the dynamic equation.

$$T_v(k) = \frac{Pv(k)}{n_{tire}(k)} \tag{23}$$

where $P_v(k)$ denotes the required power at the wheel of the tractor.

2. Obtain the required torque $T_{zy}(k)$ at the input end of the central drive.

$$T_{zy}(k) = \frac{T_v(k)}{i_0 \eta_{zy}} \tag{24}$$

where $T_{zy}(k)$ represent the required torque at the input end of the central drive.

3. Obtain the required torque T_{cvt_in} at the input end of the CVT. Firstly, the CVT speed ratio is obtained according to the central transmission demand torque and the tractor speed look-up table, and the output torque of the CVT is calculated on the basis of the known speed ratio. Then the CVT efficiency is obtained through the CVT speed ratio and torque look-up table. Accordingly, the CVT input torque is calculated from the CVT speed ratio and efficiency.
4. Obtain the required torque $T_{req}(k)$ at the input end of the torque coupler.

$$T_{req}(k) = \frac{Tcvt\_in(k)}{\eta_o(k)} + \frac{T_{PTO}(k)}{i_{PTO}(k)\eta_o(k)} \tag{25}$$

where $i_{PTO}(k)$ denotes the PTO gear ratio inside the torque coupling.

5. According to the required torque $T_{req}(k)$ and speed $n_{req}(k)$, based on the set SOC upper and lower limits, CVT speed ratio, maximum and minimum torque values of diesel engine and electric motor, diesel engine fuel consumption rate, and electric motor efficiency, interpolation calculation each stage may control variables.
6. From all the possible control variables calculated, select the speed and torque of the diesel engine and motor that satisfy the constraints of the Equation (22). Calculate the control variables and state parameter values of each component when the objective Equation (21) is the minimum value. The state change and interpolation calculation of the control variables are shown in Figure 8.

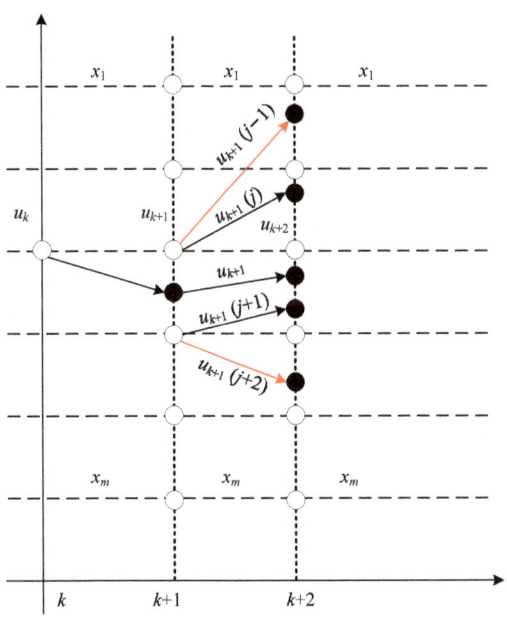

Figure 8. The solution process of dynamic programming algorithm control strategy.

When the state variables of $x(i)$ and $x(i+1)$ change, the optimal control variable at stage k is u_k. In the $k+1$ stage, the optimal control variable u_{k+1} is obtained by interpolating

$u_{k+1}(j)$ and $u_{k+1}(j+1)$. The red line in Figure 8 shows that in the $k+1$ stage, the state variable $x(i+1)$ exceeds the set upper and lower limits of the SOC.

7. Let $m = m - 1$, carry out the operation of the next stage until $k = 0$, get the optimal control parameter and SOC value, and the operation ends.

4.2. *Power following Energy Management Strategy*

The power-following energy management strategy is adopted as a comparative control strategy, which is simple in design and easy to implement. Moreover, it is a rule-based control strategy. The power-following energy management strategy employs the ratio of the diesel engine's rated power to that of the electric motor as the distribution ratio. The power demand of the whole machine is allocated based on a fixed ratio, which controls the operational status of both the diesel engine and electric motor, and give full play to the working capabilities of both.

5. Simulation Results and Analysis

During rotary tillage operations of the hybrid tractor, the PTO operates independently and is not affected by the driving conditions of the tractor. According to the field rotary tillage operation experiment, the speed, driving resistance, and PTO torque and rotational speed during the operation process are obtained. Then the measured data are input into the simulation model. Its torque speed is shown in Figure 9, and the rotary tillage operation speed is displayed in Figure 10.

Figures 11 and 12 show the MAP diagrams of diesel engine and electric motor under two control strategies during hybrid tractor rotary tillage operation. Under the global optimal energy management strategy based on dynamic programming, the working torque of the motor is large, and some negative torques appear. The diesel engine also produces a large working torque, and it is concentrated in the high-efficiency area. Under the power following energy management strategy, the motor generates a small working torque, and there is no negative torque. The working torque of the diesel engine is also small, and the efficiency of the working area is low.

Under the two control strategies, the change of CVT speed ratio is shown in Figure 13. The state change of power battery SOC is shown in Figure 14. Energy costs are shown in Figure 15.

From Figures 13–15, it can be known the CVT speed ratio is the same when the working speed is 0–3 km/h, but the CVT speed ratio is different when the working speed is 3–6 km/h under the two control strategies. Under the global optimization energy management strategy based on dynamic programming, the CVT speed ratio can not only change according to the working speed, but also make adjustments according to the required torque of the whole machine and the SOC value of the power battery. The SOC change trend of the two is basically the same, both showing a downward trend. Under the dynamic programming strategy, the SOC first drops rapidly to around 0.35, then fluctuates slowly, and some SOC rises. This corresponds to the negative torque part of the motor in Figure 13. Under the power-following control strategy, the SOC has been decreasing without increasing. In the end, the total energy consumption cost of the global optimal energy management strategy is ¥ 20.02. The total energy consumption cost of the power-following energy management strategy is ¥ 24.09.

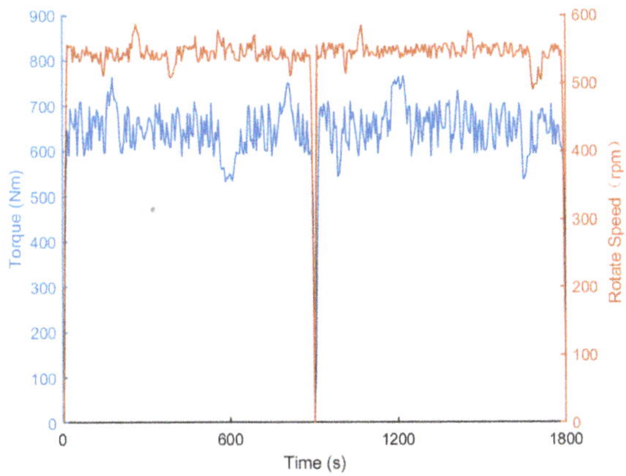

Figure 9. Torque and rotate speed of the PTO.

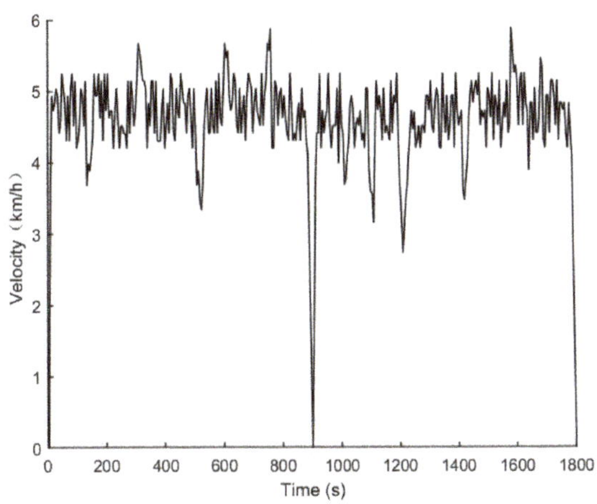

Figure 10. The speed during rotary tillage operations.

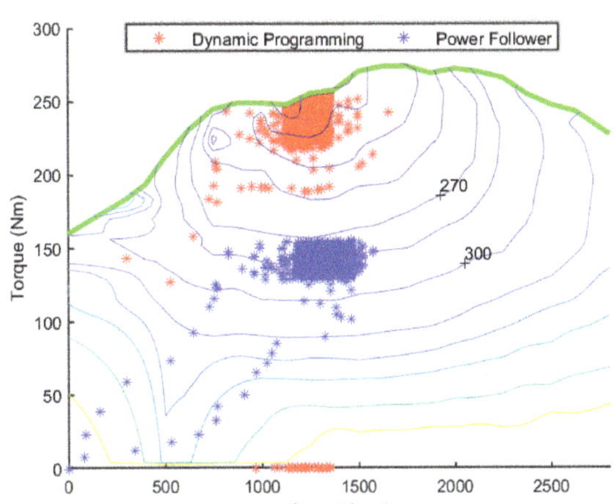

Figure 11. MAP diagram of diesel engine working.

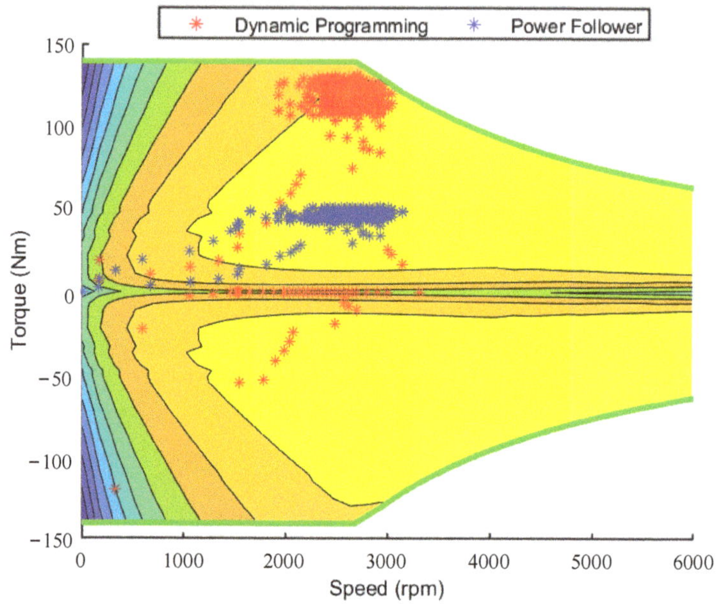

Figure 12. MAP diagram of motor working.

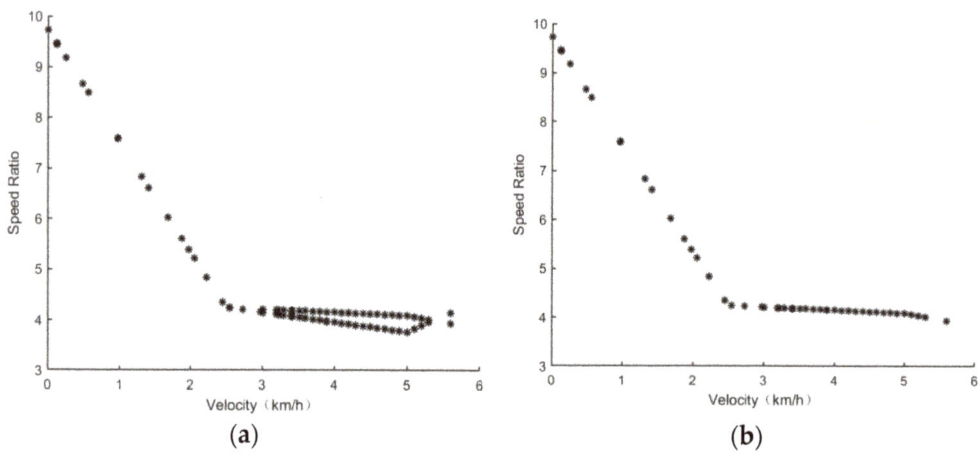

Figure 13. CVT speed ratio change diagram. (**a**) Based on dynamic programming strategy; (**b**) Based on power-following strategy.

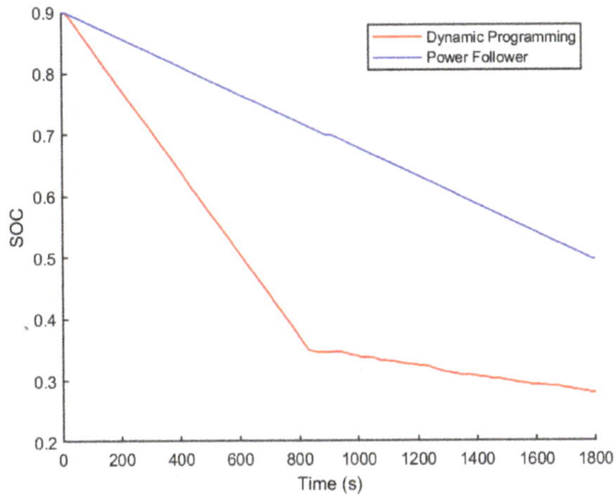

Figure 14. SOC state value changes.

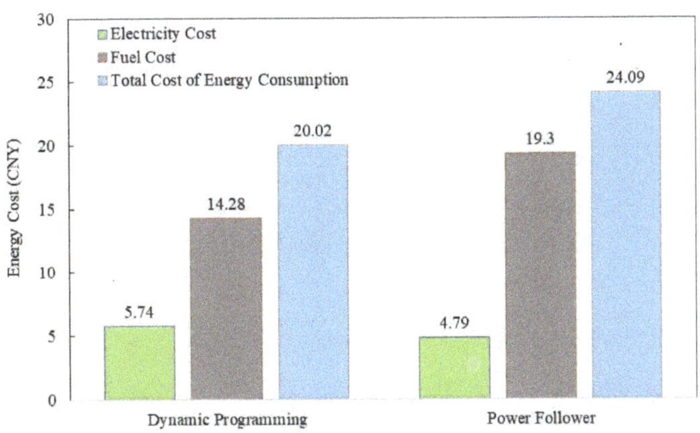

Figure 15. Energy Cost Chart.

6. Conclusions

This study describes a globally optimized energy management for an agricultural hybrid tractor equipped with a CVT. Firstly, the motor torque and speed, diesel engine torque and speed, and CVT speed ratio are used as control variables, then the SOC for the power battery is used as the state variable, and minimize the total cost of energy consumption of the whole machine is the goal. Finally, an energy management strategy based on dynamic programming is designed.

To address the issue of low energy utilization efficiency in agricultural hybrid tractors, we propose a global optimal energy management approach based on dynamic programming. Taking a hybrid tractor as the research object, based on the topology structure of the hybrid tractor equipped with CVT, the simulation model of the main components is built. Then, the control simulation model is established by adopting MATLAB. Finally, the simulation test results are obtained. Through the analysis of the simulation results of the hybrid tractor rotary tillage working conditions, it can be obtained that the control strategy based on dynamic programming can be adjusted according to specific working conditions, and the working torque of the diesel engine and electric motor and CVT speed ratio can be reasonably allocated to make the electric motor and diesel engine work in a high-efficiency area. However, the control strategy based on power following can only distribute the torque of the motor and diesel engine according to the established rules, and cannot make corresponding torque distribution according to the working conditions. In the end, the total energy consumption cost of the power-following energy management strategy and the proposed strategy are ¥ 24.09 and ¥ 20.02, respectively. Compared with the power-following energy management strategy, the total energy consumption cost of the proposed strategy is decreased by 16.89%.

A dynamic programming control strategy is designed for the agricultural hybrid tractor with CVT, and the research shows that the strategy can significantly improve the energy consumption economy of tractor. It provides new ideas for the design and research of new energy agricultural tractors. At present, we have only studied the working conditions of tractor rotary tillage. In the future, we will study the plowing and transportation conditions of the tractor, and design a control strategy framework which may contain neuro-fuzzy that can meet multiple working conditions. Additionally, readers can further improve the energy consumption economy of the whole machine by optimizing the CVT speed ratio.

Author Contributions: Conceptualization, J.Z. and G.F.; methodology, L.X.; software, G.F.; validation, J.Z., G.F. and L.X.; formal analysis, M.L. and C.S.; investigation, M.L.; resources, L.X.; data curation, X.Y. and C.S.; writing—original draft preparation, G.F.; writing—review and editing, J.Z.; visualization, X.Y.; supervision, X.Y.; funding acquisition, L.X. All authors have read and agreed to the published version of the manuscript.

Funding: This research was funded by The National Key R&D Program of China (Grant Nos. 2022YFD2001203, 2022YFD2001201B), Key Scientific and Technological Research Projects in Henan Province (Grant No. 222102240088), Key Agricultural Core Technology Research Project (Grant No. NK202216010103), Open Project of State Key Laboratory of Intelligent Agricultural Power Equipment (Grant No. SKT2023001), and Open Project of State Key Laboratory of Tractor Power System (Grant No. SKT2022001).

Data Availability Statement: Not applicable.

Conflicts of Interest: The authors declare no conflict of interest.

References

1. Fu, M.L. Characteristics of agricultural tractors emissions under real-world operating cycle. *Trans. Chin. Soc. Agric. Mach.* **2013**, *29*, 42–48.
2. Yu, W.; Shen, X.; Wu, B.; Kong, L.; Xuan, K.; Zhao, C.; Cao, X.; Hao, X.; Li, X.; Zhang, H.; et al. Real-world emission characteristics of carbonyl compounds from agricultural machines based on a portable emission measurement system. *J. Environ. Sci.* **2023**, *124*, 846–859. [CrossRef]
3. Moinfar, A.; Shahgholi, G.; Gilandeh, Y.A.; Gundoshmian, T.M. The effect of the tractor driving system on its performance and fuel consumption. *Energy* **2020**, *202*, 117803. [CrossRef]
4. Liu, M.; Lei, S.; Zhao, J.; Meng, Z.; Zhao, C.; Xu, L. Review of development process and research status of electric tractors. *Trans. Chin. Soc. Agric. Mach.* **2022**, *53*, 348–364.
5. Ghobadpour, A.; Mousazadeh, H.; Kelouwani, S.; Zioui, N.; Kandidayeni, M.; Boulon, L. An intelligent energy management strategy for an off-road plug-in hybrid electric tractor based on farm operation recognition. *IET Electr. Syst. Transp.* **2021**, *11*, 333–347. [CrossRef]
6. Steven, B.; David, H. An overview of the hybrid and electric systems R&D at the U.S.–DOE (FY 2015–2016). *World Electr. Veh. J.* **2016**, *8*, 461.
7. Mocera, F.; Somà, A. Analysis of a parallel hybrid electric tractor for agricultural applications. *Energies* **2020**, *13*, 3055. [CrossRef]
8. Si, Y.; Qian, L.J.; Qiu, L.H.; Li, H. Energy management of a 4WD hybrid electric vehicle based on ECMS. *China Mech. Eng.* **2017**, *28*, 1112–1117.
9. Li, T.; Cui, W.; Cui, N.X. Soft actor-critic algorithm-based energy management strategy for plug-in hybrid electric vehicle. *World Electr. Veh. J.* **2022**, *13*, 193. [CrossRef]
10. Rohith, J.; Mahesha, G.T. Review of energy management strategies in plug-in hybrid-electric vehicles. In *Flexible Electronics for Electric Vehicles*; Springer Nature Singapore: Singapore, 2022; pp. 83–89.
11. He, H.W.; Meng, X.F. A review on energy management technology of hybrid electric vehicles. *Trans. Beijing Inst. Technol.* **2022**, *42*, 773–783.
12. Zhang, F.Q.; Hu, X.S.; Xu, K.H.; Tang, X.L.; Cui, Y.H. Status and prospects for model predictive energy management in hybrid electric vehicles. *J. Mech. Eng.* **2019**, *55*, 86–108. [CrossRef]

13. Ma, D.B.; Zhu, F.T.; Gu, L.Q. Simulation on energy management strategy for four-wheel drive hybrid electric vehicle. *Drive Syst. Tech.* **2013**, *27*, 13–18.
14. Ming, L.; Ying, Y.; Liang, L.; Yao, L.; Zhou, W. Energy management strategy of a plug-in parallel hybrid electric vehicle using fuzzy control. *Energy Procedia* **2017**, *105*, 2660–2665. [CrossRef]
15. Luo, G.P.; Luo, Y.G.; Li, K.Q. Control strategy for plug-in hybrid electric bus based on optimal electric energy use. *Automot. Eng.* **2012**, *34*, 475–478.
16. Gao, F.Y.; Zhang, H.R.; Wang, W.X.; Li, M.M. Energy management strategy of modern tram based on the combination of rule control and driving conditions. *J. Mech. Eng.* **2022**, *58*, 1–11.
17. Geng, W.R.; Lou, D.M.; Zhang, T. Multi-objective Energy Management Strategy for Hybrid Electric Vehicle Based on Particle Swarm Optimization. *J. Tongji Univ. (Nat. Sci.)* **2020**, *48*, 1030–1039.
18. Zhang, J.J.; Feng, G.H.; Xu, L.Y.; Yan, X.; Wang, W.; Liu, M. Energy-saving control of hybrid tractors based on instantaneous optimization. *World Electr. Veh. J.* **2023**, *14*, 27. [CrossRef]
19. Chen, Z.; Mi, C.C.; Xia, B.; You, C. Energy management of power-split plug-in hybrid electric vehicles based on simulated annealing and pontryagin's minimum principle. *J. Power Sources* **2014**, *272*, 160–168. [CrossRef]
20. Li, W.; Wang, C.; Pei, H.; Xu, C.; Lin, G.; Deng, J.; Jiang, D.; Huang, Y. An improved energy management strategy of diesel-electric hybrid propulsion system based on FNN-DP strategy. *Electronics.* **2023**, *12*, 486. [CrossRef]
21. Adeleke, O.P.; Li, Y.; Chen, Q.; Zhou, W.; Xu, X.; Cui, X. Torque distribution based on dynamic programming algorithm for four in-wheel motor drive electric vehicle considering energy efficiency optimization. *World Electr. Veh. J.* **2022**, *13*, 181. [CrossRef]
22. Du, C.; Huang, S.; Jiang, Y.; Wu, D.; Li, Y. Optimization of energy management strategy for fuel cell hybrid electric vehicles based on dynamic programming. *Energies* **2022**, *15*, 4325. [CrossRef]
23. Liu, C.N. *Research on Management Strategy of Phev Based on Driving Condition Recognition and Multi-Objective Optimization*; Shang Dong University: Jinan, China, 2022.
24. Miao, Q.; Kong, F.M.; Shun, Q.; Bai, S.Z. Adaptive equivalent consumption minimization strategy based on dynamic programming. *J. South China Univ. Technol. (Nat. Sci. Ed.)* **2016**, *44*, 81–88.
25. Xia, C.G.; Zhang, Y.; Shun, Y. Research on tractor speed control of hydraulic mechanical continuously variable transmission. *Agric. Equip. Veh. Eng.* **2022**, *60*, 22–26.
26. Huazhong Agricultural University. *Tractor Cars: Tractor Car Theory*, 2nd ed.; China Agricultural Press: Beijing, China, 2001; pp. 89–97.
27. Zhou, Z.L.; Ni, Q.; Xu, L.Y. Parallel hybrid tractor drive train parameter design and performance analysis. *J. Henan Univ. Sci. Technol. (Nat. Sci.)* **2016**, *37*, 9–15.
28. Gao, S.H.; Chen, X.B.; Xue, J.; Gao, Q. Factor analysis of influencing factors of tractor operating resistance. *J. Agric. Mech. Res.* **2020**, *42*, 17–22.
29. Bin, X.; Hao, L.; Zheng-He, S.; En-Rong, M. Powertrain system design of medium-sized hybrid electric tractor. *Inf. Technol. J.* **2013**, *12*, 7228–7233. [CrossRef]
30. Liu, M.N.; Xu, L.Y.; Zhou, Z.L.; Liu, W.G. Establishment of extended range electric tractor and its rotary cultivator's simulative platforms. *China Mech. Eng.* **2016**, *27*, 413–419.
31. Likhanov, V.A.; Anfilatov, A.A. Mathematical model of the formation and burning of smoke. *IOP Conf. Ser. Mater. Sci. Eng.* **2020**, *862*, 032–050. [CrossRef]
32. Wang, S.Q. *Research on Online Energy Management Control Strategy of Phev Based on Global Optimization of Equivalent Factors*; Chong Qing University: Chongqing, China, 2021.
33. Li, S.G.; Sharkh, S.M.; Walsh, F.C.; Zhang, C.N. Energy and battery management of a plug-in series hybrid electric vehicle using fuzzy logic. *IEEE Trans. Veh. Technol.* **2011**, *60*, 3571–3585. [CrossRef]

World Electric Vehicle Journal

Article

Vehicle State Estimation Based on Sage–Husa Adaptive Unscented Kalman Filtering

Yong Chen [1,2,3,*], Hao Yan [1] and Yuecheng Li [1,2,3]

1 College of Mechanical and Electronic Engineering, Beijing Information Science and Technology University, Beijing 100192, China; 2021020066@bistu.edu.cn (H.Y.); liyc17@foxmail.com (Y.L.)
2 Collaborative Innovation Center of Electric Vehicles in Beijing, Beijing 100192, China
3 Beijing Laboratory for New Energy Vehicles, Beijing 100192, China
* Correspondence: chenyong_jz@126.com

Abstract: To combat the impacts of uncertain noise on the estimation of vehicle state parameters and the high cost of sensors, a state-observer design with an adaptive unscented Kalman filter (AUKF) is developed. The design equation of the state observer is derived by establishing the vehicle's three degrees-of-freedom (DOF) model. On this basis, the Sage–Husa algorithm and unscented Kalman filter (UKF) are combined to form the AUKF algorithm to adaptively update the statistical feature estimation of measurement noise. Finally, a co-simulation using Carsim and Matlab/Simulink confirms the algorithm is effective and reasonable. The simulation results demonstrate that the proposed algorithm, compared with the UKF algorithm, increases estimation accuracy by 19.13%, 32.8%, and 39.46% in yaw rate, side-slip angle, and longitudinal velocity, respectively. This is because the proposed algorithm adaptively adjusts the measurement noise covariance matrix, which can estimate the state parameters of the vehicle more accurately.

Keywords: vehicle state parameter estimation; Sage–Husa filtering; unscented Kalman filtering; adaptive control

Citation: Chen, Y.; Yan, H.; Li, Y. Vehicle State Estimation Based on Sage–Husa Adaptive Unscented Kalman Filtering. *World Electr. Veh. J.* **2023**, *14*, 167. https://doi.org/10.3390/wevj14070167

Academic Editor: Joeri Van Mierlo

Received: 18 May 2023
Revised: 11 June 2023
Accepted: 21 June 2023
Published: 25 June 2023

1. Introduction

Faced with the shortage of oil resources and increasing environmental pollution, electric vehicles (EVs) are considered the key to solving these problems. With the growing demand for vehicle handling characteristics and active safety in various application scenarios associated with the next-generation EVs control, more and more advanced vehicle stability control and driver assistance systems have been developed, such as active front-wheel steering (AFS), direct yaw moment control (DYC), and four-wheel steering (4WS), etc. [1]. At the core of these advanced control methods is the accurate vehicle dynamic state information, such as yaw rate, longitudinal vehicle speed, lateral and longitudinal acceleration, and side-slip angle, etc. The control effect is primarily influenced by the accuracy of the vehicle state information and the real-time availability of obtaining this information.

However, it is too expensive to equip mass-production vehicles with high-precision inertial navigation sensors (INS) and global navigation satellite systems (GNSS), and low-cost onboard sensors may fail to measure dynamic state information accurately [2]. Moreover, some of the critical state parameters are hard to measure directly, such as roll of the vehicle body. Consequently, the estimation of vehicle dynamic state is widely explored and adopted, which is the process of using sensor data and estimation schemes to estimate the dynamic state of vehicles [3].

Currently, state parameter estimation methods are broadly classified into three major categories, kinematics model-based estimation methods, data-driven-based estimation methods, and dynamics model-based estimation methods [4].

The kinematic methods, concerning the motion of vehicles, estimate the vehicle state information by directly integrating the lateral acceleration and yaw rate signals of vehicle sensors [5,6]. However, this method also integrates the noisy signals of the system, and the accumulated error increases with time, which can lead to significant deviations in the estimation results. The estimation results are also affected by unknown factors, such as sensor drift errors or calibration errors caused by temperature. Therefore, the method has some limitations.

With artificial intelligence's rapid development and application, data-driven estimation methods based on data are gradually becoming familiar. For example, neural networks estimate state parameter based on deep learning and supervised learning methods. This method does not require an accurate model. It only involves using neural networks to train the data from real-vehicle tests and verify the training effect through real-vehicle tests [7–9]. Novi et al. [9] obtained data from the Vi-Grade model, trained the artificial neural network (ANN), and combined ANN with UKF to estimate the side-slip angle. However, this training mode is mainly offline and more dependent on the existing data. In addition, changes in vehicle parameters can lead to changes in the fitted relationship derived from and based on the original data, bringing some bias to the estimation results.

For dynamic model-based methods, vehicle state information is estimated by leveraging different vehicle and tire dynamic models. Therefore, the accuracy of the model can have a significant impact on the estimation results. The vehicle dynamic models currently used are mainly 2-DOF models, 3-DOF models, and 7-DOF models [10]. Sun Wen et al. [11] used a 2-DOF model, which does not consider the effect of longitudinal motion on the state parameters. Jeong et al. [12] used the same model and established a linear tire model. However, a linear model will affect the estimation performance of the algorithm when the vehicle is running in a nonlinear region. The 7-DOF model, on the other hand, has a complex structure, which undoubtedly increases the computational effort of the algorithm. Therefore, the 3-DOF model, which considers lateral, yaw, and longitudinal motions, becomes the preferred choice. Based on this, the estimation methods can be further divided into filter-based vehicle state estimation, such as the Kalman filter (KF), extended Kalman filter (EKF), unscented Kalman filter (UKF), particle filter (PF), unscented particle filter (UPF), etc., and observer-based vehicle state estimation, such as sliding mode observer [13], nonlinear observer, etc.

KF is a linear optimal filter that obtains a posterior optimal estimate based on the system's state equation, measurement equation, and the statistical properties of noise between them, using observations and prior forecast. References [14,15] used KF to estimate vehicle parameters. However, systems are often nonlinear, and the linear model-based KF estimator cannot be applied in all cases. So, the KF needs to be improved. Currently, EKF and UKF are relatively popular. The EKF expands the nonlinear equations in a Taylor series around priori state estimates, and then uses the KF to handle the linear problem [16,17]. For the EKF algorithm, the accuracy of the vehicle model parameters significantly impacts the estimation accuracy. Compared to the EKF, UKF uses sigma points instead of the Taylor series expansion. In most cases, sigma points approximate nonlinear transformation better than linearization. It can eliminate the process of calculating complex Jacobi matrices, and its accuracy is higher in dealing with more complex nonlinear problems [18,19]. Huang Yuhao et al. [20] compared UKF with EKF, proving that UKF has a higher accuracy. In addition, many scholars have also studied state estimation based on PF [21,22]. Chu Wenbo et al. [23] proposed an information-fusion observer based on the UPF algorithm. Although the UPF algorithm can handle nonlinear and non-Gaussian distributions, it is computationally intensive, affecting its efficiency and real-time requirement.

Regarding the UKF algorithm, matrices Q and R are both noise covariance matrices, and their values usually require experience or trial and error to determine. However, the noise constantly changes when a vehicle is driving, and the original UKF cannot achieve sound filtering effects under different working conditions. Therefore, some scholars have proposed a filtering method for adaptive adjustment of the noise covariance matrix, which enables better adaptability and robustness for the filtering algorithm. Wang Zhenpo et al. [24] combined UKF with fuzzy control to realize the process's adaptive adjustment of measuring the noise covariance matrix. Fan Tiane et al. [25] used shallow long short-term memory networks (LSTM) to optimize UKF. The estimation error of battery state of charge (SOC) and state of energy (SOE) is reduced to 0.43% and 0.46%, respectively. Li Gang et al. [26] introduced a Sage–Husa adaptive EKF algorithm and enhanced the adaptive rules built upon it, which increased the estimation accuracy of the center of mass's side-slip angle. Zhou Bing et al. [27] suggested a double adaptive unscented Kalman filter algorithm based on the fuzzy control adaptive adjustment of slip rate to address the issue of starting value sensitivity in semi-trailer state estimation. Xu Daxing et al. [28] proposed an improved algorithm based on Sage–Husa for process noise and provided an accurate estimation method for process noise statistics. Luo Zeyuan et al. [29] researched the transient interference problem of the Sage–Husa algorithm and proposed the divergence calculation technique. The simulation results show that the improved algorithm has strong robustness.

When the process noise variance is known or small, Yang Rui et al. [30] used the Sage–Husa algorithm to estimate the noise variance. However, in the Sage–Husa algorithm, compensating for both Q and R at the same time can easily lead to a divergence in the filtering results, and the process noise covariance matrix Q has a certain degree of robustness after compensation [31–33]. In comparison, the size of the measurement noise covariance matrix R has a more significant impact on the filtering effect.

Through research on existing achievements, this paper proposes an adaptive unscented Kalman filter method based on the Sage–Husa algorithm to obtain vehicle state parameters under uncertain noise interference accurately. This algorithm takes the front wheel angle and vehicle lateral acceleration as input signals, ignoring the process noise covariance matrix Q update, and adaptively adjusts the statistical feature estimation of the system measurement noise. Moreover, we improve the estimator to ensure that the covariance of the measurement noise is always positive to prevent the filtering results from diverging. Finally, we verified the rationality of the algorithm through joint simulation with Carsim and Matlab/Simulink, and typical operating condition experiments were conducted.

The main contributions of this paper are as follows:

(1) The Sage–Husa algorithm is improved to avoid non-positive definiteness of the covariance matrix and to ensure its positivity.
(2) The improved Sage–Husa algorithms are adopted to dynamically update the mean and covariance matrices of the measurement noise, which effectively improves the filtering accuracy and prevents its divergence.
(3) Sage–Husa algorithm is integrated with the UKF algorithm to form the AUKF algorithm for dynamic vehicle state estimation. The simulation results demonstrate that AUKF increases estimation accuracy by 19.13%, 32.8%, and 39.46% in yaw rate, side-slip angle, and longitudinal velocity, respectively, proving the algorithm's validity in providing accurate vehicle state information for active vehicle safety control.

The structure of this paper is as follows.

In Section 2, we establish the vehicle dynamics model and the tire model. Section 3 describes the principle and implementation of UKF and AUKF. Section 4 presents the comparative simulation results of UKF and AUKF. Section 5 provides a summary of this work.

2. Vehicle State Parameter Estimation Model

2.1. 3-DOF Vehicle Dynamics Model

A non-linear 3-DOF vehicle dynamic model with lateral, longitudinal, and yaw is built, based on a linear 2-DOF model [34] as shown in Figure 1, to simulate the vehicle motion state more accurately and reliably under various road circumstances. This model treats the car as if it were a rigid body, ignoring the pitch, roll, and vertical motions, as well as air resistance and tire-rolling resistance. Meanwhile, this model assumes that the steering angles of the two front wheels are equal.

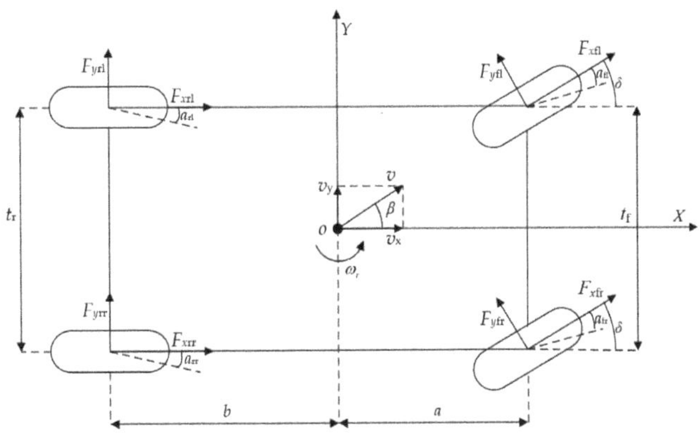

Figure 1. 3-DOF vehicle dynamics model.

The 3-DOF's vehicle dynamics model equations are as follows
The longitudinal dynamic equation is

$$ma_x = (F_{xfl} + F_{xfr})\cos\delta - (F_{yfl} + F_{yfr})\sin\delta + F_{xrl} + F_{xrr} \tag{1}$$

The lateral dynamic equation is

$$ma_y = (F_{xfl} + F_{xfr})\sin\delta - (F_{yfl} + F_{yfr})\cos\delta + F_{yrl} + F_{yrr} \tag{2}$$

The yaw dynamics equation is

$$I_z\dot{\omega}_r = \frac{t_f}{2}[(F_{xfr} - F_{xfl})\cos\delta + (F_{yfl} - F_{yfr})\sin\delta] + \frac{t_r}{2}(F_{xrr} - F_{xrl}) + a[(F_{xfl} + F_{xfr})\sin\delta + (F_{yfl} + F_{yfr})\cos\delta] - b(F_{yfl} + F_{yfr}) \tag{3}$$

where m is the vehicle mass; F_x and F_y are the longitudinal and lateral forces of the wheels, respectively; fl, fr, rl and rr represent the left front wheel, right front wheel, left rear wheel and right rear wheel, respectively; t_f and t_r are wheelbases of the front and rear wheels of automobiles; $\dot{\omega}_r$ is the yaw acceleration of the vehicle; I_z is the moment of inertia of the vehicle around the Z axis; and a and b are the distance between the center of gravity of the vehicle and the front and rear axles, respectively.

From the vehicle's dynamics model equations, the state equation and measurement equation of the 3-DOF vehicle can be derived as follows

$$\begin{cases} \dot{\omega}_r = \frac{(a^2k_1+b^2k_2)}{I_z v_x}\omega_r + \frac{ak_1-bk_2}{I_z}\beta - \frac{ak_1}{I_z}\delta \\ \dot{\beta} = (\frac{ak_1-bk_2}{mv_x^2} - 1)\omega_r + \frac{k_1+k_2}{mv_x}\beta - \frac{k_1}{mv_x}\delta \\ \dot{v}_x = \omega_r\beta v_x + a_x \end{cases} \tag{4}$$

$$a_y = \frac{ak_1 - bk_2}{mv_x}\omega_r + \frac{k_1 + k_2}{m}\beta - \frac{k_1}{m}\delta \tag{5}$$

where k_1 and k_2 are the cornering stiffness of the front and rear axle, respectively; v_x is the vehicle longitudinal speed; δ is the front wheel steering angle; β is side-slip angle; ω_r is yaw rate; and a_x and a_y are the longitudinal and lateral acceleration of the vehicle, respectively.

The mechanism for maintaining vehicle stability relies heavily on sensors. This paper estimates the side-slip angle, yaw rate, and longitudinal vehicle speed from the vehicle's lower-cost longitudinal acceleration and steering-angle sensor outputs. For nonlinear systems, the standard state space formulae are

$$\begin{cases} X(k) = f(X(k-1), u(k)) + W(k) \\ Z(k) = h(X(k), u(k)) + V(k) \end{cases} \tag{6}$$

where f and h are nonlinear state-equation functions and observation-equation functions, respectively; $X(k)$ is the state vector that cannot be observed directly; $u(k)$ is the control-input vector; $W(k)$ is the system Gauss white noise and its covariance matrix is Q; and $V(k)$ is the measurement of Gauss white noise and its covariance matrix is R.

The two-dimensional system-input vector is defined as

$$u = [\delta \ a_x]^{\mathrm{T}} \tag{7}$$

The three-dimensional system-state vector is defined as

$$X = [\omega_r \ \beta \ v_x]^{\mathrm{T}} \tag{8}$$

The system-observation vector is defined as

$$Z = [a_y]^{\mathrm{T}} \tag{9}$$

2.2. Tire Model

Tires are one of the most essential components of automobiles. In the simulation analysis, the precision of the selected tire model should match the accuracy of the built 3-DOF vehicle model, and the tires have the structural complexity and nonlinearity of mechanical properties. Therefore, it is imperative to select the appropriate tire model.

The most often used formula in vehicle dynamics studies is H.B. Pacejka's "magic tire". It is a model of a tire constructed by a unique sine function. One set of straightforward formulas with excellent simulation accuracy thoroughly explain the mechanical properties of tires under various operating circumstances [35]. It can be provided as follows

$$y = D\sin\{C\arctan[Bx - E(Bx - \arctan(Bx))]\} \tag{10}$$

$$Y(X) = y(x) + \Delta S_{\mathrm{v}} \tag{11}$$

$$x = X + \Delta S_{\mathrm{h}} \tag{12}$$

where Y denotes the lateral or longitudinal force; x represents the side-slip angle β or slip rate s; y is the tire-roll angle; D is the peak value, representing the maximum of the curve; C is the shape factor of the curve, i.e., whether the curve symbolizes longitudinal force, lateral force or correction moment; B is the stiffness factor; E is the curvature factor and represents the shape near the maximum value of the curve. S_{v} represents the offset in the vertical direction of the curve, and S_{h} represents the offset in the horizontal direction of the curve.

Then, the longitudinal and lateral forces of the tire can be expressed as

$$F_{xij} = D_{ij}\sin\Big\{C_{ij}\tan^{-1}[B_{ij}(1 - E_{ij})s_{ij} + E_{ij}\tan^{-1}(B_{ij}s_{ij})]\Big\} \tag{13}$$

$$F_{yij} = D_{ij}\sin\Big\{C_{ij}\tan^{-1}[B_{ij}(1 - E_{ij})a_{ij} + E_{ij}\tan^{-1}(B_{ij}a_{ij})]\Big\} \tag{14}$$

3. Vehicle State Parameter Estimation Based on AUKF

3.1. UKF Algorithm

The main idea of UKF is to transform a nonlinear system into a linear system for processing by unscented transformation (UT). The UKF algorithm is expressed as follows [36]:

(1) Obtain a set of sampling points (sigma points) and calculate the corresponding weights of these sampling points It is assumed that the $\overline{X}$ and variance P of the n-dimensional random variable state vector X are known. Then, by obtaining $2n + 1$ sigma points

X and the appropriate weights via the subsequent unscented transformation, the statistical properties of $f(x)$ may be computed.

$$\begin{cases} X^{(0)} = \overline{X},\ i = 0 \\ X_{k-1}{}^{(i)} = \overline{X} + (\sqrt{(n+\lambda)P})_i,\ i = 1 \sim n \\ X_{k-q}{}^{(i)} = \overline{X} - (\sqrt{(n+\lambda)P})_i,\ i = n+1 \sim 2n \end{cases} \tag{15}$$

These sampling sites' related weights are determined as

$$\begin{cases} \omega_m{}^{(0)} = \omega_c{}^{(0)} = \frac{\lambda}{n+\lambda},\ i = 0 \\ \omega_m{}^{(i)} = \omega_c{}^{(i)} = \frac{1}{2(n+\lambda)},\ i = 1 \sim 2n \end{cases} \tag{16}$$

where λ is used to reduce the overall prediction error, which can be chosen based on experience but should ensure that the matrix $(n + \lambda)\ P$ is a positive semi-definite. The subscripts m and c stand for the mean and covariance, respectively.

(2) According to Equations (15) and (16), a set of sampling points and their corresponding weights are calculated

$$X^{(i)}(k|k) = [\hat{X}(k|k),\ \hat{X}(k|k) + \sqrt{(n+\lambda)P(k|k)},\ \hat{X}(k|k) - \sqrt{(n+\lambda)P(k|k)}] \tag{17}$$

(3) One-step prediction of the set of $2n + 1$ sigma points using the state equation

$$X^{(i)}(k+1|k) = f[X^{(i)}(k|k),\ k] \tag{18}$$

(4) Calculation of one-step prediction and covariance matrix of the system state variables

$$\widehat{X}(k+1|k) = \sum_{i=0}^{2n} \omega^{(i)} X^{(i)}(k+1|k) \tag{19}$$

$$P_{x_k} = \sum_{i=0}^{2n} \omega^{(i)} [X^{(i)}(k+1|k) - \widehat{X}(k+1|k)][X^{(i)}(k+1|k) - \widehat{X}(k+1|k)]^{\mathrm{T}} + Q \tag{20}$$

(5) The predicted observations are calculated by bringing sigma points into the observation equation

$$Z^{(i)}(k+1|k) = h\,[X^{(i)}(k+1|k),\ k+1] \tag{21}$$

(6) The mean, covariance and cross-covariance are calculated analogously to Equations (19) and (20)

$$\overline{Z}(k+1|k) = \sum_{i=0}^{2n} \omega^{(i)} Z^{(i)}(k+1|k) \tag{22}$$

$$P_{z_k} = \sum_{i=0}^{2n} \omega^{(i)} [Z^{(i)}(k+1|k) - \overline{Z}(k+1|k)][Z^{(i)}(k+1|k) - \overline{Z}(k+1|k)]^{\mathrm{T}} + R \tag{23}$$

$$P_{x_k z_k} = \sum_{i=0}^{2n} \omega^{(i)} [X^{(i)}(k+1|k) - \widehat{X}(k+1|k)][Z^{(i)}(k+1|k) - \overline{Z}(k+1|k)]^{\mathrm{T}} \tag{24}$$

(7) Kalman gain matrix is calculated

$$K(k+1) = P_{x_k z_k} P_{z_k}^{-1} \tag{25}$$

(8) System status and covariance matrix are updated

$$\overset{\frown}{X}(k+1|k+1) = \overset{\frown}{X}(k+1|k) + K(k+1)[Z(k+1) - \overline{Z}(k+1|k)] \tag{26}$$

$$P(k+1|k+1) = P(k+1|k) - K(k+1)P_{z_k}K^{\mathrm{T}}(k+1) \tag{27}$$

3.2. AUKF Algorithm

The Sage–Husa algorithm is a maximum a posteriori (MAP) estimation algorithm [37]. When the noise means and covariance matrix are unknown, the magnitudes of q, r, Q, and R are approximated online from the observations. In the estimate phase, the Sage–Husa method optimizes the filtering performance by raising the weighting coefficients of the fresh data by an asymptotic elimination factor, reducing the influence of time-old data on the present estimation results.

For the uncertainty of measurement noise in the actual process, this paper proposes an adaptive unscented Kalman filter algorithm based on the Sage–Husa theory and updates R_k online to reduce the algorithm's complexity without affecting the accuracy. Its flow chart is shown in Figure 2. The AUKF algorithm, in contrast to the conventional UKF algorithm, continuously corrects and estimates the model parameters, as well as the statistical characteristics of the noise based on its quantitative data, which can decrease the estimation error, suppress filtering divergence, and increase filtering accuracy.

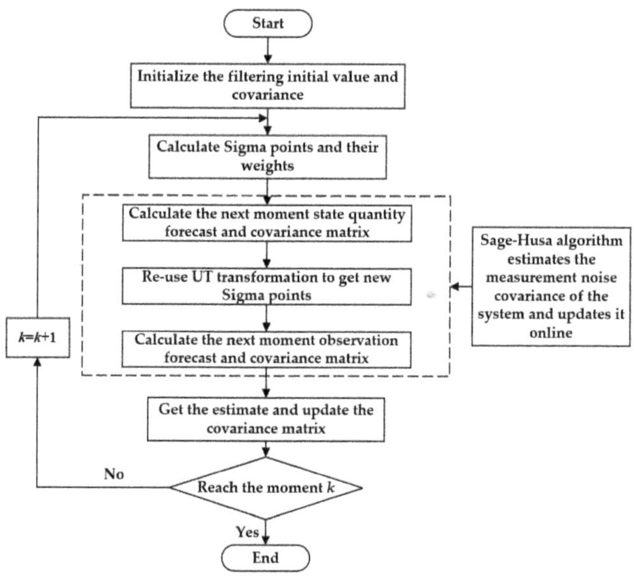

Figure 2. The block diagram of AUKF.

The steps are as follows

$$d_k = \frac{1-\gamma}{1-\gamma^k} \tag{28}$$

$$\varepsilon_k = Z_k - h[\overline{X}(k|k-1)] - r_k \tag{29}$$

$$\hat{r}_k = (1-d_k)\hat{r}_{k-1} + d_k[Z_k - \sum_{i=0}^{2n} \omega^{(i)} Z^{(i)}(k|k-1)] \tag{30}$$

$$\hat{R}_k = (1-d_k)\hat{R}_{k-1} + d_k[\varepsilon_k\varepsilon_k^{\mathrm{T}} - \sum_{i=0}^{2n}\omega^{(i)}[Z^{(i)}(k+1|k) - \overline{Z}(k+1|k)][Z^{(i)}(k+1|k) - \overline{Z}(k+1|k)]^{\mathrm{T}}] \tag{31}$$

where r_k represents the observed noise's mean, while R_k represents its covariance; ε_k is a new information sequence, representing the discrepancy between the actual and expected observations; γ is a gradual elimination factor, and the range of values is 0~1.

Sage–Husa adaptive filtering is suboptimal filtering, so the filtering results sometimes diverge. The primary cause is that filter divergence may result from subtraction in a solution of Equation (31) for R_k because it has a non-positive definite condition. To guarantee that R_k is positive, the following techniques are employed in this paper

$$\hat{R}_k = (1-d_k)\hat{R}_{k-1} + d_k(\varepsilon_k\varepsilon_k^{\mathrm{T}}),\ \hat{R}_k \le 0 \tag{32}$$

4. Simulation Results and Analyses

In order to verify the accuracy and feasibility of the proposed vehicle state observer based on the Sage–Husa adaptive UKF algorithm, co-joint simulation performs in Carsim and MATLAB/Simulink. Finally, the estimation results of AUKF and UKF are compared. All simulations below are carried out in Matlab R2020a and Carsim2019 running on a laptop computer with AMD R7-5800H CPU @3.2 GHz and 16 GB RAM.

The simulation environment in Carsim is set to a typical steering angle step-input condition, sinusoidal steering condition, and double-lane change condition. These driving conditions can cover various situations in daily driving, such as turning, overtaking, and continuous overtaking. In addition, when the noise covariance matrix is time-varying, the algorithm's performance is tested by altering the statistical properties of measurement noise under the double-lane change condition. The road adhesion coefficient is set to 0.85, the vehicle's initial speed is 40 km/h, the transmission ratio of the steering wheel to the front wheel is 20, and the sampling time is 0.001 s. In Carsim, the car is chosen as a C-class hatchback with the fundamental parameters in Table 1.

Table 1. Vehicle Parameter Settings.

Parameter	Value
m	1410 kg
a	1.015 m
b	1.895 m
k_1	−122,540 N·rad^{-1}
k_2	−100,500 N·rad^{-1}
I_z	1536.7 kg·m^2

Assigning the basic parameters of the algorithm, the initial value of the error covariance matrix P is set to eye (3), the initial value of the system process noise covariance matrix Q is set to eye (3) $\times$ 0.001, and the initial value of the measurement noise covariance matrix R is set to 100.

4.1. Steering Angle Step-Input Condition

We set the initial value of the system-state vector $X(t_0)$ to $[0, 0, 40/3.6]^{\mathrm{T}}$. The initial value of the system-input vector $u(t)$ is $[\delta, a_x]^{\mathrm{T}}$, and its input waveform is shown in Figure 3. The observed quantity Z is a_y, and the waveform after adding Gaussian white noise is shown in Figure 4.

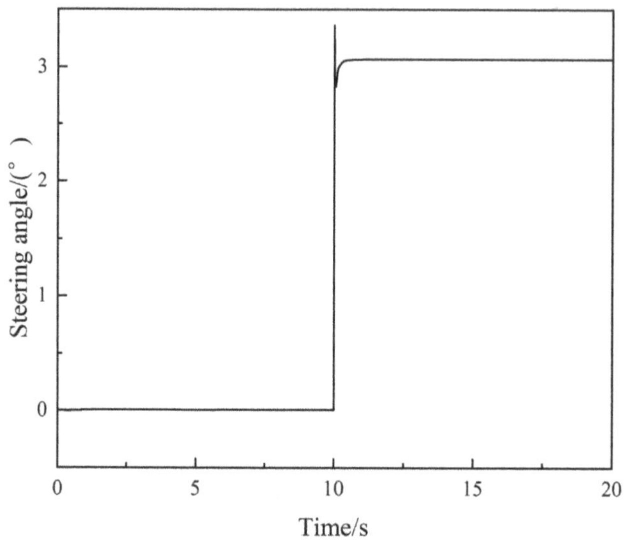

Figure 3. Steering angle variation under the step-input condition.

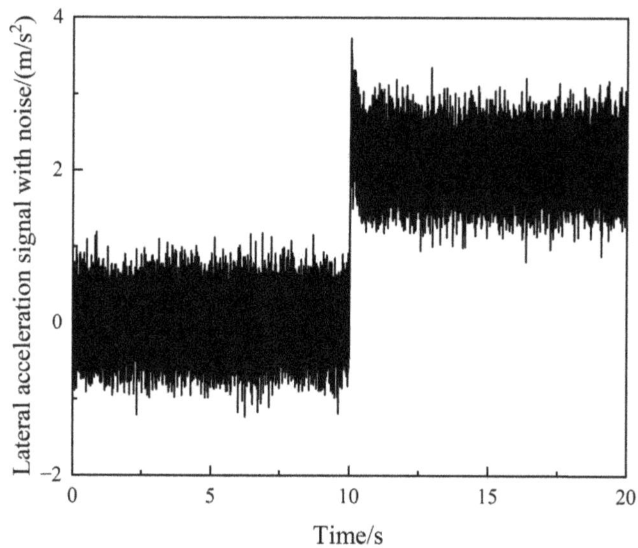

Figure 4. Observation signal with noise under the steering angle step-input condition.

Figures 5–7 depict the simulation results, where Carsim's output represents the ideal value, UKF and AUKF originating from the estimated values obtained by the UKF and AUKF algorithms, respectively. Figures 5 and 6 show how the standard UKF estimation are significantly impacted when the steering wheel is twisted violently at 10 s. They demonstrate that UKF cannot be precisely predicted when there is unknown noise interference. The AUKF algorithm's estimated value is more accurately calculated than the classic UKF algorithm and is roughly consistent with the ideal value, which can effectively lessen the influence of unidentified noise.

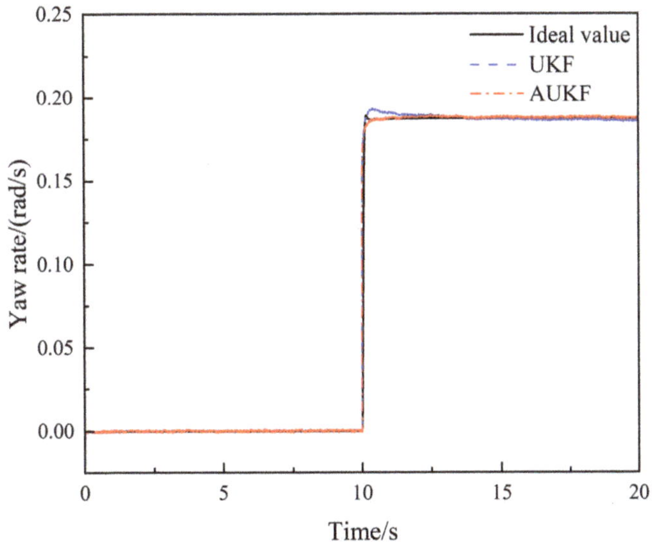

Figure 5. Comparison of yaw rate estimates under the steering angle step-input condition.

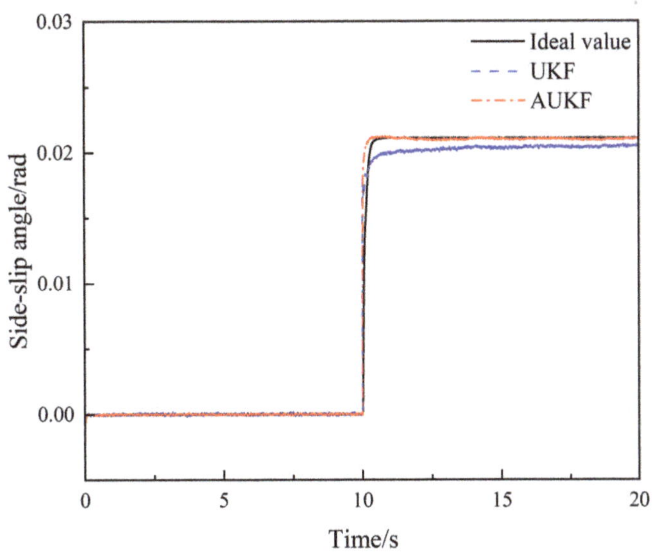

Figure 6. Comparison of side-slip angle estimates under the steering angle step-input condition.

Figure 7 shows that there is a certain deviation in the estimation of UKF and AUKF within 0–3 s when calculating the longitudinal vehicle speed. Then, within 3–10 s, both UKF and AUKF are approaching the actual value. At 10 s, due to step input of the steering wheel, the input value of the observer undergoes a significant change in an instant, and the estimated value of UKF suddenly increases and deviates from the actual value. At the same time, AUKF is less affected and can still accurately estimate the longitudinal speed of the vehicle within 10–20 s.

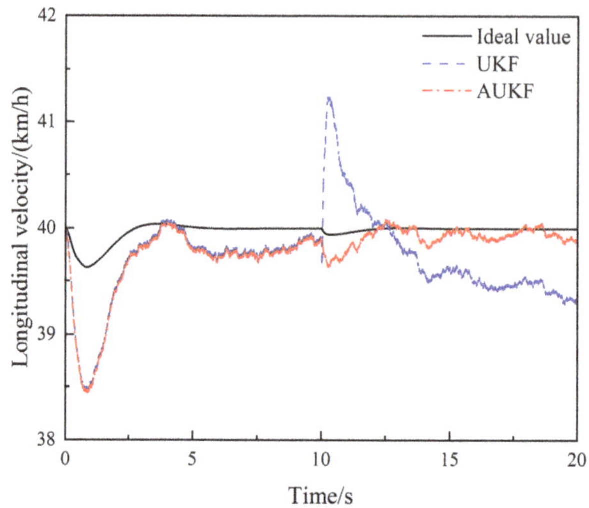

Figure 7. Comparison of longitudinal velocity estimates under the steering angle step-input condition.

The mean absolute error (MAE) and root mean square error (RMSE) are indicators used in this research to quantify further and examine the estimation values. The root means square error and means absolute error can be used to assess the precision and tracking abilities of the estimation results, respectively. The expressions are defined as

$$\text{MAE} = \frac{1}{n}\sum_{i=1}^{n} |y_i - \hat{y}_i| \tag{33}$$

$$\text{RMSE} = \sqrt{\frac{1}{n}\sum_{i=1}^{n} (y_i - \hat{y}_i)^2} \tag{34}$$

As shown in Tables 2 and 3, the estimation accuracy and stability of the AUKF algorithm are better than those of the UKF algorithm in the MAE indicators and the RMSE indicators. The accuracy of yaw rate estimation under the steering angle step-input condition has been improved by 3.2%, the accuracy of side-slip angle estimation has been improved by 9.5%, and the estimation of longitudinal velocity has been improved by 32.5%.

Table 2. MAE between estimated value and true value under the steering angle step-input condition.

Algorithm	Yaw Rate (rad/s)	Side-Slip Angle (rad)	Longitudinal Velocity (km/h)
UKF	0.0012	0.000462	0.3726
AUKF	0.0009	0.000140	0.1982

Table 3. RMSE between estimated value and true value under the steering angle step-input condition.

Algorithm	Yaw Rate (rad/s)	Side-Slip Angle (rad)	Longitudinal Velocity (km/h)
UKF	0.0063	0.000907	0.4644
AUKF	0.0061	0.000821	0.3133

4.2. Sinusoidal Steering Condition

We set the initial value of the system state vector $X(t_0)$ to $[0, 0, 40/3.6]^T$. The initial value of the system input vector $u(t)$ is $[\delta, a_x]^T$, and its input waveform is shown in Figure 8.

The observed quantity Z is a_y, and the waveform after adding Gaussian white noise is shown in Figure 9.

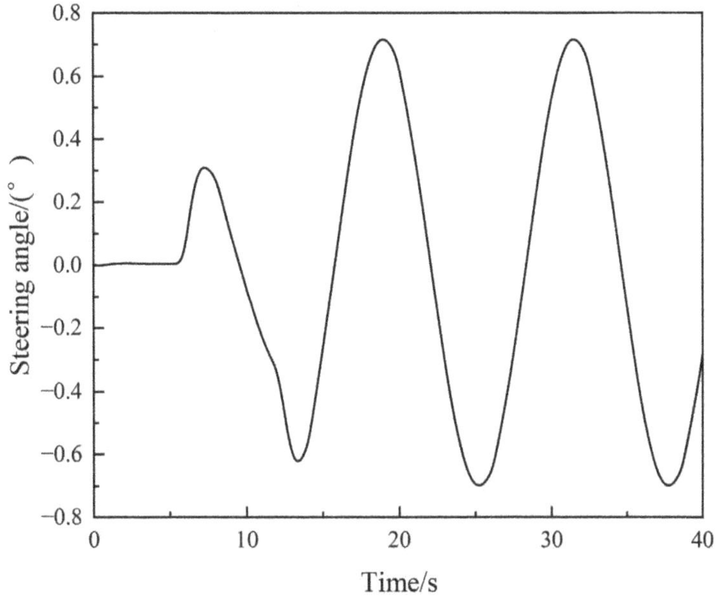

Figure 8. Steering-angle variation under the sinusoidal steering condition.

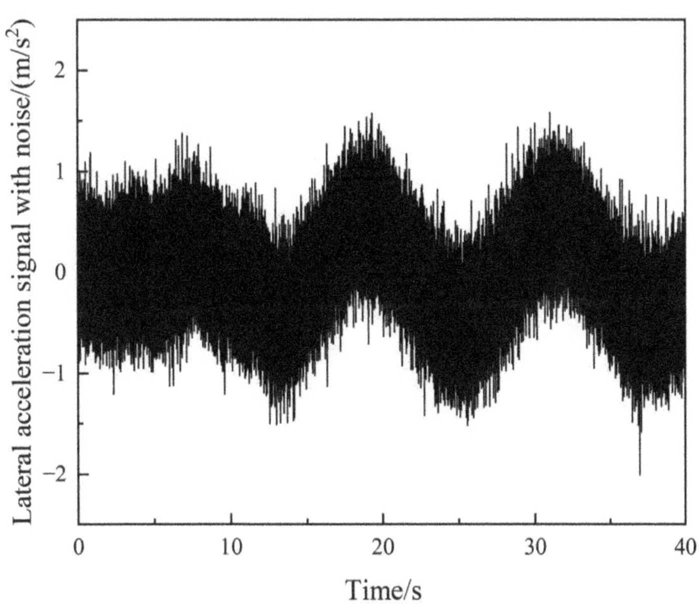

Figure 9. Observation Signal with Gaussian white noise.

The simulation results are shown in Figures 10–15. Figures 10 and 12 show that under the sinusoidal steering condition, the car changes lanes after 5 s. The original UKF deviates from the ideal value when the steering wheel angle experiences a quick shift, but the proposed AUKF algorithm can successfully track the ideal value. Figures 11 and 13 show that the error and diffusion of AUKF are smaller than those of UKF, indicating that AUKF can better reduce the impact of unknown noise. It can be seen from Figures 14 and 15 that the divergence of UKF is larger than AUKF because UKF cannot dynamically adjust the measurement noise covariance matrix. The error of UKF becomes larger and larger as time goes on, and the error of AUKF is relatively stable, which proves that the algorithm can effectively reduce the influence of unknown noise and curb the divergence of the filtering results.

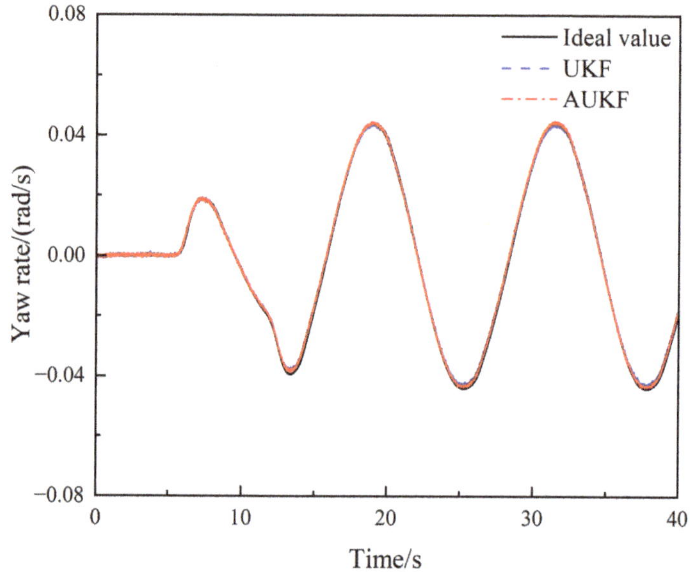

Figure 10. Comparison of yaw rate estimates under the sinusoidal steering condition.

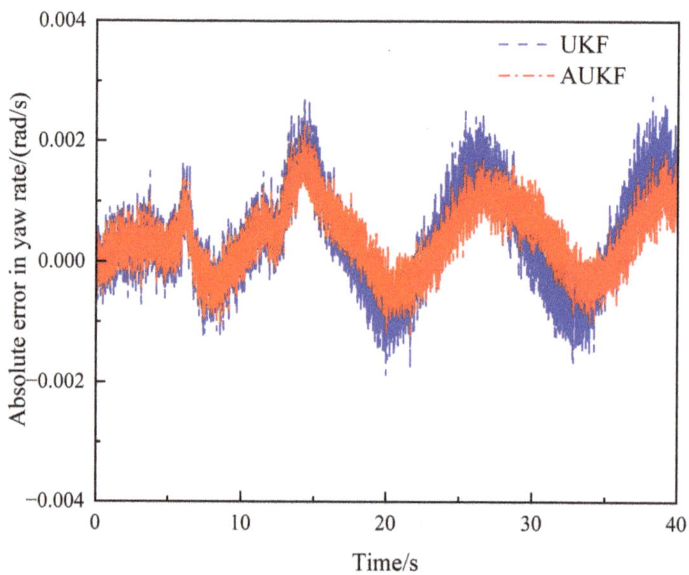

Figure 11. Absolute error of yaw rate estimates under the sinusoidal steering condition.

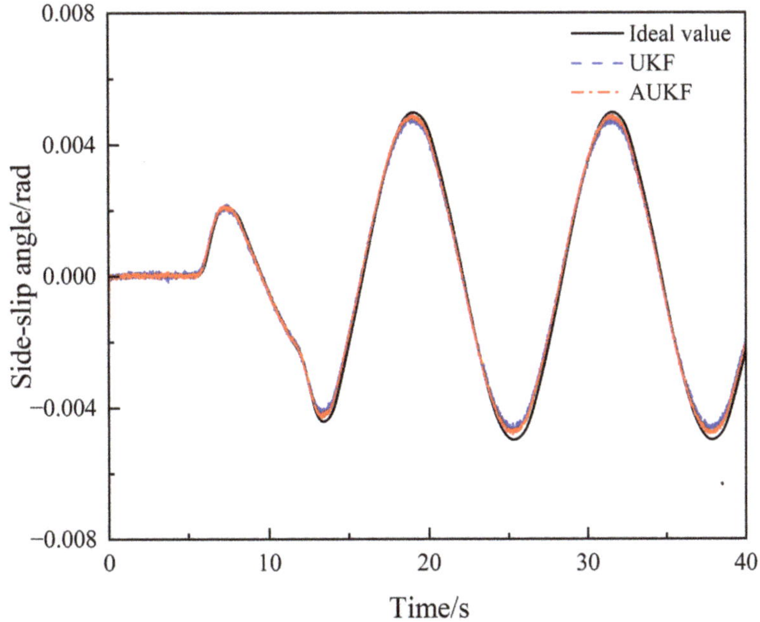

Figure 12. Comparison of side-slip angle estimates under the sinusoidal steering condition.

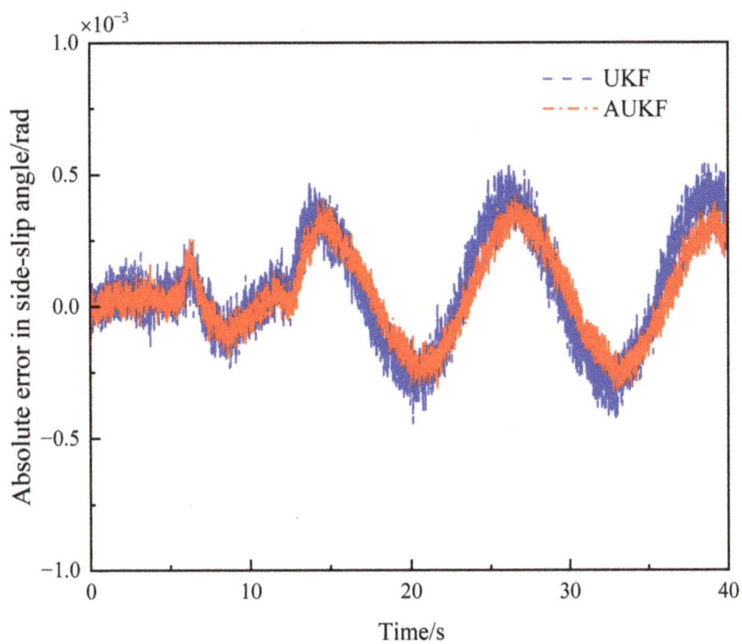

Figure 13. Absolute error of side-slip angle estimates under the sinusoidal steering condition.

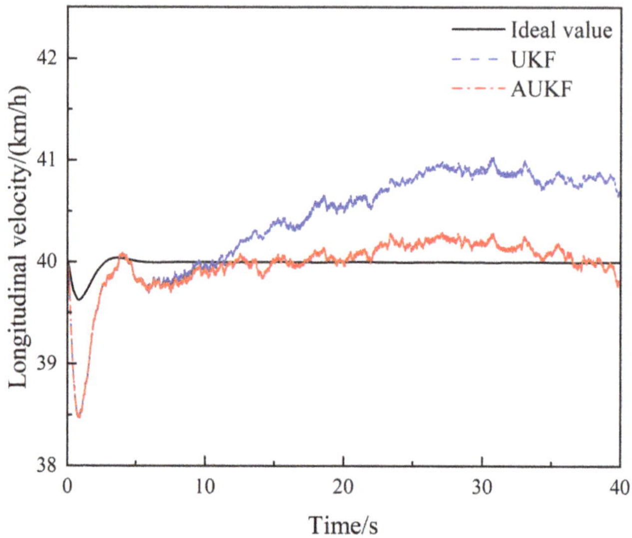

Figure 14. Comparison of longitudinal velocity estimates under the sinusoidal steering condition.

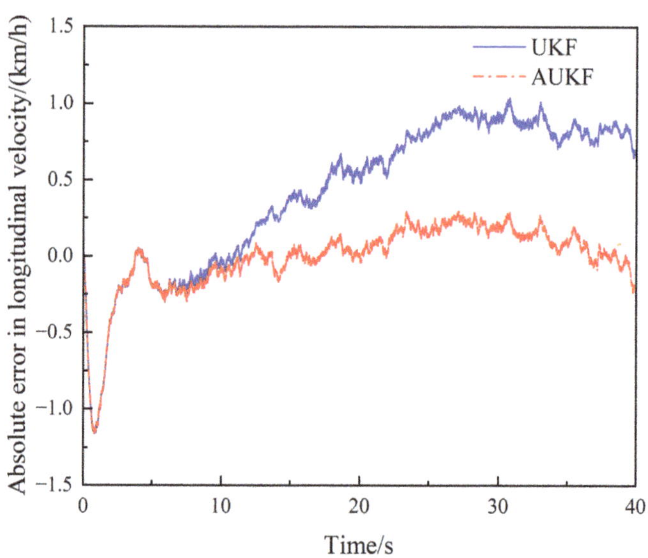

Figure 15. Absolute error of longitudinal velocity estimates under the sinusoidal steering condition.

As shown in Tables 4 and 5, the accuracy of yaw rate estimation has been improved by 24.8%, the accuracy of side-slip angle estimation has been improved by 19.6%, and the estimation of longitudinal velocity by AUKF has been improved by 63.4% under the sinusoidal steering condition.

Table 4. MAE between estimated value and true value under the sinusoidal steering condition.

Algorithm	Yaw Rate (rad/s)	Side-Slip Angle (rad)	Longitudinal Velocity (km/h)
UKF	0.000726	0.000176	0.557
AUKF	0.000551	0.000141	0.154

Table 5. RMSE between estimated value and true value under the sinusoidal steering condition.

Algorithm	Yaw Rate (rad/s)	Side-Slip Angle (rad)	Longitudinal Velocity (km/h)
UKF	0.000916	0.000219	0.6434
AUKF	0.000689	0.000176	0.2356

4.3. Double-Lane Change Condition

We set the initial value of the system state vector $X(t_0)$ to $[0, 0, 40/3.6]^T$. The initial value of the system input vector $u(t)$ is $[\delta, a_x]^T$, and its input waveform is shown in Figure 16. The observed quantity Z is a_y, and the waveform after adding time-varying Gaussian white noise is shown in Figure 17, in which the statistical characteristics of the measurement noise becomes 10 times that of the first 10 s at 10–20 s, i.e., $10R_k$.

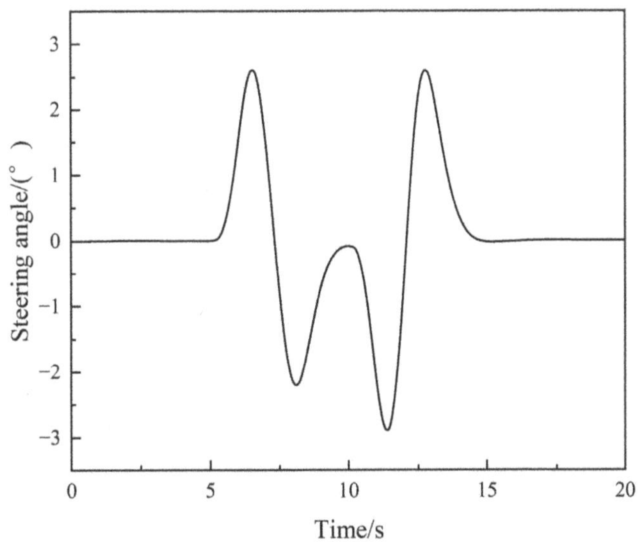

Figure 16. Steering-angle variation under the double-lane change condition.

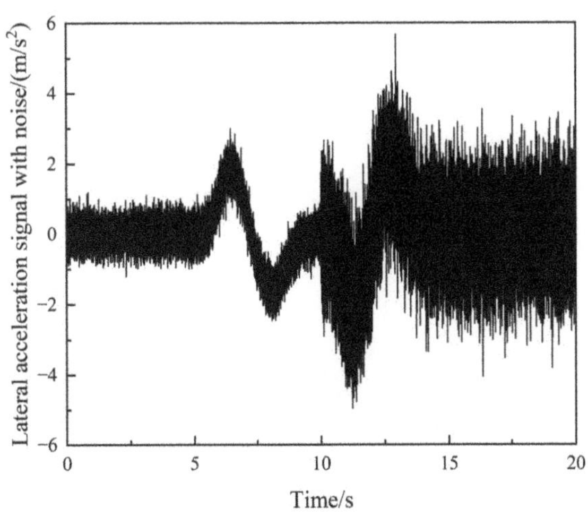

Figure 17. Observation Signal with time-varying Gaussian white noise.

The simulation results are shown in Figures 18–23. Figures 18 and 20 show that under the double-lane change condition, the car changes lanes between 5–15 s. The original UKF deviates from the ideal value when the steering wheel angle experiences a quick shift, but the proposed AUKF algorithm can successfully track the ideal value. Figures 19 and 21 show that the error of AUKF is smaller than that of UKF, indicating that AUKF can better reduce the impact of unknown noise. And after 10 s, when the statistical characteristics of the measurement noise covariance increase by ten times, the divergence of UKF significantly increases, proving that UKF cannot accurately estimate state parameters when the noise covariance is time-varying. The fluctuation of AUKF is relatively small, and the error is within an acceptable range.

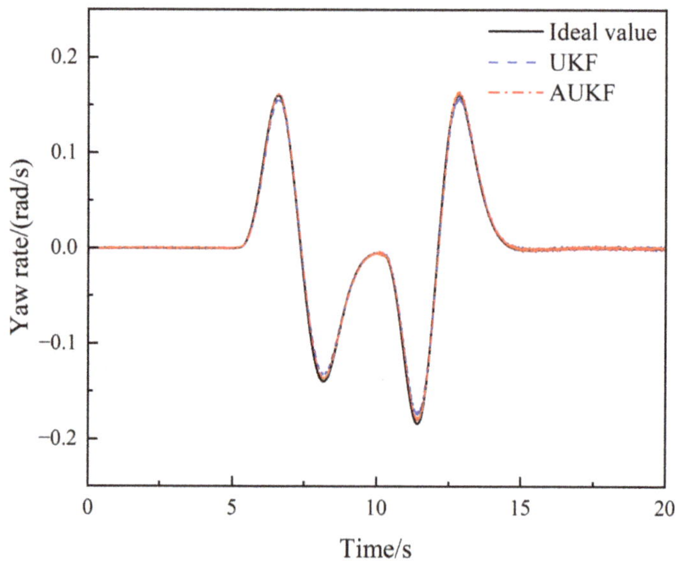

Figure 18. Comparison of yaw rate estimates under the double-lane change condition.

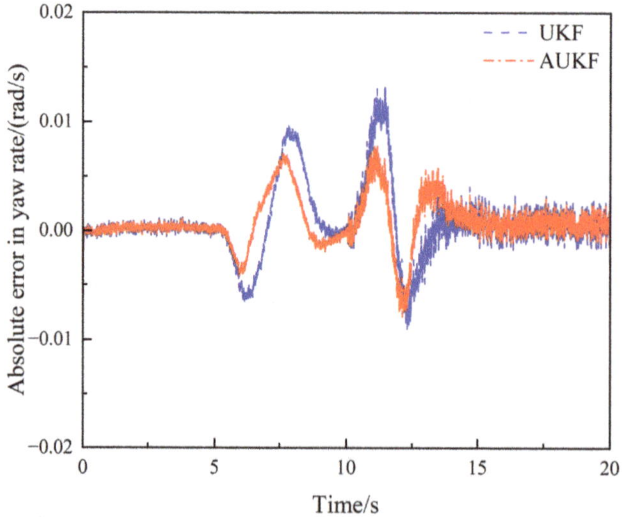

Figure 19. Absolute error of yaw rate estimates under the double-lane change condition.

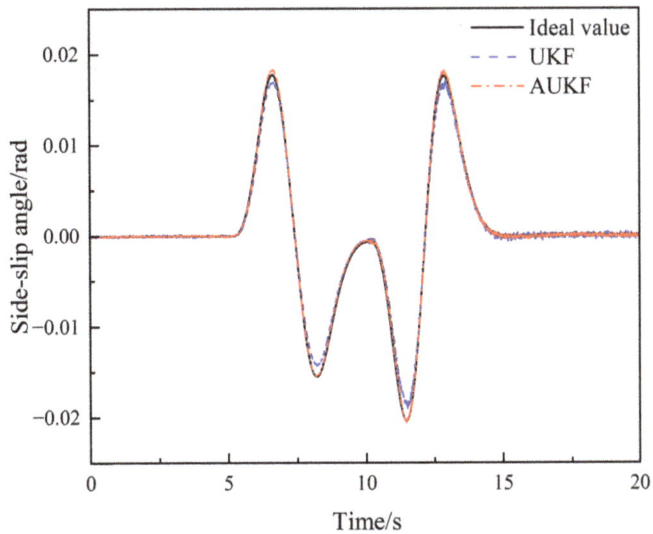

Figure 20. Comparison of side-slip angle estimates under the double-lane change condition.

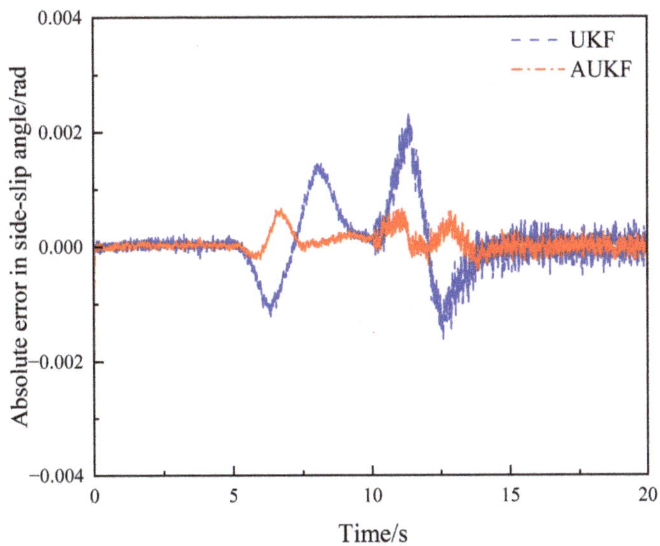

Figure 21. Absolute error of side-slip angle estimates under the double-lane change condition.

As shown in Figure 22, the estimates of UKF and AUKF are the same within 0–5 s, and both show some deviation. Whereas, after 5 s, when the driver starts to turn the steering wheel, the UFK is unable to provide real-time updates in calculating the longitudinal speed and deviates as the error increases. As can be seen in Figure 23, both UKF and AUKF are affected after the statistical characteristics of the noise is expanded by a factor of 10, but the fluctuation of AUKF is smaller than that of UKF, and the estimation results are still more accurate than those of UKF. The simulation results show that the proposed algorithm has a higher accuracy and can cope with the time-varying noise covariance characteristic values.

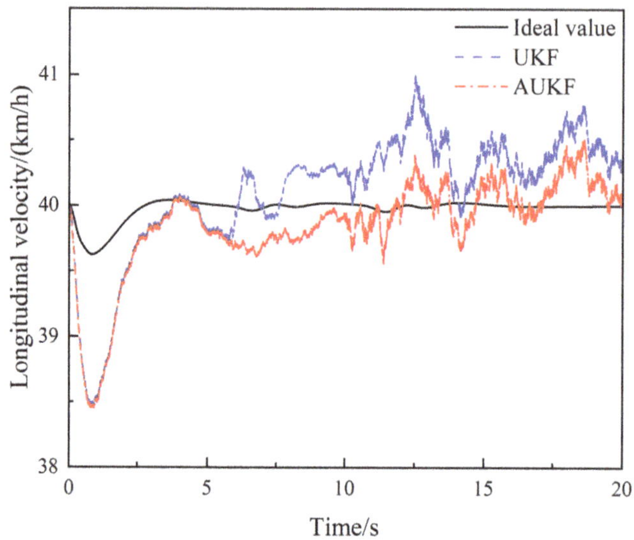

Figure 22. Comparison of longitudinal velocity estimates under the double-lane change condition.

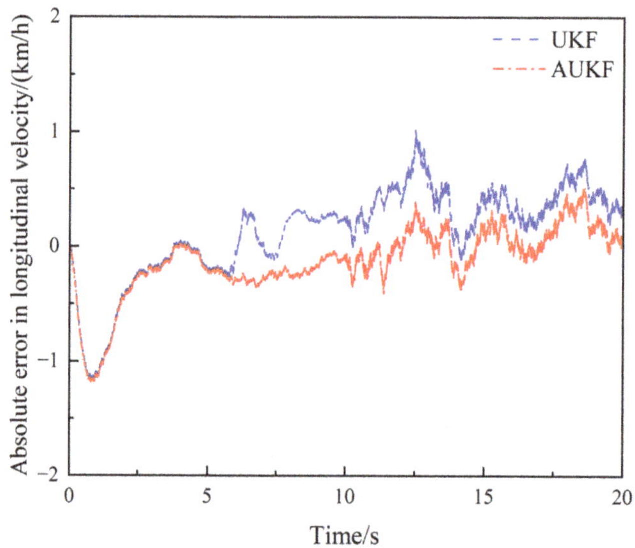

Figure 23. Absolute error of longitudinal velocity estimates under the double-lane change condition.

As shown in Tables 6 and 7, the accuracy of yaw rate estimation has been improved by 29.4%, the accuracy of side-slip angle estimation has been improved by 69.3%, and the estimation of longitudinal velocity by AUKF has been improved by 22.5% under the double-lane change condition.

Table 6. MAE between estimated value and true value under the double-lane change condition.

Algorithm	Yaw Rate (rad/s)	Side-Slip Angle (rad)	Longitudinal Velocity (km/h)
UKF	0.0020	0.000363	0.348
AUKF	0.0016	0.000116	0.234

Table 7. RMSE between estimated value and true value under the double-lane change condition.

Algorithm	Yaw Rate (rad/s)	Side-Slip Angle (rad)	Longitudinal Velocity (km/h)
UKF	0.0034	0.000578	0.4289
AUKF	0.0024	0.000177	0.3321

As shown in Tables 8–10, the AUKF algorithm has a longer single-step running time than the UKF algorithm. This is because the AUKF algorithm needs to calculate the weight factors and check whether the covariance matrix is positively definite. The computational complexity is higher than that of the UKF algorithm, and therefore, the total running time is longer.

Table 8. Comparison of algorithm complexity in steering angle step-input condition.

Algorithm	Total Running Time (s)	Single-Step Running Time (s)	Computational Complexity
UKF	23.433	0.00117	$O(n^2)$
AUKF	24.026	0.00120	$O(n^2)$

Table 9. Comparison of algorithm complexity in sinusoidal steering condition.

Algorithm	Total Running Time (s)	Single-Step Running Time (s)	Computational Complexity
UKF	52.716	0.00132	$O(n^2)$
AUKF	56.439	0.00141	$O(n^2)$

Table 10. Comparison of algorithm complexity in double-lane change condition.

Algorithm	Total Running Time (s)	Single-Step Running Time (s)	Computational Complexity
UKF	24.892	0.00124	$O(n^2)$
AUKF	26.214	0.00131	$O(n^2)$

5. Conclusions

When the vehicle is disturbed by uncertain noises, the traditional vehicle state estimation methods will appear to reduce the accuracy or even diverge. To solve this problem, an AUKF algorithm based on the Sage–Husa algorithm is proposed in this paper. By ignoring the update of the process noise covariance matrix Q and adjusting the measurement noise covariance matrix R online, the estimation of vehicle state parameters under unknown noise disturbance is solved.

This article uses Carsim and Matlab/Simulink for joint simulation to estimate the vehicle's side-slip angle, yaw rate, and longitudinal speed. The simulation results show that the accuracy of the AUKF algorithm is improved by 3.2%, 9.5%, and 32.5% under the steering angle step-input condition, 24.8%, 19.6%, and 63.4% under the sinusoidal steering condition, and 29.4%, 69.3%, and 22.5% under the double-lane change condition. Compared with the UKF algorithm, although the single-step running time of the AUKF algorithm is slightly increased, its accuracy is greatly improved, which can effectively filter out unknown noise. In practical applications, the three working conditions proposed in this article can cover various situations in daily driving. The proposed method eliminates the need to obtain critical state parameters from high-precision sensors, reduces the cost of the vehicle, and can provide more accurate parameters for vehicle decision-making and control systems, which significantly helps to improve the vehicle's handling characteristics.

However, when estimating the longitudinal speed of the vehicle, both UKF and AUKF have significant errors at the beginning. This may be caused by errors in the tire model. In our future research, we will build more accurate models to reduce the uncertainty error further and consider the vehicle state estimation under different road surface adhesion coefficients and road slopes. Then, we will improve our proposed algorithm by incorporating more complex working conditions verification and conduct a real-vehicle test.

Author Contributions: Conceptualization, Y.C. and H.Y.; Data curation, H.Y.; Formal analysis, H.Y.; Methodology, Y.C. and H.Y.; Project administration, Y.C.; Software, H.Y.; Validation, Y.C. and H.Y.; Visualization, H.Y.; Writing—original draft, H.Y.; Writing—review and editing, Y.C. and Y.L. All authors have read and agreed to the published version of the manuscript.

Funding: This research is funded by Project of Beijing Municipal Education Commission for Beijing laboratory construction—Beijing Laboratory for New Energy Vehicles (PXM2020_014224).

Data Availability Statement: Not applicable.

Conflicts of Interest: The authors declare no conflict of interest.

References

1. Indu, K.; Aswatha Kumar, M. Electric Vehicle Control and Driving Safety Systems: A Review. *IETE J. Res.* **2023**, *69*, 482–498. [CrossRef]
2. Guo, H.; Cao, D.; Chen, H.; Lv, C.; Wang, H.; Yang, S. Vehicle dynamic state estimation: State of the art schemes and perspectives. *IEEE/CAA J. Autom. Sin.* **2018**, *5*, 418–431. [CrossRef]
3. Mazzilli, V.; Ivone, D.; De Pinto, S.; Pascali, L.; Contrino, M.; Tarquinio, G.; Gruber, P.; Sorniotti, A. On the benefit of smart tyre technology on vehicle state estimation. *Veh. Syst. Dyn.* **2022**, *60*, 3694–3719. [CrossRef]
4. Jin, X.; Yin, G.; Chen, N. Advanced estimation techniques for vehicle system dynamic state: A survey. *Sensors* **2019**, *19*, 4289. [CrossRef]
5. Chen, B.C.; Hsieh, F.C. Sideslip angle estimation using extended Kalman filter. *Veh. Syst. Dyn.* **2008**, *46*, 353–364. [CrossRef]
6. Piyabongkarn, D.; Rajamani, R.; Grogg, J.A.; Lew, J.Y. Development and experimental evaluation of a slip angle estimator for vehicle stability control. *IEEE Trans. Control Syst. Technol.* **2009**, *17*, 78–88. [CrossRef]
7. Viehweger, M.; Vaseur, C.; Aalst, S.; Acosta, M.; Regolin, E.; Alatorre, A.; Desmet, W.; Naets, F.; Ivanov, V.; Ferrara, A.; et al. Vehicle state and tyre force estimation: Demonstrations and guidelines. *Veh. Syst. Dyn.* **2021**, *59*, 675–702. [CrossRef]
8. González, L.P.P.; Sánchez, S.S.S.; Garcia-Guzman, J.; Boada, M.J.L.; Boada, B.L. Simultaneous Estimation of Vehicle Roll and Sideslip Angles through a Deep Learning Approach. *Sensors* **2020**, *20*, 3679. [CrossRef]
9. Novi, T.; Capitani, R.; Annicchiarico, C. An integrated artificial neural network–unscented Kalman filter vehicle sideslip angle estimation based on inertial measurement unit measurements. *Proc. Inst. Mech. Eng. Part D J. Automob. Eng.* **2019**, *233*, 1864–1878. [CrossRef]
10. Yang, S.; Lu, Y.; Li, S. An overview on vehicle dynamics. *Int. J. Dyn. Control* **2013**, *1*, 385–395. [CrossRef]
11. Sun, W.; Wang, Z.; Wang, J.; Wang, X.; Liu, L. Research on a Real-Time Estimation Method of Vehicle Sideslip Angle Based on EKF. *Sensors* **2022**, *22*, 3386. [CrossRef]
12. Jeong, D.; Ko, G.; Choi, S.B. Estimation of sideslip angle and cornering stiffness of an articulated vehicle using a constrained lateral dynamics model. *Mechatronics* **2022**, *85*, 102810. [CrossRef]
13. Song, R.; Fang, Y. Estimation of Vehicle Sideslip Angle based on Modified Sliding Mode Observer and Recurrent Neural Network. In Proceedings of the 2022 7th Asia-Pacific Conference on Intelligent Robot Systems (ACIRS), Tianjin, China, 1–3 July 2022; pp. 135–139.
14. Zhang, F.; Wang, Y.; Hu, J.; Yin, G.; Chen, S.; Zhang, H.; Zhou, D. A novel comprehensive scheme for vehicle state estimation using dual extended H-infinity kalman filter. *Electronics* **2021**, *10*, 1526. [CrossRef]
15. Venhovens, P.J.T.; Naab, K. Vehicle dynamics estimation using Kalman filters. *Veh. Syst. Dyn.* **1999**, *32*, 171–184. [CrossRef]
16. Zong, C.F.; Hu, D.; Yang, X.; Pan, Z.; Xu, Y. Vehicle driving state estimation based on extended Kalman filter. *J. Jilin Univ. (Eng. Technol. Ed.)* **2009**, *39*, 7–11.
17. Singh, K.B.; Arat, M.A.; Taheri, S. Literature review and fundamental approaches for vehicle and tire state estimation. *Veh. Syst. Dyn.* **2019**, *57*, 1643–1665. [CrossRef]
18. Heidfeld, H.; Schünemann, M.; Kasper, R. Experimental Validation of a GPS-Aided Model-Based UKF Vehicle State Estimator. In Proceedings of the 2019 IEEE International Conference on Mechatronics (ICM), Ilmenau, Germany, 18–20 March 2019; pp. 537–543.
19. Villano, E.; Lenzo, B.; Sakhnevych, A. Cross-combined UKF for vehicle sideslip angle estimation with a modified Dugoff tire model: Design and experimental results. *Meccanica* **2021**, *56*, 2653–2668. [CrossRef]
20. Huang, Y. Estimation of Vehicle Status and Parameters Based on Nonlinear Kalman Filtering. In Proceedings of the 2022 6th International Conference on Robotics and Automation Sciences (ICRAS), Wuhan, China, 9–11 June 2022; pp. 200–205.
21. Xiao, Z.; Xiao, D.; Havyarimana, V.; Jiang, H.; Liu, D.; Wang, D.; Zeng, F. Toward accurate vehicle state estimation under non-Gaussian noises. *IEEE Internet Things J.* **2019**, *6*, 10652–10664. [CrossRef]

22. Wang, Q.; Zhao, Y.; Lin, F.; Zhang, C.; Deng, H. Integrated control for distributed in-wheel motor drive electric vehicle based on states estimation and nonlinear MPC. *Proc. Inst. Mech. Eng. Part D J. Automob. Eng.* **2022**, *236*, 893–906. [CrossRef]
23. Chu, W.; Luo, Y.; Dai, Y.; Li, K. In–wheel motor electric vehicle state estimation by using unscented particle filter. *Int. J. Veh. Des.* **2015**, *67*, 115–136. [CrossRef]
24. Wang, Z.P.; Xue, X.; Wang, Y.C. State parameter estimation of distributed drive electric vehicle based on adaptive unscented Kalman filter. *J. Beijing Inst. Technol.* **2018**, *38*, 698–702.
25. Fan, T.E.; Liu, S.M.; Tang, X.; Qu, B.H. Simultaneously estimating two battery states by combining a long short-term memory network with an adaptive unscented Kalman filter. *J. Energy Storage* **2022**, *50*, 104553. [CrossRef]
26. Li, G.; Zhao, D.; Xie, R.; Han, H.; Zong, C. Vehicle State Estimation Based on Improved Sage–Husa Adaptive Extended Kalman Filtering. *Automot. Eng.* **2015**, *37*, 1426–1432.
27. Zhou, B.; Li, T.; Wu, X.; Lei, F. Semi-trailer State Estimation Based on Double Adaptive Unscented Kalman Filter. *J. Hunan Univ. (Nat. Sci.)* **2022**, *49*, 63–73.
28. Xu, D.; Wang, B.; Zhang, L.; Chen, Z. A New Adaptive High-Degree Unscented Kalman Filter with Unknown Process Noise. *Electronics* **2022**, *11*, 1863. [CrossRef]
29. Luo, Z.; Fu, Z.; Xu, Q. An Adaptive multi-dimensional vehicle driving state observer based on modified Sage–Husa UKF algorithm. *Sensors* **2020**, *20*, 6889. [CrossRef]
30. Yang, R.; Zhang, A.; Zhang, L. A novel adaptive H-Infinity cubature Kalman filter algorithm based on Sage-Husa estimator for unmanned underwater vehicle. *Math. Probl. Eng.* **2020**, *9*, 456–463. [CrossRef]
31. Bian, H.; Jin, Z.; Wang, J. The innovation-based estimation adaptive Kalman filter algorithm for INS/GPS integrated navigation system. *J. Shanghai Jiaotong Univ.* **2006**, *40*, 1000.
32. Dey, A.; Sadhu, S.; Ghoshal, T.K. Adaptive Gauss–Hermite filter for non-linear systems with unknown measurement noise covariance. *IET Sci. Meas. Technol.* **2015**, *9*, 1007–1015. [CrossRef]
33. Narasimhappa, M.; Mahindrakar, A.D.; Guizilini, V.C.; Terra, M.H.; Sabat, S.L. MEMS-based IMU drift minimization: Sage Husa adaptive robust Kalman filtering. *IEEE Sens. J.* **2019**, *20*, 250–260. [CrossRef]
34. Yu, Z. *Automobile Theory*, 5th ed.; China Machine Press: Beijing, China, 2009; pp. 144–146.
35. Pacejka, H. *Tire and Vehicle Dynamics*, 3rd ed.; Butterworth-Heinemann: Oxford, UK, 2012; pp. 593–601.
36. Wan, E.A.; Van Der Merwe, R. The unscented Kalman filter. In *Kalman Filtering and Neural Networks*; John Wiley & Sons, Inc.: New York, NY, USA, 2001; pp. 221–280.
37. Liu, K.; Zhao, W.; Sun, B.; Wu, P.; Zhu, D.; Zhang, P. Application of updated Sage–Husa adaptive Kalman filter in the navigation of a translational sprinkler irrigation machine. *Water* **2019**, *11*, 1269. [CrossRef]

World Electric Vehicle Journal

MDPI

Article

Control Strategy of Torque Distribution for Hybrid Four-Wheel Drive Tractor

Xianghai Yan [1,2,3], Hui Zhang [1], Xianzhe Li [1], Yanying Li [1] and Liyou Xu [1,3,*]

1 Vehicle and Traffic Engineering College, Henan University of Science and Technology, Luoyang 471003, China
2 Henan Province Collaborative Innovation Center for Advanced Manufacturing of Mechanical Equipment, Luoyang 471003, China
3 State Key Laboratory of Intelligent Agricultural Power Equipment, Luoyang 471003, China
* Correspondence: xlyou@haust.edu.cn

Abstract: Based on the analysis of the operating conditions of the tractor, a Hybrid four-wheel drive tractor is proposed, and formulate the torque distribution control strategy based on fuzzy control, to control the driving wheel slip rate of the Hybrid four-wheel drive tractor in the high traction efficiency operating range of the tractor. The vehicle model of the Hybrid four-wheel drive tractor is established in AVL-CRUISE software, and the torque distribution control strategy based on fuzzy control is established in MATLAB/Simulink software. The AVL-CRUISE and MATLAB/Simulink co-simulation was carried out based on the plowing condition of the tractor. The simulation results show that the torque distribution control strategy based on fuzzy control can control the driving wheel slip rate of the Hybrid four-wheel drive tractor in the high traction efficiency operating range, the power performance of the Hybrid four-wheel drive tractor is improved, while the engine runs smoothly and is always in the high-efficiency range of engine operation, and the economy is better.

Keywords: parallel hybrid; hybrid tractor; slip rate; torque distribution; fuzzy control; four-wheel drive

Citation: Yan, X.; Zhang, H.; Li, X.; Li, Y.; Xu, L. Control Strategy of Torque Distribution for Hybrid Four-Wheel Drive Tractor. *World Electr. Veh. J.* **2023**, *14*, 190. https://doi.org/10.3390/wevj14070190

Academic Editors: Fachao Jiang, Yongyu Li and Weiwei Kong

Received: 18 June 2023
Revised: 10 July 2023
Accepted: 13 July 2023
Published: 18 July 2023

1. Introduction

The tractor is the most widely used agricultural machinery. Its main job is to mount agricultural implements for operation in the field and for transportation on the road [1,2]. With the change in tractive force, the slip rate of the tractor also constantly varies [3,4]. With the increase in tractive force, the slip rate will also increase. Excessive slip rate on the one hand will affect the tractor's tractive force, wasting engine power; on the other hand, excessive wheel slip will damage the soil structure and cause increased tires wear, reducing efficiency [5,6]. Related research shows that the wheel tractor will have high traction efficiency if the slip rate is controlled between 10% and 15% [7], so the slip rate of the tractor drive wheel needs to be controlled to improve the tractor's traction efficiency. The current research on changing the slip rate of tractor drive wheels is mainly focused on changing the tillage depth [8,9]. The change in tillage depth will have some effect on the growth of crops [10,11]. The torque distribution of the tractor drive wheels directly affects the slip rate of the drive wheels [12–14], and the four-wheel drive tractor can effectively reduce the slip rate of the drive wheels [15]. The traditional four-wheel drive tractor uses a splitter to distribute the torque of the front- and rear-drive wheels, and the arrangement of the mechanical components is not convenient, at the same time, the four-wheel drive tractor with the splitter is not able to distribute the front- and rear-drive wheel torque flexibly, and the working condition is poorly adapted. The electric drive system can conveniently arrange the power source, realize the front- and rear-independent drive, and flexibly distribute the torque of the front- and rear-drive wheels [16–18]. Some researchers in the vehicle field have proposed a compound hybrid electric vehicle configuration, which adds a front-drive motor to drive the front wheels based on the power-split hybrid electric vehicle configuration, and Toyota's Highlander vehicle uses three motors and an engine to improve the fuel economy and power of the vehicle [19]. However, both compound hybrid electric vehicle configurations and power-split hybrid electric vehicle configurations have complex control systems and high costs. Cong Guo et al. proposed coordinated control of torque distribution and drive anti-skid for front- and rear-independent drive electric vehicles in the vehicle field, which achieved good results and improved the dynamics of four-wheel drive Electric vehicles [20]. However, the current Electric vehicle has the problem of short range, which is not conducive to agricultural production. Liyou Xu et al. proposed an Extended range four-wheel drive electric tractor [21], which distributes the torque of the front- and rear-drive wheels of the tractor and achieved good results. However, the Extended range electric vehicle configuration has the problem of too many energy conversions and low energy utilization.

The four-wheel drive tractor with a splitter has the problem of difficult component arrangement and not being able to flexibly distribute the torque of the front- and rear-drive wheels; the compound hybrid configuration has the problem of complex structure and difficult design of the control system; the pure electric system has the problem of short-range time; and the Extended range electric vehicle configuration has the problem of low energy utilization with a high number of energy conversions. In this paper, by analyzing the defects of the existing tractor power system, a Hybrid four-wheel drive tractor is proposed by adding the front-drive motor to the parallel hybrid tractor, the parameters of the main components are calculated and matched, and a torque distribution control strategy is developed to distribute the torque to the front-drive motor, the rear-drive motor and the engine. The effectiveness of the torque distribution control strategy was verified by joint simulation of AVL-CRUISE and Matlab/Simulink. A theoretical basis is provided for the design and development of the Hybrid four-wheel drive tractor.

2. Hybrid Four-Wheel Drive Tractor Working Condition Analysis

The main work of the tractor is in the field mounted agricultural equipment for plowing, harrowing, fertilization, seeding, and other operations, as well as road transport operations. Plowing condition is the highest frequency of tractor working condition, but also the most basic and heavy working condition, the tractor plowing condition analysis, can fully reflect the performance of the tractor, and this paper to plowing condition as an example of the tractor analysis.

The longitudinal mechanics of the tractor during operation is modeled as

$$F_{qt} = F_f + F_i + F_w + F_j + F_g = F_{qr} + F_{qf}, \tag{1}$$

where, F_{qt}, total tractor tractive force; F_f, rolling resistance; F_i, gradient resistance; F_w, air resistance; F_j, acceleration resistance; F_g, plowing resistance; F_{qr}, rear-drive wheel tractive force; F_{qf}, front-drive wheel tractive force.

Air resistance and acceleration resistance are negligible during tractor operation. The resistance of tractor operation mainly comes from plowing resistance, rolling resistance, and gradient resistance.

$$F_{qt} = F_f + F_i + F_g = F_{qr} + F_{qf}, \tag{2}$$

$$F_i = Gi \tag{3}$$

$$F_f = Gf \tag{4}$$

where, G, tractor gravity; i, road gradient; f, rolling resistance coefficient.

Tractor operation will produce slip, reducing the efficiency of the tractor. Research shows that as the tractive force increases, the slip rate will also increase.

The relationship between tractive force and adhesion coefficient can be expressed as:

$$F_q = \varphi F_z, \tag{5}$$

$$\varphi = \varphi_{\max}\left(1 - e^{-\frac{\delta}{\delta^*}}\right) \tag{6}$$

where, φ, drive wheel tractive force coefficient; $\varphi_{\max}$, maximum adhesion coefficient; δ, slip rate; δ^*, characteristic slip rate; F_q, drive wheel tractive force; F_z, drive wheel vertical load.

Therefore, by reasonably distributing the tractor front- and rear-drive wheel, tractive forces can effectively reduce the slip rate and improve the traction efficiency of the tractor.

The tire model describes the ground forces on the tires as a mathematical function [22]. The selection of the tire model, which directly affects the prediction of the traction performance of the drive wheels, is crucial for the drive system modeling. In this paper, the tire model uses the Duggof tire model [23], and the tractive force expression of the drive wheels is:

$$F_q = \begin{cases} F_z\left[\varphi - \varphi^2\frac{F_z(1-\delta)}{4c\delta}\right], & \frac{c\delta}{1-\delta} \geq \frac{\varphi F_z}{2} \\ \frac{c\delta}{1-\delta} & , \frac{c\delta}{1-\delta} < \frac{\varphi F_z}{2} \end{cases}, \tag{7}$$

where, c, tire longitudinal stiffness.

The tractor distributes the front- and rear-drive wheel tractive forces according to the tractive force distribution coefficient (k) to obtain:

$$F_{qf} = F_{qt}k, \tag{8}$$

$$F_{qr} = F_{qt}(1-k) \tag{9}$$

From Formulas (7) and (8), the front-drive wheel tractive force is

$$F_{qf} = \begin{cases} \left(F_{zf}\left[\varphi_f - \varphi_f^2\frac{F_{zf}(1-\delta)}{4c\delta}\right] + F_{qr}\right)k, & \frac{c\delta}{1-\delta} \geq \frac{\varphi F_{zf}}{2} \\ \left(\frac{c\delta}{1-\delta} + F_{qr}\right)k & , \frac{c\delta}{1-\delta} < \frac{\varphi F_{zf}}{2} \end{cases}, \tag{10}$$

where, k, tractive force distribution coefficient; F_{zf}, front-drive wheel vertical load; φ_f, front-drive wheel tractive force coefficient.

From Formulas (7) and (9), the rear-drive wheel tractive force is

$$F_{qr} = \begin{cases} \left(F_{zr}\left[\varphi_r - \varphi_r^2\frac{F_{zr}(1-\delta)}{4c\delta}\right] + F_{qf}\right)(1-k), & \frac{c\delta}{1-\delta} \geq \frac{\varphi F_{zr}}{2} \\ \left(\frac{c\delta}{1-\delta} + F_{qf}\right)(1-k) & , \frac{c\delta}{1-\delta} < \frac{\varphi F_{zr}}{2} \end{cases}, \tag{11}$$

where, F_{zr}, the rear-drive wheel vertical load; φ_r, the rear-drive wheel tractive force coefficient.

From Formulas (10) and (11), it can be obtained that there is a high degree of non-linearity between the front-drive wheel tractive force, the rear-drive wheel tractive force, and the tractive force distribution coefficient of the tractor.

3. Hybrid Four-Wheel Drive Tractor Model Analysis

3.1. Hybrid Four-Wheel Drive Tractor Drive System Model Analysis

The drive configuration of the tractor has an important impact on the dynamics of the tractor. The traditional four-wheel drive tractor uses a mechanical splitter to distribute the torque of the front- and rear-drive wheels, which has a complex structure and difficult arrangement of mechanical components, and cannot adjust the tractive force distribution of the front- and rear-drive wheels according to the road conditions, and has a low traction efficiency [24]. The power system of the Hybrid four-wheel drive tractor in this paper includes the front-drive system and the rear-drive system. The front-drive system includes the front-drive motor, which is responsible for driving the front-drive wheels. The rear-drive system includes the engine and the rear-drive motor, the engine is the main power source and the rear-drive motor is the auxiliary power source, the power of the engine and the rear-drive motor output power to the rear-drive wheels through the coupling action of the power coupler. The engine and the rear-drive motor can work separately or jointly to provide power to the rear-drive wheels. No mechanical connection between the front-drive system and the rear-drive system, convenient for the arrangement of components, and the front- and rear-drive wheel tractive force can be flexibly adjusted according to the road conditions, to develop a flexible front- and rear wheel torque distribution control strategy, in order to make the tractor slip rate control in a more reasonable range, to improve the tractor traction efficiency, the Hybrid four-wheel drive tractor power system used in this paper is shown in Figure 1.

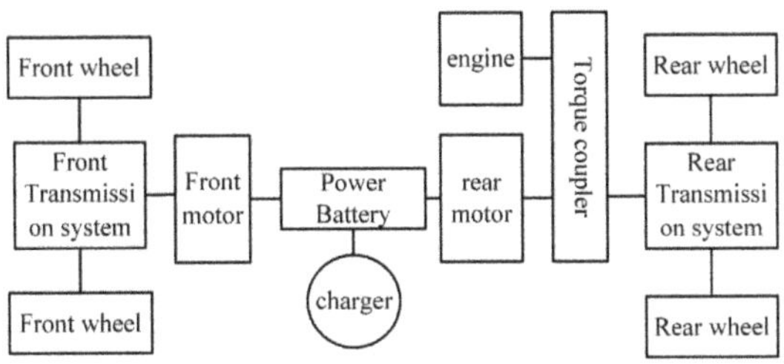

Figure 1. Hybrid four-wheel drive tractor power system.

3.2. *Power System Matching Design*

3.2.1. Determination of Rated Tractive Force

Rated tractive force is the tractor in the horizontal lot with the basic plowing speed and drive wheel slip rate in the specified value or engine in the standard working condition, the maximum tractive force can be issued (take the smaller of the two values). Plowing is the most basic and heavy tractor operation in agricultural production, rated tractive force determination, the first should meet the plowing operation requirements.

$$F_g = zbhk, \tag{12}$$

where, F_g, plowing resistance; z, the number of plowshares, according to the selection of agricultural machinery, $z = 6$; b, the width of the single plowshare, $b = 0.45$m; h, plowing depth, $h = 0.25$m; k, soil specific resistance, take $7\text{N}/\text{cm}^2$.

The driving resistance of the tractor will fluctuate due to changes in operating conditions and farm equipment, and should generally have a reserve tractive force of 10~20%. Therefore, the rated tractive force is:

$$F_T = (1.1 \sim 1.2)F_g. \tag{13}$$

3.2.2. Power System Matching Design

In conventional tractor design, the rated tractive force is used to determine the engine power. In the design of a Hybrid four-wheel drive tractor, the rear-drive system proposed in this paper can be considered as a traditional tractor engine for power selection compared to the traditional tractor engine power selection. And in the rear-drive system power distribution, the traditional parallel hybrid tractor compared to the traditional tractor in the structure of only a small power motor used to improve the power output of the engine, cannot play well with the electric drive system drive efficiency high environmental pollution low characteristics, in recent years with the development of electric drive system, electric drive system costs continue to reduce, so this paper uses a small power engine with high power electric motor. At the same time, in order to achieve a better anti-skid effect, the front-drive motor also uses a high-power motor.

Engine power:

$$P_e = \frac{0.65 \times F_T \times v_l}{3.6\eta_T}, \tag{14}$$

Rear-drive motor power:

$$P_{rm} = \frac{0.35 \times F_T \times v_l}{3.6\eta_T}, \tag{15}$$

Front-drive motor power:

$$P_{fm} = \frac{0.35 \times F_T \times v_l}{3.6\eta_T}, \tag{16}$$

where, v_l, the driving speed of the tractor during plowing operation, take 7 km/h; η_T, the efficiency of the transmission system, take 0.86.

In summary, the engine is an inline 4-cylinder supercharged inter-cooled high-pressure common rail diesel engine, model LR4A3LRP-T4-U3, with a rated power of 85 kw and rated speed of 2300 r/min. The rear-drive motor is a permanent magnet synchronous motor with a rated power of 45 kw, rated speed of 2500 r/min and rated voltage of 320 v. The front-drive motor is a Permanent magnet synchronous motor, its rated power is 45 kw, rated speed 2500 r/min, rated voltage 320 v.

3.2.3. Torque Coupler Design

For the selection of coupling device of parallel hybrid vehicle configuration, it can be divided into torque coupling type, speed coupling type, and power coupling type. Torque-coupled power sources are coupled in such a way that the output torques of the engine and motor are independent of each other, and the torque after coupling is the algebraic sum of the output torques of the two power sources. Speed coupling refers to the engine and motor power in the coupling, the two output speeds independent of each other, the final synthesis of the speed is a number of power sources each speed of the algebraic sum, and the output torque into a fixed proportion, the actual use of the tractor need to output a larger force, so the speed coupling for the tractor is not suitable for the design. Power-coupled engine and motor torque and speed are independent of each other, the output torque and output speed are the algebraic sums of the engine and motor [25], but the control system design is more complex. The torque-coupled type has a simple structure and the control system design is simple, so the torque-coupled type is used in this paper. The torque coupler is shown in Figure 2.

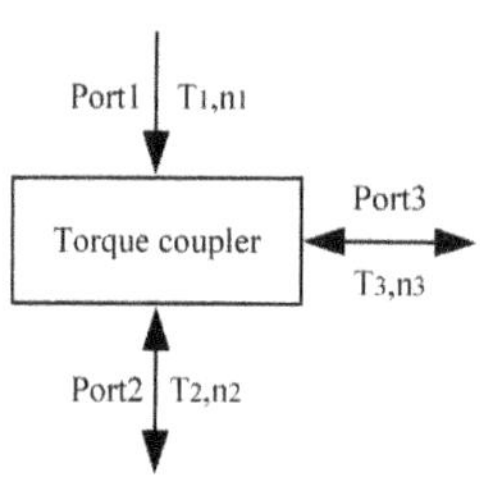

Figure 2. Torque coupler.

The torque coupler port 1 herein is a unidirectional torque and speed input, and ports 2 and 3 are bidirectional torque and speed inputs or outputs. In a hybrid tractor application, port 1 is connected to the engine via a transmission; port 2 is connected to the rear-drive motor via a transmission; and port 3 is connected to the drive wheels via a transmission.

4. Torque Distribution Control Strategy for Hybrid Four-Wheel Drive Tractor

The torque distribution control strategy of the Hybrid four-wheel drive tractor proposed in this paper includes the front- and rear-drive wheel torque distribution control strategy and the rear-drive system torque distribution control strategy, and the two torque distribution control strategies will be described separately below.

4.1. Front- and Rear-Drive Wheel Torque Distribution Control Strategy

From Formulas (2)–(4), the tractor in the plowing operation, plowing resistance, gradient resistance, and rolling resistance are important factors affecting the drive wheel slip rotation, air resistance and acceleration resistance value is small, neglected. Therefore, the concept of road resistance coefficient is introduced, the road resistance coefficient is the sum of the gradient and rolling resistance coefficient, and it can reflect the road resistance situation. The tractor's drive wheel slip rate is controlled at about 10~15%, and the tractor has the highest traction efficiency.

During tractor plowing, the tractor center of mass will move backwards, the tractor axle load distribution will change, the rear axle load will become larger, and the rear-drive wheel rolling resistance will also become larger. The change in rear-drive wheel rolling resistance will have an effect on the distribution of tractive force between the front- and rear-drive wheels of the tractor. From the joint simulation data of AVL-CRUISE and MATLAB/Simulink, the real-time rolling resistance of the rear-drive wheel and the real-time axle load distribution of the tractor are obtained, and the rolling resistance coefficients of the front-drive wheel, the rear-drive wheel and the whole vehicle are calculated. Comparing each rolling resistance coefficient, it can be seen that the rear-drive wheel rolling resistance coefficient is larger than the whole vehicle rolling resistance coefficient than the front-drive wheel rolling resistance coefficient. The rear-drive wheel of the tractor carries more tractive force compared with the front-drive theory, so the rear-drive wheel rolling resistance coefficient is substituted into the calculation of road resistance coefficient, which can better reflect the resistance of the tractor.

The tractor operating conditions are complex and variable, and it is known from Formulas (10) and (11) that there is a high degree of nonlinearity between the tractor's front-drive wheel tractive force, rear-drive wheel tractive force, and tractive force distribution coefficient k. The fuzzy controller has a better control effect for the nonlinear system. The plowing resistance and road resistance coefficient are selected as the input variables of the fuzzy controller, and the tractive force distribution factor k is selected as the output variable of the fuzzy controller.

The front- and rear-torque distribution control strategy of a Hybrid four-wheel drive tractor is shown in Figure 3. In order to verify the superiority of the front- and rear-drive wheel torque distribution control strategy based on fuzzy control proposed in this paper, the front- and rear-torque distribution control strategy based on fixed ratio distribution is set as the comparison control strategy, and the front- and rear-torque distribution control strategy based on fixed ratio distribution is shown in Figure 4.

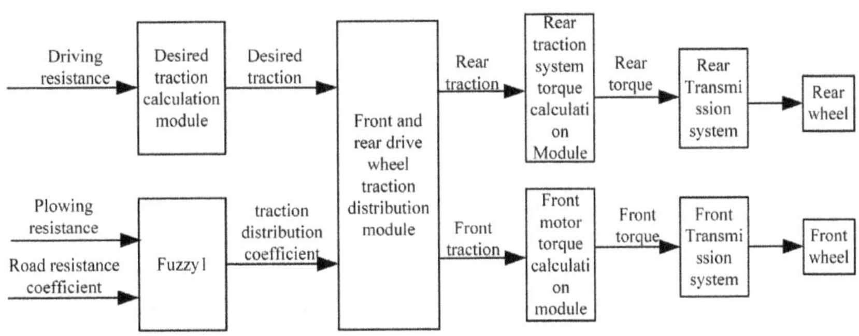

Figure 3. Control strategy for torque distribution of front- and rear-drive wheels based on fuzzy control.

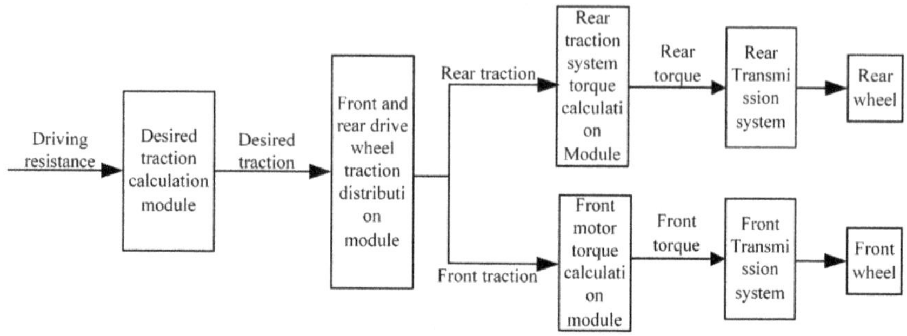

Figure 4. Control strategy for torque distribution of front- and rear-drive wheels based on fixed ratio distribution.

In Figure 3, the AVL-CRUISE software inputs the driving resistance and outputs the desired tractive force through the desired tractive force calculation module. The plowing resistance and road resistance coefficient are used as input variables of fuzzy controller 1, and the output variable tractive force distribution coefficient k is obtained by fuzzification, fuzzy inference and anti-fuzzification of fuzzy controller 1, and then the k value is transmitted to the front- and rear-drive wheel tractive force distribution module. The front- and rear-drive wheel tractive force distribution module distributes the tractive force of the front-drive wheel and the rear-drive wheel through the demand tractive force and the tractive force distribution coefficient. The front-drive wheel tractive force and rear-drive wheel tractive force are calculated to obtain the front-drive system torque and rear-drive system torque, which are transmitted to the front-drive wheels and rear-drive wheels through the transmission system. In Figure 4, the torque distribution control strategy based on fixed ratio distribution is the same as the strategy proposed in this paper except that the tractive force distribution coefficient is different from the strategy proposed in this paper.

4.1.1. Affiliation Function Design

The exact values of plowing resistance were converted into proportional values with the theoretical domain of [0.8, 1] for plowing resistance, [−0.05, 0.22] for road resistance coefficient, and [0.12, 0.22] for tractive force distribution coefficient.

The formula for converting the exact value of plowing resistance into a proportional value is:

$$F_{gp} = \frac{F_g}{F_{g\max}}, \tag{17}$$

where, F_{gp}, the proportional value of the plowing resistance; F_g, the plowing resistance at a certain moment; $F_{g\max}$, the maximum value of the plowing resistance.

And, respectively, stipulate the following fuzzy subsets: the fuzzy subset of proportional values of plowing resistance is E(F) = {NB,NM,Z,PM,PB}. The fuzzy subset of road resistance coefficient is E(R) = {NH,NL,Z,PL,PH}. The fuzzy subset of the tractive force distribution coefficient is E(k) = {ND,NB,NM,NS,Z,PS,PM,PB,PD}. The fuzzy subsets {NH,NL,Z,PL,PH} represent negative high, negative low, medium, positive low, and positive high, respectively. The fuzzy subsets {ND,NB,NM,NS,Z,PS,PM,PB,PD} represent negative pole, negative large, negative medium, negative small, medium, positive small, positive medium, positive large, and positive pole, respectively.

In summary, the affiliation function of plowing resistance is shown in Figure 5, the affiliation function of road resistance coefficient is shown in Figure 6, and the affiliation function of tractive force distribution coefficient is shown in Figure 7.

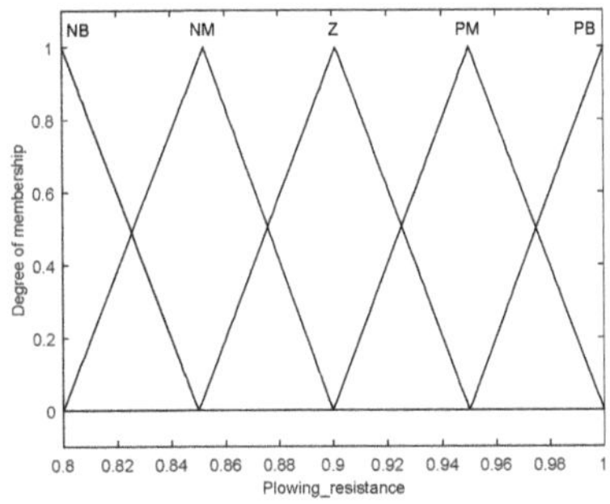

Figure 5. Affiliation function of plowing resistance.

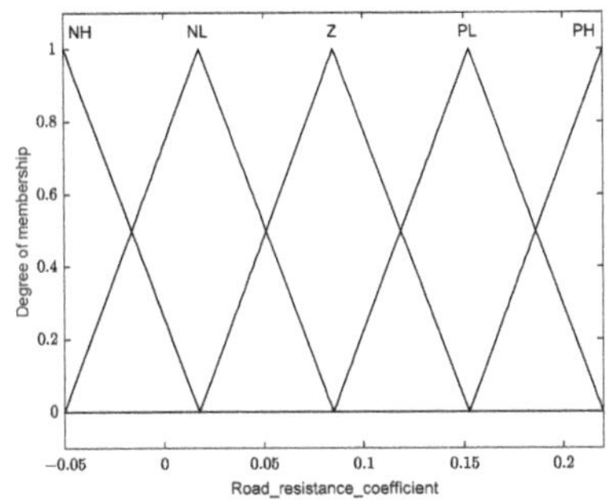

Figure 6. Affiliation function of road resistance coefficient.

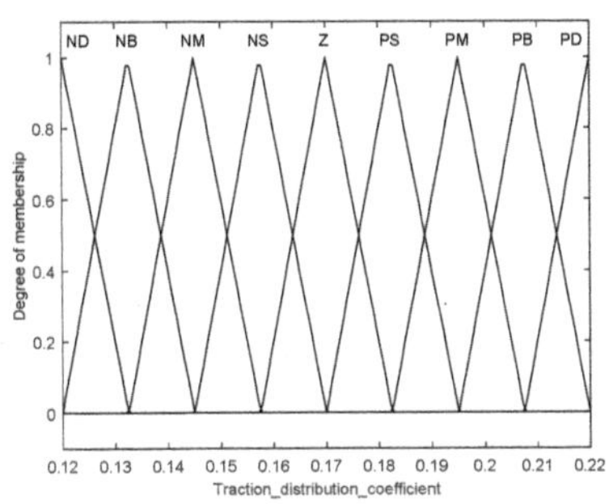

Figure 7. Affiliation function of tractive force distribution coefficient.

4.1.2. Fuzzy Rules Design

The design of fuzzy rules needs to fully consider the operating conditions of the tractor, the design of fuzzy rules as shown in Table 1.

Table 1. Fuzzy control rules library.

k		Plowing Resistance				
		NB	NM	Z	PM	PB
Road resistance coefficient	NH	ND	ND	ND	NB	NM
	NL	ND	NB	NM	NS	Z
	Z	NM	NS	Z	PS	PM
	PL	Z	PS	PM	PB	PD
	PH	PS	PM	PB	PD	PD

where, k, tractive force distribution coefficient.

4.2. Control Strategy for Torque Distribution of Rear-Drive System

The torque distribution of the parallel hybrid tractor is highly nonlinear, and the fuzzy control strategy is well suited for the hybrid tractor control system. The rear-drive system desired torque and battery SOC are used as input variables, and the motor torque distribution coefficient is used as the output variable to establish the fuzzy control-based rear-drive system torque distribution control strategy. The rear-drive system torque distribution control strategy is shown in Figure 8.

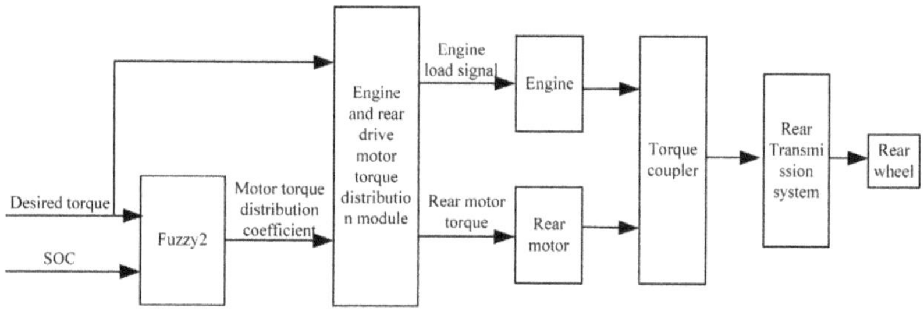

Figure 8. Control strategy for torque distribution of rear-drive system.

4.2.1. Affiliation Function Design

The exact value of the desired torque of the rear-drive system is converted into a proportional value, and the domain of the desired torque of the rear-drive system is [0, 1], the domain of the battery SOC is [0, 1], and the domain of the motor torque distribution coefficient is [−0.2, 1].

Motor torque distribution coefficient is the ratio of the rear motor torque to the desired torque of the rear-drive system.

$$\lambda = \frac{T_m}{T_r}, \tag{18}$$

where, T_r, the desired torque of the rear-drive system at a certain moment; T_m, rear motor torque; λ, motor torque distribution coefficient.

The formula for converting the exact value of the desired torque of the rear-drive system into a proportional value is:

$$T_{rp} = \frac{T}{T_{\max}}, \tag{19}$$

where, T_{rp}, the proportional value of the desired torque of the rear-drive system; $T_{\max}$, the maximum desired torque of the rear-drive system.

And, respectively, stipulate the following fuzzy subsets: E(T_{rp}) = {NB,NM,Z,PM,PB}, E(SOC) = {NH,NL,Z,PL,PH}, E(λ) = {TS,S,M,B,TB}. The fuzzy subsets {NB,NM,Z,PM,PB} represent negative large, negative medium, medium, medium, and large, respectively. The fuzzy subsets {NH,NL, Z,PL,PH}stand for negative high, negative low, medium, positive low, and positive high, respectively. Fuzzy subsets {TS,S,M,B,TB} represent very small, small, medium, large, and very large, respectively.

In summary, the affiliation function of the desired torque is shown in Figure 9, the affiliation function of the SOC is shown in Figure 10, and the affiliation function of the motor torque distribution coefficient is shown in Figure 11.

4.2.2. Fuzzy Rules Design

The design of fuzzy rules needs to fully consider the operating characteristics of the parallel hybrid configuration. The designed fuzzy rules are shown in Table 2.

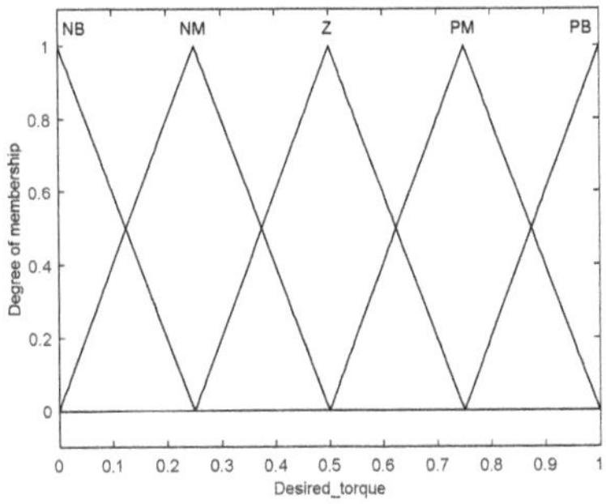

Figure 9. Affiliation function of desired torque.

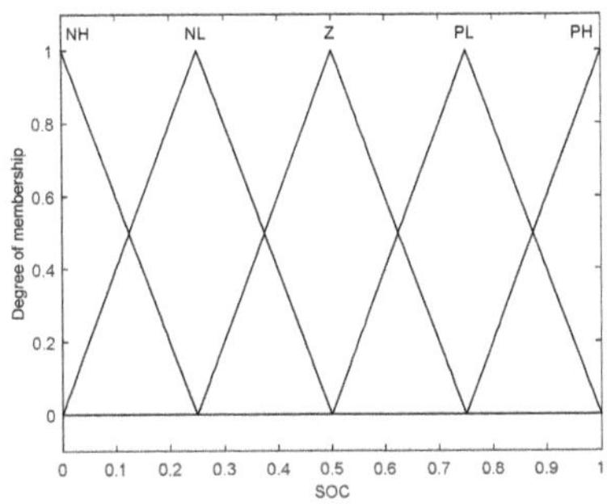

Figure 10. Affiliation function of SOC.

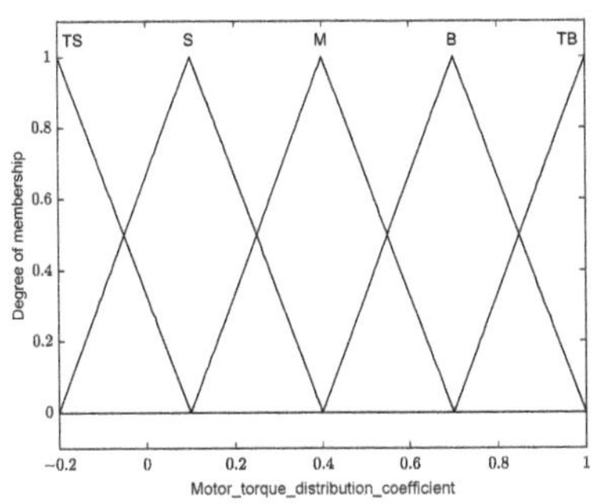

Figure 11. Affiliation function of motor torque distribution coefficient.

Table 2. Fuzzy control rules library.

λ		Desired Torque				
		NB	NM	Z	PM	PB
SOC	NH	TS	TS	S	S	M
	NL	TS	S	S	M	M
	Z	S	S	M	B	B
	PL	TB	B	B	M	B
	PH	TB	TB	B	M	B

where, λ, motor torque distribution coefficient.

5. Comparative Analysis

In this paper, AVL-CRUISE software is used to establish a vehicle model of a Hybrid four-wheel drive tractor [26–28], the main technical parameters of the Hybrid four-wheel drive tractor are shown in Table 3, and MATLAB/Simulink software is used to build a torque distribution control strategy based on fuzzy control, and the joint simulation analysis of AVL-CRUISE and MATLAB/Simulink is carried out.

Table 3. Hybrid four-wheel drive tractor main technical parameters.

Project	Parameter	Value
Vehicle	Quality/kg Wheel base/mm	7350 2800
Engine	Rated power/kw Rated speed/(r/min)	85 2300
Front motor	Rated power/kw Rated speed/(r/min)	45 2500
Rear motor	Rated power/kw Rated speed/(r/min)	45 2500
Gearbox	4th gear transmission ratio	3.6
Front final reduction drive	transmission ratio	5.5
Rear final reduction drive	transmission ratio	5.0

The vehicle model of the Hybrid four-wheel drive tractor built by AVL-CRUISE software is shown in Figure 12. The force1 module and force2 module in the figure are the two sets of plowing resistance modules established.

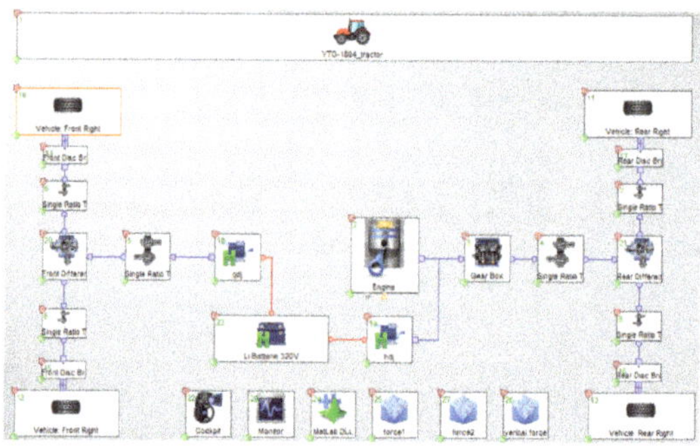

Figure 12. Hybrid four-wheel drive tractor vehicle model.

The torque distribution control strategy based on fuzzy control established by MATLAB/Simulink software is shown in Figure 13.

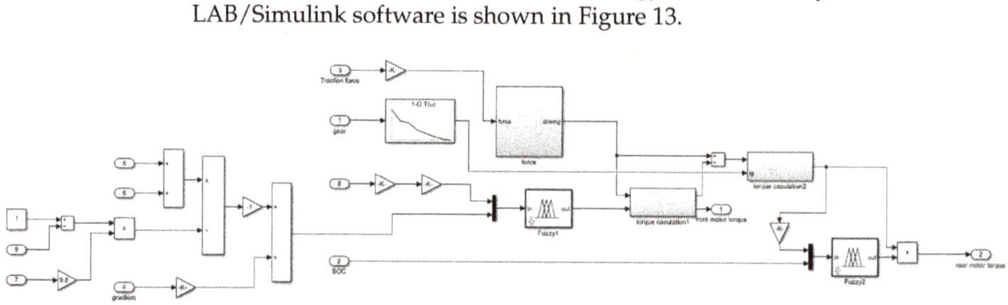

Figure 13. Torque distribution control strategy based on fuzzy control.

When the tractor is actually working, the resistance of different operating conditions is different. According to different operating conditions, tractor operation mode can be divided into three: heavy load mode (80% of the rated tractive effort above), medium load mode (about 20% to 80% of the rated tractive effort), and light load mode (20% of the rated tractive effort below). Tractor plowing working condition operation, resistance is higher, belong to the heavy load mode.

The actual plowing operation of the tractor is mostly flat, and the gradient is generally less than 10%. In order to fully verify the effect of the two torque distribution control strategies on the performance of the Hybrid four-wheel drive tractor, a certain road gradient was set along with the plowing resistance. Considering the effect of rolling friction resistance and gradient resistance set plowing resistance. At the same time, in order to verify the effect of two torque distribution control strategies on the working stability of a Hybrid four-wheel drive tractor set two groups of plowing resistance, one group for small plowing resistance group, and one group for large plowing resistance group.

The established small plowing resistance group is shown in Figure 14 and the established large plowing resistance group is shown in Figure 15.

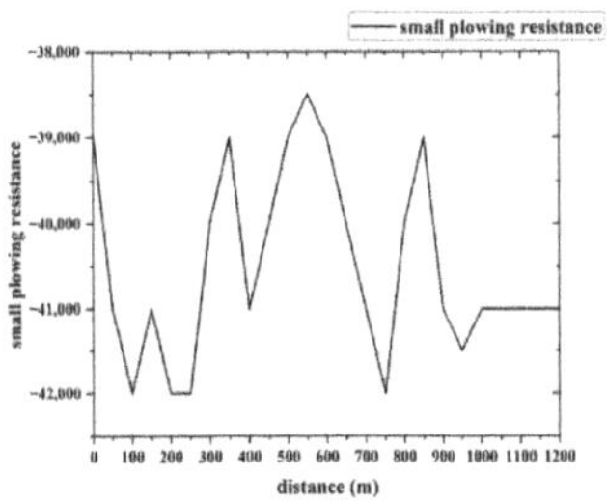

Figure 14. Small plowing resistance.

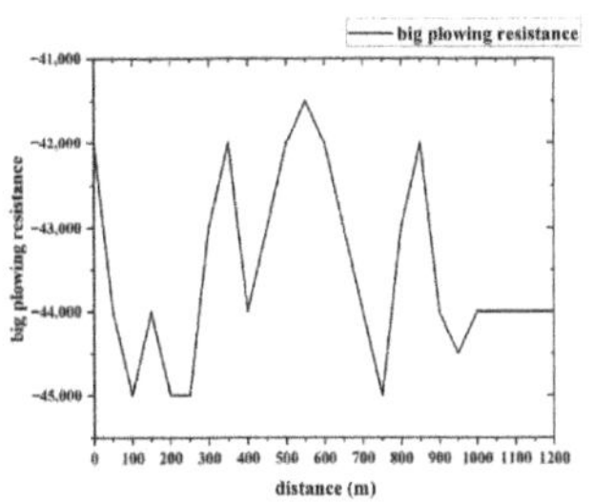

Figure 15. Big plowing resistance.

The set road slope is shown in Figure 16.

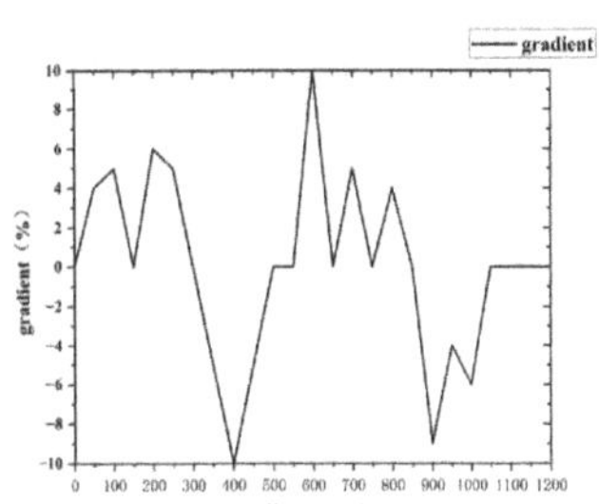

Figure 16. Gradient.

As shown in Figures 17 and 18, when the Hybrid four-wheel drive tractor was loaded with a small plow resistance group, the fuzzy control-based torque distribution control strategy reduced the tractor rear wheel slip rate by 4.5% on average and the maximum slip rate by 18.7% compared to the fixed ratio-based torque distribution control strategy. The fuzzy control-based torque distribution control strategy can basically control the rear-drive wheel slip rate in the range of 10% to 15%, and the variance of the rear-drive wheel slip rate is only 1.16 for the fuzzy control-based torque distribution control strategy and 3.87 for the fixed ratio-based torque distribution control strategy, which shows that the fuzzy control strategy based torque distribution control strategy can not only control the rear-drive wheel slip rate. This shows that the torque distribution control strategy based on the fuzzy control strategy can not only control the slip rate of the rear-drive wheel at 10%~15%, but also the slip rate fluctuation is small, and the Hybrid four-wheel drive tractor can obtain high traction efficiency. At the same time, the fuzzy control-based torque distribution control strategy has a small difference in the slip rate of the front-drive wheel compared to the fixed ratio-based torque distribution control strategy.

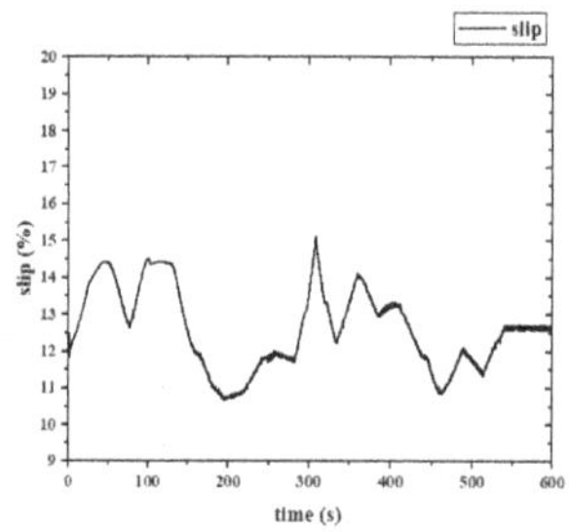

Figure 17. Slip rate of rear-drive wheels with fuzzy control.

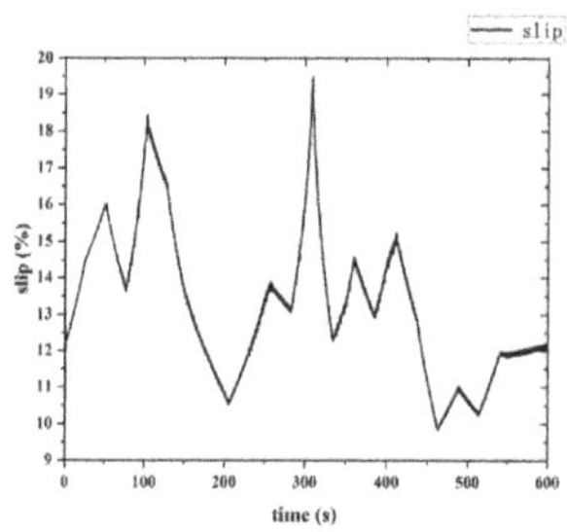

Figure 18. Slip rate of rear-drive wheels with fixed ratio distribution.

When the Hybrid four-wheel drive tractor is loaded with a large plowing resistance group, as shown in Figure 19, the Hybrid four-wheel drive tractor with fuzzy control-based torque distribution control strategy can complete the operation task, and as shown in Figure 20, the Hybrid four-wheel drive tractor with fixed ratio distribution based torque distribution control strategy cannot complete the operation, which shows that the Hybrid four-wheel drive tractor with fuzzy control-based torque distribution control strategy has better adaptability to the working conditions and can cope with the complex agricultural production environment.

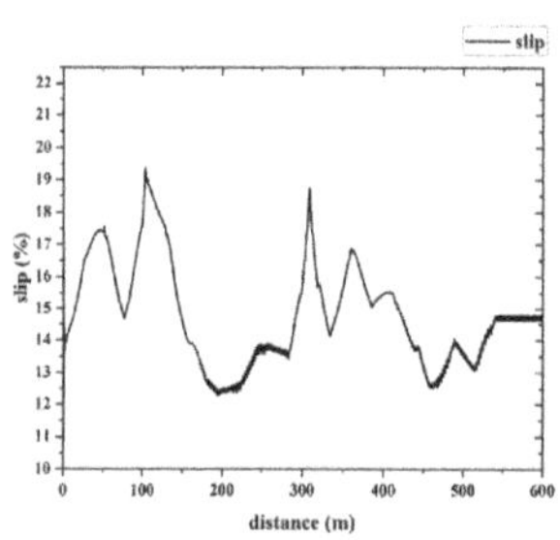

Figure 19. Slip rate of rear-drive wheels with fuzzy control.

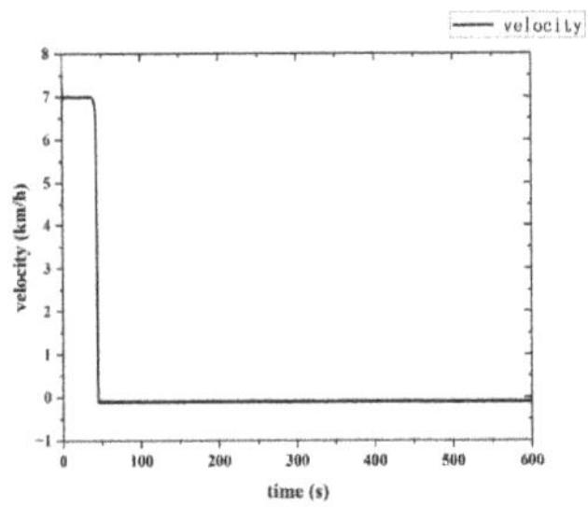

Figure 20. Tractor velocity with fixed ratio distribution.

Through the comparative analysis of two torque distribution control strategies, it can be seen that the fuzzy control-based torque distribution control strategy can significantly reduce the drive wheel slip rate, and the Hybrid four-wheel drive tractor can obtain higher traction efficiency. And good adaptability to working conditions, more widely used.

The rear-drive system adopts the fuzzy control-based torque distribution control strategy, based on the analysis of the tractor plowing conditions, as shown in Figure 21, the tractor plowing operation engine speed fluctuations are not large, smooth operation, while the engine torque in the engine high efficiency torque operating range, the economy is better.

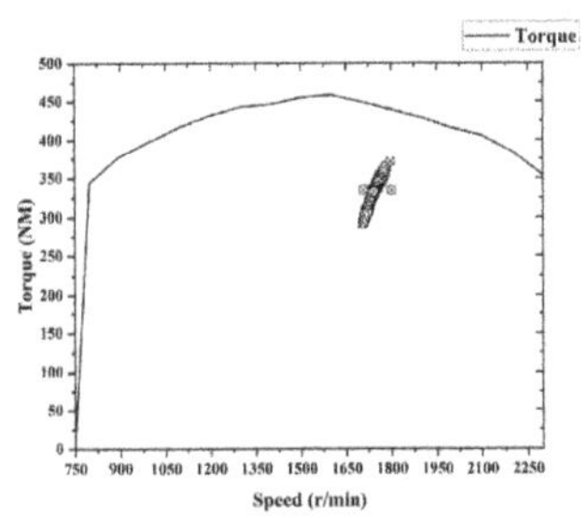

Figure 21. Engine working point distribution map.

6. Conclusions

(1). By analyzing the defects of the existing tractor power system, this paper, a Hybrid four-wheel drive tractor is proposed, and the parameters of the main components are calculated and matched, while a torque distribution control strategy is developed.

(2). The vehicle model of the Hybrid four-wheel drive tractor was established in AVL/CRUISE software, and the vehicle torque distribution control strategy based on fuzzy control and the vehicle torque distribution control strategy based on fixed ratio distribution were established in MATLAB/SIMULINK software. The joint simulation was carried out with the tractor's plowing operation as an example.

(3). The simulation results show that the fuzzy control-based front- and rear-drive wheel torque distribution control strategy can control the slip rate of the tractor rear-drive wheel at 10%~15% compared with the fixed ratio distribution based front- and rear-drive wheel torque distribution control strategy, the average slip rate of the tractor rear wheels has been reduced by 4.5%, the maximum slip rate of the rear-drive wheels has been reduced by 18.7%, and the variance has been reduced by 70%, and the traction performance has been greatly improved.

(4). The simulation results show that when the Hybrid four-wheel drive tractor was loaded with a large plowing resistance group, the Hybrid four-wheel drive tractor with the fuzzy control-based front- and rear-drive wheel torque distribution control strategy was able to complete the operation, and the Hybrid four-wheel drive tractor with the fixed ratio distribution based front- and rear-drive wheel torque distribution control strategy was not able to complete the operation. The results show that the Hybrid four-wheel drive tractor with fuzzy control-based front- and rear-drive wheel torque distribution control strategy has better adaptability to working conditions and can cope with complex agricultural production environments.

(5). The simulation results show that the established torque distribution control strategy for the rear-drive system is able to control the engine operating point within the period of efficient engine operation with better economy.

Author Contributions: Conceptualization, X.Y., H.Z. and L.X.; methodology, X.Y., H.Z. and X.L.; software, H.Z. and Y.L.; validation, X.Y.; data curation, X.Y. and H.Z.; writing—original draft preparation, H.Z.; writing—review and editing, H.Z. All authors have read and agreed to the published version of the manuscript.

Funding: This research was funded by Key Research Programs of Higher Education Institutions in Henan Province, grant number [22B416001]; and National Key Research and Development Program of the 14th Five-Year Plan, grant number [2022YFD2001203, 2022YFD2001201B]; and GG project on key core technologies in agriculture, grant number [NK202216010103]; and State Key Laboratory of Tractor Power System Open Subjects, grant number [SKT2022001]. And The APC was funded by [2022YFD2001203, 2022YFD2001201B].

Data Availability Statement: Not applicable.

Conflicts of Interest: The authors declare no conflict of interest.

References

1. Luo, C.; Wen, C.; Meng, Z.; Liu, H.; Li, G.; Fu, W.; Zhao, C. Research on the Slip Rate Control of a Power Shift Tractor Based on Wheel Speed and Tillage Depth Adjustment. *Agronomy* **2023**, *13*, 281. [CrossRef]
2. Mocera, F.; Martini, V.; Somà, A. Comparative Analysis of Hybrid Electric Architectures for Specialized Agricultural Tractors. *Energies* **2022**, *15*, 1944. [CrossRef]
3. Shao, X.; Yang, Z.; Mowafy, S.; Zheng, B.; Song, Z.; Luo, Z.; Guo, W. Load Characteristics Analysis of Tractor Drivetrain under Field Plowing Operation Considering Tire-Soil Interaction. *Soil Tillage Res.* **2023**, *227*, 105620. [CrossRef]
4. Xia, G.; Xia, Y.; Tang, X.; Gao, J.; Wang, S.; Sun, B. Speed Regulation Control of the Dual-Flow Transmission System for a Tractorusing Slip Rate-Resistance Interval Division. *Trans. Chin. Soc. Agric. Eng.* **2021**, *37*, 47–55. [CrossRef]
5. Bulgakov, V.; Aboltins, A.; Beloev, H.; Nadykto, V.; Kyurchev, V.; Adamchuk, V.; Kaminskiy, V. Maximum Admissible Slip of Tractor Wheels without Disturbing the Soil Structure. *Appl. Sci.* **2021**, *11*, 6893. [CrossRef]
6. Zhu, S.; Wang, L.; Zhu, Z.; Mao, E.; Chen, Y.; Liu, Y.; Du, X. Measuring Method of Slip Ratio for Tractor Driving Wheels Based on Machine Vision. *Agriculture* **2022**, *12*, 292. [CrossRef]
7. Zhang, S.; Du, Y.; Zhu, Z.; Mao, E.; Liu, J.; Shi, J. Integrated Control Method of Traction & Slip Ratio for Rear-Driving High-Power Tractors. *Trans. Chin. Soc. Agric. Eng.* **2016**, *32*, 47–53. [CrossRef]
8. Ma, Y.; Li, R.; Liu, Y.; Xu, J. Research on Plowing Control Method of Agricultural Tractor Based on Slip Rate. *J. Agric. Mech. Res.* **2020**, *42*, 259–263. [CrossRef]
9. Zhao, G.; Xia, C. SimulationX-Based Tractor Slip Rate Control Research. *J. Agric. Mech. Res.* **2021**, *43*, 240–245. [CrossRef]
10. Zhai, Z.; Li, Y.; Guo, J.; Wang, J.; Dong, G.; Guo, Z.; Pang, H. Effect of Tillage Depth on Soil Physical Properties and Yield of Winter Wheat-Summer Maize. *Trans. Chin. Soc. Agric. Eng.* **2017**, *33*, 115–123. [CrossRef]
11. Zhai, C.; Yang, S.; Wang, X.; Zhang, C.; Song, J. Status and Prospect of Intelligent Measurement and Control Technology for Agricultural Equipment. *Trans. Chin. Soc. Agric.* **2022**, *53*, 1–20. [CrossRef]
12. Yu, Y.; Hao, S.; Guo, S.; Tang, Z.; Chen, S. Motor Torque Distribution Strategy for Different Tillage Modes of Agricultural Electric Tractors. *Agriculture* **2022**, *12*, 1373. [CrossRef]
13. Zhang, S.; Wen, C.; Ren, W.; Luo, Z.; Xie, B.; Zhu, Z.; Chen, Z. A Joint Control Method Considering Travel Speed and Slip for Reducing Energy Consumption of Rear Wheel Independent Drive Electric Tractor in Ploughing. *Energy* **2023**, *263*, 126008. [CrossRef]
14. Khalid, M.; Smith, J.L. Axle Torque Distribution in 4WD Tractors. *J. Terramechanics* **1981**, *18*, 157–167. [CrossRef]
15. Li, Y.; Jia, Y.; Sun, P.; Wei, T.; Wang, Y.; Zhou, Y. Discussion of Four-Wheel Drive Tractor Traction Efficiency. *Tract. Farm Transp.* **2014**, *41*, 1–4.
16. Huan-huan, Z.; Xu'ai, X.; Kebao, Y.; Guoping, Y. Research on Torque Distribution for an Electric Vehicle with In-Wheel Motors. In Proceedings of the 2014 IEEE Conference and Expo Transportation Electrification Asia-Pacific (ITEC Asia-Pacific), Beijing, China, 31 August–3 September 2014; pp. 1–6.
17. Wu, S.; Li, Y.; Guan, Y.; Liu, T.; Che, C. Distribution Method of Automotive Torque for Hub Motor Considering Energy Consumption Optimization. *Int. J. Automot. Technol.* **2023**, *24*, 913–928. [CrossRef]
18. Zhai, L.; Sun, T.; Wang, J. Electronic Stability Control Based on Motor Driving and Braking Torque Distribution for a Four In-Wheel Motor Drive Electric Vehicle. *IEEE Trans. Veh. Technol.* **2016**, *65*, 4726–4739. [CrossRef]

19. Zhang, X.; Mi, C. *Vehicle Power Management: Modeling, Control and Optimization*; China Machine Press: Beijing, China, 2013; ISBN 978-7-111-41689-0.
20. Guo, C.; Fu, C.; Zhai, J.; Cao, K.; Luo, R.; Liu, Y.; Pan, H.; Qiao, S. Coordinated Control of Torque Distribution and Acceleration Slip Regulation for Front- and Rear-Independent-Drive Electric Vehicles. *J. Chongqing Univ.* **2022**, *45*, 97–112. [CrossRef]
21. Xu, L.; Zhang, J.; Liu, M. Torque Distribution Strategy of Extended Range Electric Tractor. *J. Henan Univ. Sci. Technol. (Nat. Sci.)* **2017**, *38*, 80–86. [CrossRef]
22. Yin, X.; Lu, Z. Simulation Research on Acceleration Slip Regulation System for Four-Wheel Drive Tractor Using Fuzzy Control Method. *Agric. Equip. Veh. Eng.* **2010**, *12*, 6–10. [CrossRef]
23. Dugoff, H.; Fancher, P.S.; Segel, L. *An Analysis of Tire Traction Properties and Their Influence on Vehicle Dynamic Performance*; SAE International: Warrendale, PA, USA, 1970.
24. Zhang, J. *Research of Control Strategy for Extended-Range Electric Tractor*; Henan University Of Science and Technology: Luoyang, China, 2017.
25. Kang, J. *Design of Dynamic Coupling System for Series-Parallel Hybrid Electric Tractor*; Henan University Of Science and Technology: Luoyang, China, 2022.
26. Li, Y.; Liu, M.; Xu, L.; Lei, S. Control Strategy of Series Hybrid Tractor Based on Nonlinear Program Genetic Algorithm. *J. Jiangsu Univ. (Nat. Sci. Ed.)* **2023**, *44*, 166–172. [CrossRef]
27. Fang, S.; Wang, N.; Yi, K.; Hou, R.; Xu, L.; Xia, X. Research on Operation Performance of Pure Electric Tractor Based on CRUISE. *J. Shandong Univ. Technol. (Nat. Sci. Ed.)* **2019**, *33*, 20–25. [CrossRef]
28. Liu, M.; Xu, L.; Zhou, Z.; Liu, W. Establishment of Extended Range Electric Tractor and Its Rotary Cultivator's Simulative Platforms. *China Mech. Eng.* **2016**, *27*, 413–419. [CrossRef]

World Electric Vehicle Journal

MDPI

Article

Research on Parameter Optimization Design Method for Dual-Motor Coupled Drive System

Tonghui Li *, Nan Zhang, Xiaoyu Gao and Daqian Pang

China North Vehicle Research Institute, Beijing 100072, China; zheshizn@outlook.com (N.Z.); gaoxiaoyunannan@outlook.com (X.G.); pangdaqian@hotmail.com (D.P.)
* Correspondence: litonghui0908@gmail.com

Abstract: To improve energy utilization efficiency and extend the driving range of electric vehicles, this paper proposes a Dual-Motor Coupled Drive System (DMCDS) with a simple structure and establishes a dynamic mathematical model to analyze power flow characteristics under different driving modes. Considering the interdependence between the optimization of component sizes and system control in multi-motor drive systems, a two-layer hybrid optimization method is proposed to determine the optimal component sizes, balancing vehicle performance with minimal system energy losses. To evaluate the effectiveness of the proposed optimization design method, extensive simulation analysis was carried out in MATLAB. The results demonstrate that the optimization of motor sizes and gear ratios can enhance the energy efficiency of the drive system. In comparison with prototype scheme before optimization, the high-efficiency region utilization of motors EM_R and EM_S increased by 45% and 48%, respectively. Compared with the prototype and single-motor drive system, the average drive efficiency after optimization increased by 2.5% and 4.2%, respectively, and the energy consumption per 100 km decreased by 3.6% and 6.8%, respectively. These results confirm the efficacy of the proposed optimization design method in achieving an energy-saving effect.

Keywords: electric vehicles; dual-motor coupled drive; two-layer hybrid optimization

Citation: Li, T.; Zhang, N.; Gao, X.; Pang, D. Research on Parameter Optimization Design Method for Dual-Motor Coupled Drive System. *World Electr. Veh. J.* **2023**, *14*, 282. https://doi.org/10.3390/wevj14100282

Academic Editors: Fachao Jiang, Yongyu Li and Weiwei Kong

Received: 5 August 2023
Revised: 12 September 2023
Accepted: 13 September 2023
Published: 8 October 2023

1. Introduction

As global petroleum resources continue to deplete rapidly and air quality worsens, electric vehicles have been experiencing rapid advancements [1]. Pure electric vehicles have diverse energy sources, such as wind, solar, and hydro power, providing advantages like simple structures and zero emissions in comparison to conventional vehicles [2]. However, the main challenge for electric vehicles remains their limited energy density, long charging times, and restricted driving range due to current battery technology [3]. To mitigate these challenges, besides technological breakthroughs in batteries, the widely embraced and most effective solution lies in reducing the energy losses in the drive system [4]. In pursuit of this objective, a plethora of methods have been proposed, primarily focusing on powertrain configurations, energy management strategies (EMSs), and component size optimization.

At present, the powertrains of EVs on the market are mostly driven by a single motor coupled to a single-speed transmission. The utilization of a single-speed transmission offers a cost-effective solution by effectively reducing the mass, volume, and cost [5–7]. However, the single-motor drive systems exhibit a lower efficiency at low torques and low speeds, leading to a higher probability of the motor operating within the low-efficiency region. Therefore, it is necessary to find other powertrain structures that can improve drive efficiency. Many studies have shown that dual-input coupling powertrain systems have been widely used in electric vehicles to reduce energy consumption and improve efficiency [8–11]. Utilizing two smaller motors instead of a single high-power source allows for a reduction in the torque capacity of each individual motor, thereby facilitating the development of high-speed motors and increasing the power density of the drive system.

Moreover, the operating points of the two motors can be adjusted to optimize the efficiency of the drive system [12].

After determining the powertrain configuration, the subsequent task involves optimizing the component sizes and designing the EMS for the drive system [13]. Regarding EMS design, stochastic dynamic programming (SDP) [14], Pontryagin's Maximum Principle (PMP) [15], and the dynamic programming (DP) algorithm [16] are widely employed optimization algorithms. The DP algorithm can provide a global optimal strategy when the entire driving cycle information is available. Nevertheless, its real-time online application is limited as it necessitates knowledge of future road grade and vehicle speed information [17].

Currently, some research progress has been made in the optimization of drive system parameters. References [18–20] utilized genetic algorithms and particle swarm optimization (PSO) algorithms to optimize the component parameters of hybrid drive systems, resulting in reduced energy consumption. However, these studies only performed static optimization of the objective function during parameter optimization, without considering the impact of component parameter variations on the high-efficiency region and the coordinated control of power sources. Focusing on a single aspect alone cannot achieve optimal system performance; instead, an integrated optimization of both component parameters and system control strategies is required.

In recent years, researchers around the world have made notable progress in cooperative optimization methods and double-layer control strategies [21–25]. Angelo et al. proposed a novel double-layer control architecture designed to drive the longitudinal motion of electric vehicles. The control architecture, by combining the two control strategies, can reduce the overall energy consumption of electric vehicles [26]. Fathy et al. demonstrated the significant influence of control strategy optimization on the effectiveness of parameter optimization, affirming the existence of coupling between parameter optimization and control strategy optimization [27]. Fang et al. identified Pareto optimal solutions using a comprehensive optimization approach to concurrently optimize powertrain components and control systems [28]. In the context of multi-mode hybrid electric vehicles, Zhuang integrated energy management control strategy optimization, topology configuration optimization, and component parameter matching, proposing both cage optimization and iterative optimization architectures. The results demonstrated that the iterative optimization architecture efficiently converged to the global optimal solution [29]. Nguyen et al. proposed a two-loop optimization algorithm, combined with the global search method and non-dominated sorting genetic algorithm-II to find optimal motor sizes and transmission ratios for the powertrain of electric vehicles equipped with two motors and multi-gear ratios. The simulation results showed that the optimization of both motor sizes and gear ratios considerably enhances the energy efficiency of the powertrain system [30].

The objective of this research is to develop a dual-motor coupled drive system for electric vehicles that enhances the vehicle's energy efficiency while ensuring dynamic performance. To further improve the energy utilization efficiency of the DMCDS, a two-layer hybrid optimization method is proposed to synergistically optimize the system component sizes and control strategies. Through simulation experiments, the optimal control parameters are determined, and the most suitable parameter configuration for the driving cycle conditions is identified.

The rest of this paper is organized as follows: Section 2.1 introduces the overall configuration of the DMCDS. The modeling and driving modes of the DMCDS are displayed in Section 2.2. In Section 2.3, a two-layer hybrid optimization method is presented, which aims to determine the optimal parameters of the system components. The simulation results and discussion are given in Section 3, where two typical driving cycles are used to evaluate the effectiveness of the proposed method. Finally, the conclusions are presented in Section 4.

2. Materials and Methods

2.1. Configuration of DMCDS

As the central component of electric vehicles, the Dual-Motor Coupled Drive System exerts a direct impact on the dynamic performance of the vehicle. By conducting a comparative analysis of multiple configuration schemes, this study proposes a dual-motor coupled drive system configuration that incorporates a planetary gear mechanism as the power coupling device, as illustrated in Figure 1. It is mainly composed of two motors, EM_R and EM_S, two electromagnetic brakes, B1 and B2, and a power coupling gearbox. Notably, motors EM_S and EM_R are, respectively, linked to the sun gear and ring gear, while the carrier establishes a connection to the power output of the main reducer.

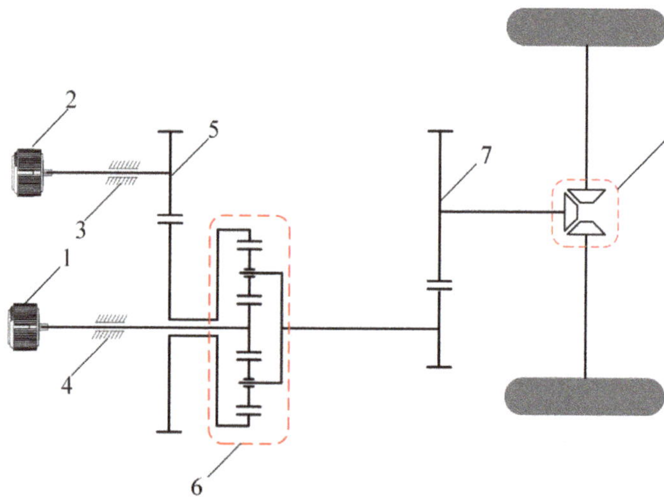

Figure 1. Schematic diagram of DMCDS. 1. EM_S; 2. EM_R; 3. B1; 4. B2; 5. ring gear; 6. planetary gear mechanism; 7. main reduction gear; 8. differential mechanism.

2.2. Modeling and Mode Analysis

2.2.1. Dynamics Modeling of DMCDS

The primary focus of this paper is to investigate a dual-motor coupled drive system, consisting of two motors, which achieve power coupling through the utilization of a planetary gear mechanism. In situations where the vehicle speed and required torque are known, the operational states of EM_S and EM_R are not uniquely determined for the DMCDS. By combining the static kinematic equations of the planetary gear mechanism with the dynamic model of the planetary gear set, we derive the dynamic model of the dual-motor coupled drive system as shown in the following equation.

$$\begin{cases} \dot{\omega}_{ms} = \dfrac{a(T_{ms} - \frac{T_{mr}}{k+1}) - b(T_{ms} - \frac{T_{mr} - 2T_c}{k+1})}{ad - bc} \\ \dot{\omega}_{mr} = \dfrac{c(T_{ms} - \frac{T_{mr}}{k+1}) - d(T_{ms} - \frac{T_{mr} - 2T_c}{k+1})}{bc - ad} \\ \dot{\omega}_c = \frac{\dot{\omega}_{ms} + k\dot{\omega}_{mr}}{1+k} = \frac{i_m \dot{v}_a}{r_w} \\ a = \frac{J_r}{k+1} + \frac{2kJ_c}{(k+1)^2}, b = -(\frac{J_r}{k+1} + \frac{kJ_p}{(k-1)^2}) \\ c = J_s + \frac{2J_c}{(k+1)^2}, d = J_s + \frac{2J_p}{(k-1)^2} \end{cases} \quad (1)$$

where ω_{ms}, ω_{mr}, and ω_c are the output speeds of motors EM_S and EM_R, and the carrier, respectively; T_{ms}, T_{mr}, and T_c are the output torque of motors EM_S and EM_R, and the planet carrier; J_s, J_r, J_c, and J_p represent the equivalent moments of inertia of the sun gear, ring gear, carrier, and planetary gear, respectively; i_m, r_w, and k denote the main reduction ratio, rolling radius of the wheels, and the planetary gear ratio, respectively; and v_a denotes the vehicle speed.

2.2.2. Driving Mode Analysis

The operational states of components in the DMCDS vary as it operates in different driving modes, leading to distinct dynamic mathematical models. Therefore, determining the operational states for different driving modes is crucial for the efficient functioning of the DMCDS. Through effective coordination of motor and brake controls, the DMCDS can seamlessly switch between three driving modes: Motor EM_S independent drive (M1S), Motor EM_R independent drive (M1R), and Dual-Motor Coupled Drive (DMC). Table 1 provides the working states of each component under different driving modes.

Table 1. System operation status.

Working States	Driving Modes	M1	M2	B1	B2
Park/Neutral	N/P	○	○	○	○
Driving states	M1S	•	○	○	•
	M1R	○	•	•	○
	DMC	•	•	○	○

• signifies the activation of the motor or engagement of the brake, while ○ denotes the deactivation of the motor or disengagement of the brake.

When operating in the M1S mode, the DMCDS disengages brake B1, engages brake B2, and deactivates motor EM_R. The power from motor EM_S is transmitted through the sun gear and output by the planet carrier, resulting in a higher transmission ratio for the system. This mode exhibits its advantage in situations where the vehicle requires significant torque at low speeds, allowing EM_S to operate efficiently in its high-efficiency region. It proves particularly effective for low-speed, high-torque scenarios, such as rapid acceleration and uphill driving. Figure 2 illustrates the equivalent lever model of the DMCDS.

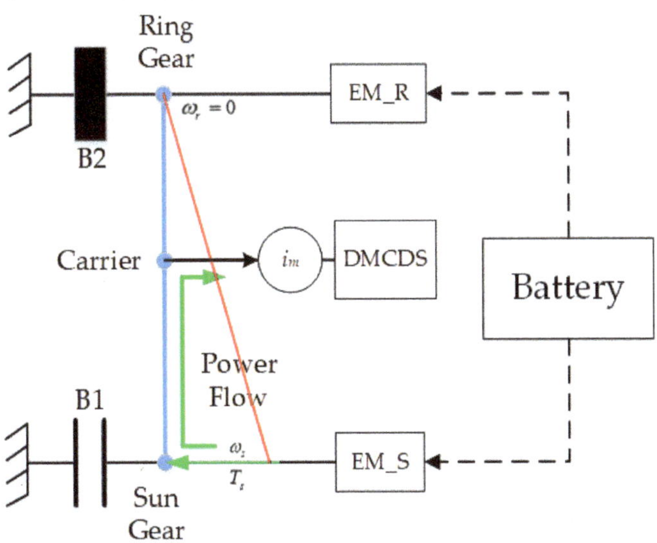

Figure 2. Equivalent lever model in M1S mode. The red line in the figure represents the lever and the green line represents the power flow.

Utilizing the equivalent lever model in conjunction with the dynamic model described by Equation (1), the dynamic mathematical model for the M1S mode can be deduced.

$$\begin{cases} (T_{ms} - J_{ms}\dot{\omega}_{ms})\cdot(1 + k) - \frac{T_w}{i_m} = J_c\dot{\omega}_c \\ J_c\dot{\omega}_c = \frac{m_s r_w^2}{i_m^2}\cdot\frac{\dot{v}_a i_m}{r_w} = \frac{m_s r_w \dot{v}_a}{i_m} \\ \omega_{ms} = \frac{v_a i_m}{r_w}(1 + k) \end{cases} \quad (2)$$

where J_{ms} denotes the equivalent moment of inertia of motor EM_S, T_w denotes the load torque on the driving wheel, and m_s denotes the vehicle weight.

When operating in the M1R mode, the DMCDS deactivates motor EM_S and engages brake B1 to apply a braking force on the sun gear, effectively transforming the DMCDS into a single-degree-of-freedom system. The output torque of EM_R is transmitted through the ring gear and planet carrier, delivering power to the wheels via the main reducer. Therefore, selecting the M1R mode when the vehicle demands higher power allows the system to leverage its advantages, with motor EM_R operating more efficiently within its high-efficiency range. Figure 3 illustrates the equivalent lever model of the DMCDS.

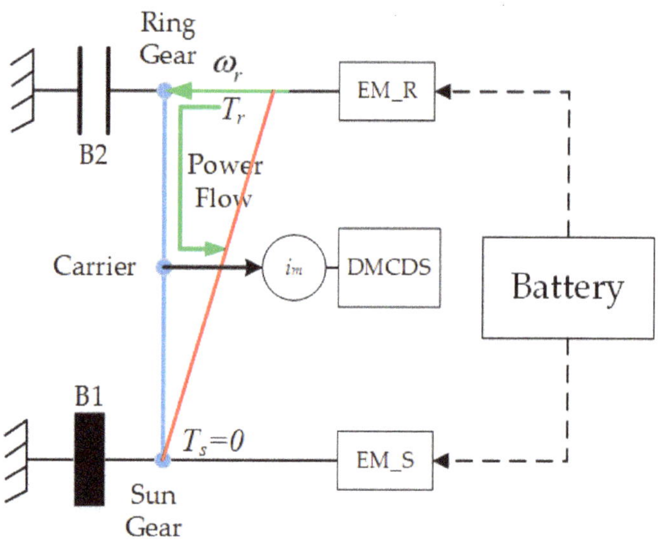

Figure 3. Equivalent lever model in M1R mode. The red line in the figure represents the lever and the green line represents the power flow.

Utilizing the equivalent lever model in conjunction with the dynamic model described by Equation (1), the dynamic mathematical model for the M1S mode can be deduced.

$$\begin{cases} (T_{mr} - J_{mr}\dot{\omega}_{mr})\frac{1+k}{k} - \frac{T_w}{i_m} = \frac{m_s r_w \dot{v}_a}{i_m} \\ \omega_{mr} = \frac{v_a i_m}{r_w}\cdot\frac{1+k}{k} \end{cases} \quad (3)$$

where J_{mr} denotes the equivalent moment of inertia of motor EM_R.

When operating in the Dual-Motor Coupled Drive mode, the DMCDS disengages both brakes B1 and B2. Motors EM_S and EM_R operate simultaneously, with both motors providing power to the planetary gear mechanism. The equivalent lever model of the DMCDS in this mode is illustrated in Figure 4.

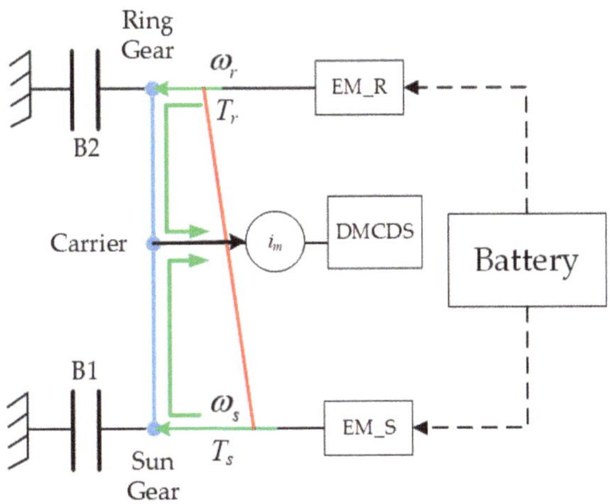

Figure 4. Equivalent lever model in DMC mode. The red line in the figure represents the lever and the green line represents the power flow.

When the single-motor drive mode falls short of satisfying the high-power requirements or when the motor cannot operate within the high-efficiency range, the Dual-Motor Coupled Drive mode can be utilized. By adjusting the output speeds of both motors, continuous speed regulation is achieved, contributing to the improved operational efficiency of both motors. This mode proves advantageous for driving scenarios with substantial power demands or higher vehicle speeds. The dynamic model under the DMC drive mode can be represented as follows:

$$\begin{cases} (\min(T_{ms}, T_{mr}/k) - J_{ms}\dot{\omega}_{ms} - J_{mr}\dot{\omega}_{mr}/k)(1+k) = \frac{m_s r_w \dot{v}_a + T_w}{i_m} \\ \omega_{ms} + k\omega_{mr} = (1+k)\frac{v_a i_m}{r_w} \end{cases} \quad (4)$$

2.3. Parameter Optimization of DMCDS

2.3.1. Mathematical Models

(1) Vehicle model

To examine the relationship between variations in vehicle speed and the output characteristics of the Dual-Motor Coupled Drive System, it is essential to establish a longitudinal vehicle dynamics model that accounts for slip ratio. Leveraging the system configuration and vehicle dynamics, the longitudinal dynamic model of the vehicle can be expressed as follows:

$$\begin{cases} T_c i_m \eta_t - T_w = J_w \dot{\omega}_w \\ T_w = \left(m_s g f_r + C_D A_f {v_a}^2/21.15\right) r_w \\ v_a = \frac{\omega_w r_w}{1+\lambda} \\ \omega_w i_m (k+1) = \omega_{ms} + k\omega_{mr} \end{cases} \quad (5)$$

where ω_w, T_w, and J_w are the speed, torque, and moment of inertia of the wheel, respectively; f_r is the tire rolling resistance coefficient; C_D is the aerodynamic drag coefficient; A_f is the vehicular frontal area; and λ is the slip ratio of the driving wheel. The values of the vehicle parameters are displayed in Table 2.

Table 2. Basic parameters of the vehicle.

Parameter	Meaning	Value
m_s (kg)	Mass of vehicle	1949
A_f (m^2)	Frontal area	2.66
C_D	Air resistance coefficient	0.4
fr	Tire rolling friction coefficient	0.015
r_w (m)	Tire radius	0.343
$v_{\max}$ (km/h)	Maximum velocity	150
t_{acc} (s)	0–100 km/h acceleration time	9

(2) Motor model

In the domain of electric vehicles, the peak power of the motor is commonly determined by the acceleration demands. Consequently, if the acceleration time is pre-established, the total peak power of the two motors remains constant and can be expressed as

$$P_{\text{maxs}} + P_{\text{max}r} = 2P_{\text{max}b} \tag{6}$$

where P_{maxs} and P_{maxr} are the peak power of motors EM_S and EM_R, respectively; and P_{maxb} denotes the peak power of the baseline motor EMB, which equals half of the maximum vehicle power.

To maximize the average efficiency of the DMCDS under various driving cycles, selection of the power levels for the motors is essential. To streamline the model, the efficiency map of both motors has the same shape as motor EMB. As a result, EM_S and EM_R share the same speed range as EMB, while the torque range is proportional to their respective power, which can be expressed as

$$\begin{cases} P_{\text{maxs}} = 2\alpha P_{\text{max}b}, P_{\text{max}r} = 2(1-\alpha)P_{\text{max}b} \\ T_{\text{maxs}} = 2\alpha T_{\text{max}b}, T_{\text{max}r} = 2(1-\alpha)T_{\text{max}b} \\ \alpha \in (0,1) \end{cases} \tag{7}$$

where α is the power scaling factor between the motors; and T_{maxs}, T_{maxr}, and T_{maxb} are the peak output torque of motors EM_S, EM_R, and EMB, respectively.

The case of α = 1 corresponds to EM_S and EM_R being the baseline motors in the efficiency map shown in Figure 5.

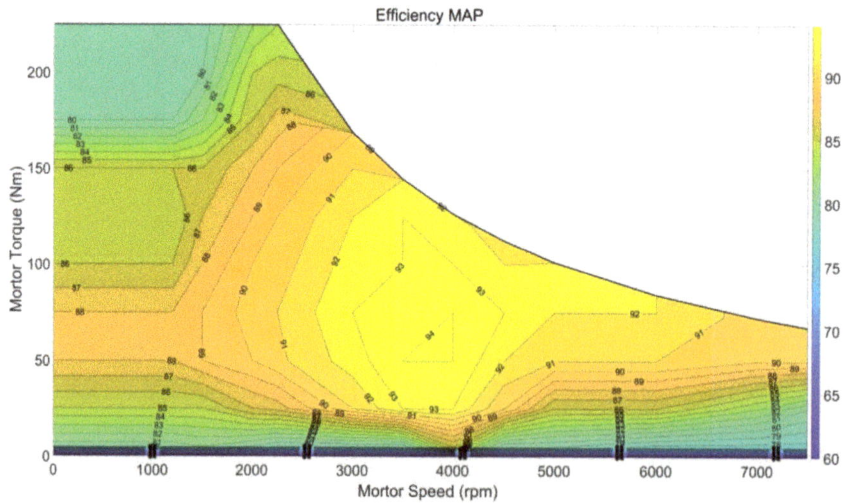

Figure 5. Efficiency map of motor EMB.

(3) Battery model

The thermal temperature effect and battery life are ignored in this paper, and the basic physical model of the battery is simplified as a voltage source with open-circuit voltage and internal resistance, each of which depend on the battery *SOC*, so the mathematical model of the battery can be expressed as

$$SOC(t+1) = -\frac{U_{oc} - \sqrt{U_{oc}^2 - 4R_{bat} \cdot P_{bat}}}{2R_{bat} \cdot Q_{bat}} + SOC(t) \quad (8)$$

where U_{oc} is the battery open circuit voltage; R_{bat} is the internal resistance; P_{bat} is the output power of the battery, which is also the electric power consumed by the two motors; Q_{bat} is the capacity of battery; and the index t represents any time instant.

It is worth noting that both U_{oc} and R_{bat} can be obtained from the look-up table of the battery *SOC*, and then the instantaneous internal energy of battery P_{ele} can be obtained from

$$P_{ele} = -U_{oc}Q_{bat}(SOC(t+1) - SOC(t)) \quad (9)$$

(4) Efficiency model

Since this paper is mainly focused on the influence of the selection of the DMCDS on the economic performance of electric vehicles, it is necessary to establish efficiency models of the DMCDS, including a motor efficiency model and a transmission efficiency model.

The motor efficiency can be defined as a function of output speed and torque, and the efficiency in a certain state can be obtained by the interpolation method. The efficiency map of the motors has the same shape as motor EMB.

Here, we only consider the gear efficiency in the coupling box, including the dynamic efficiency of the planetary gear set and the efficiency of the main reduction gears. For a pair of gears, one of the most widely used efficiency models is as follows [31]:

$$\eta = 1 - \left| \frac{1}{5}\left(\frac{1}{z_a} \pm \frac{1}{z_b}\right) \right| \quad (10)$$

where Z_a and Z_b represent the number of teeth, and the $\pm$ symbol indicates external (+) and internal gear pairs (−). In terms of the planetary gear set, there are different control strategies and energy losses for different design parameters and driving modes, so the efficiency model should be established for each of the three driving modes.

$$\begin{cases} \eta_{r(s-c)} = 1 - \frac{k \cdot (1 - \eta_{c(s-r)})}{1+k} & M1S \\ \eta_{s,r-c} = 1 - \left| \frac{\omega_{ms} - \omega_c}{\omega_{ms} + k\omega_{mr}} \right| (1 - \eta_{c(s-r)}) & DMC \\ \eta_{s(r-c)} = 1 - \frac{1 - \eta_{c(s-r)}}{1+k} & M1R \end{cases} \quad (11)$$

where $\eta_{r(s-c)}$ indicates the efficiency when the ring gear is fixed and power is input into the sun gear and output from the planet carrier; $\eta_{c(s-r)}$ is the efficiency when the planet carrier is fixed with power input into the sun gear and output from the ring gear; $\eta_{s(r-c)}$ denotes the efficiency when the sun gear is fixed and power is input into the ring gear and output from the planet carrier; and $\eta_{s,r-c}$ denotes the efficiency when power is input into the ring gear and sun gear and output from the planet carrier.

2.3.2. Optimization Problem

The performance of the DMCDS is closely related to the parameter design of components such as the motors and the power-coupled gearbox. To achieve the optimal match between the DMCDS output characteristics and the vehicle load requirements, it is crucial to determine the external characteristic parameters of the motors based on the power constraints. Furthermore, considering the driving cycle conditions, the rated parameters and transmission ratio should be determined, and the high-efficiency region should be optimized to enhance the energy utilization efficiency.

(5) Inner-layer optimization

Once the system configuration and component parameters have been established, the state transition equation of the DMCDS can be formulated based on Bellman's optimization theory. The optimal control problem of the system can be formulated as follows:

$$\begin{cases} x(t+1) = f(x(t), u(t)) \\ J^*(x(t)) = \min\limits_{u(t)}\{J^*(x(t+1)) + L(x(t), u(t))\} \end{cases} \tag{12}$$

where x_t is the current state variables, u_t is the current decision variables, x_{t+1} denotes the state variables at the next time step, $J^*(x(t))$ denotes the optimal value function from stage t to the terminal state, and $L(x(t), u(t))$ is the stage cost function of the system.

For the dual-motor coupled drive system, determining the optimal power allocation ratio at each moment under specific driving cycles is essential to enhance the driving performance and improve the energy-economic efficiency. Consequently, the state variables primarily consist of the power allocation ratios of the two motors, denoted as x_1, and the current operating mode of the drive system, denoted as x_2. The power allocation ratio increment is denoted as the decision variable $u1$, while the command for mode switching serves as the decision variable u_2. The power allocation ratio of the two motors represents the ratio of the output power of motor EM_R to the total required power and can be expressed as follows:

$$\text{PAR} = \frac{P_{mr} \cdot \eta_c}{\omega_c \cdot T_c} \tag{13}$$

where PAR denotes the power allocation ratio of the two motors, P_{mr} is the output power of motor EM_R, and η_c represents the transmission efficiency of the planetary gear.

When PAR = 0, the DMCDS operates in the M1S mode. Conversely, when PAR = 1, the DMCDS operates in the M1R mode. For the range 0 < PAR < 1, the DMCDS operates in the DMC mode. The state transition equations for state variables x_1 and x_2 are as follows:

$$\begin{aligned} x_1(t+1) &= x_1(t) + u_1(t) \\ x_2(t+1) &= \begin{cases} -1 & x_2(t) + u_2(t) < -1 \\ x_2(t) + u_2(t) & \text{other} \\ 1 & x_2(t) + u_2(t) > 1 \end{cases} \end{aligned} \tag{14}$$

where x_2 takes values from the set {−1, 0, 1}, corresponding to the M1S mode, DMC mode, and M1R mode, respectively; and u_2 is restricted to the set $\{-1, 0, 1\}$, signifying the downshift, neutral, and upshift, respectively.

During the adoption of the dynamic programming optimization process, the primary objective of the system is to minimize the energy losses within the DMCDS, which can be expressed as

$$\begin{cases} J = \sum\limits_{t=0}^{N-1} L(x(t), u(t)) = \sum\limits_{t=0}^{N-1} L_{mR}(t) + L_{mS}(t) + L_{gear}(t) \\ L_{mR}(t) = \frac{T_{mr}(t) \cdot \omega_{mr}(t)}{1000 \cdot \eta_{mr}(t)^{sign(T_{mr})}} \cdot \left(1 - \eta_{mr}(t)^{sign(T_{mr})}\right) \\ L_{mS}(t) = \frac{T_{ms}(t) \cdot \omega_{ms}(t)}{1000 \cdot \eta_{ms}(t)^{sign(T_{ms})}} \cdot \left(1 - \eta_{ms}(t)^{sign(T_{ms})}\right) \\ L_{gear}(t) = \frac{T_{mr}(t) \cdot \omega_{mr}(t) + T_{ms}(t) \cdot \omega_{ms}(t)}{1000} (1 - \eta_C(t)) \end{cases} \tag{15}$$

where L_{mR}, L_{mS}, and L_{gear} represent the power losses of motors EM_R and EM_S, and the planetary gear mechanism, respectively; and η_{mr} and η_{ms} denote the operational efficiencies of motors EM_R and EM_S, respectively.

To ensure the safety and efficient operation of the DMCDS during the optimization process, the following constraints need to be applied:

$$\begin{cases} T_{ms\_\min}(n_{ms}(t)) \le T_{ms}(t) \le T_{ms\_\max}(n_{ms}(t)) \\ T_{mr\_\min}(n_{mr}(t)) \le T_{mr}(t) \le T_{mr\_\max}(n_{mr}(t)) \\ n_{ms\_\min} \le n_{ms}(t) \le n_{ms\_\max} \\ n_{mr\_\min} \le n_{mr}(t) \le n_{mr\_\max} \end{cases} \tag{16}$$

where $T_{ms\_\min}$, $T_{ms\_\max}$ and $T_{mr\_\min}$, $T_{mr\_\max}$ denote the minimum and maximum torque of motors EM_S and EM_R at the current speed, respectively; and $n_{ms\_\min}$, $n_{ms\_\max}$ and $n_{mr\_\min}$, $n_{mr\_\max}$ denote the minimum and maximum speed of motors EM_S and EM_R, respectively.

(6) Outer-layer component parameter optimization

When optimizing the parameters of the DMCDS, it is necessary to consider the joint minimization of energy losses and component costs. Therefore, the objective function *J(p)* is defined as a weighted sum of energy losses and component costs:

$$\begin{cases} J(p) = \gamma_1 \cdot J_1(p)/J_1{}^N + \gamma_2 \cdot J_2(p)/J_2{}^N \\ J_2(p) = C_{mot} + C_{pe} \\ C_{mot} = -779.1 + 450.8 \cdot \ln(T_{\max}) \\ C_{pe} = 3160 + 30.1 \cdot T_{\max} \end{cases} \tag{17}$$

where $J_1(p)$ is the energy losses of the DMCDS, corresponding to the inner optimization objective; $J_2(p)$ is the overall cost of the electric system; C_{mot} and C_{pe} are the costs associated with the motor and controller, respectively; and $T_{\max}$ is the peak torque of the motors. γ_1 and γ_2 are the weighting factors, $\gamma_1, \gamma_2 \in [0,1]$, and $\gamma_1 + \gamma_2 = 1$. To achieve a more reasonable weight distribution, and recognizing that the two objective sub-functions carry distinct physical meanings, we introduce the objective expectations $J_1{}^N$ and $J_2{}^N$ as normalization factors.

Since the efficient performance of the DMCDS primarily relies on the rated parameters of the motors and the planetary gear ratio, these five parameters are chosen as the optimization variables. These optimization variables need to satisfy the dynamic performance requirements, including maximum vehicle speed, maximum climbing gradient, and 0–100 km/h acceleration time. When the vehicle is operating at the maximum speed, the DMCDS component parameters must satisfy the following equation:

$$\begin{cases} P_{ms} + P_{mr} \ge \dfrac{(m_s g f_r + C_D A_f v_{\max}{}^2/21.15) v_{\max}}{3600 \eta_{sys}} \\ T_c(v_{\max}) \ge \dfrac{(m_s g f_r + C_D A_f v_{\max}{}^2/21.15) r_w}{\eta_{sys} i_m} \end{cases} \tag{18}$$

To meet the requirements for the maximum climbing gradient, the component parameters must satisfy the following equation:

$$\begin{cases} P_{ms} \ge \dfrac{(m_s g f_r \cos(\theta_{\max}) + m_s g \sin(\theta_{\max}) + C_D A_f v_{10}{}^2/21.15) v_{10}}{3600 \eta_{sys}} \\ T_{\max s}(k+1) \ge \dfrac{(m_s g f_r \cos(\theta_{\max}) + m_s g \sin(\theta_{\max}) + C_D A_f v_{10}{}^2/21.15) r_w}{\eta_{sys} i_m} \end{cases} \tag{19}$$

To meet the requirements for the 0–100 km/h acceleration time, the component parameters need to satisfy the following equation:

$$\left\{ P_{ms} + P_{mr} \ge \frac{1}{3600 \eta_{sys} t_{acc}} \left((m_s g f_r + C_D A_f v_{acc}{}^2/21.15) \int_0^{t_{acc}} v_{acc} \left(\frac{t}{t_{acc}} \right)^{0.5} dt \right) \right. \tag{20}$$

where η_{sys} is the system efficiency, $\theta_{\max}$ is the maximum gradient, v_{acc} is the vehicle speed at the end of acceleration, and t_{acc} is the acceleration time.

Considering vehicle dynamic performance indicators, the feasible range of the optimization variables is obtained and shown in Table 3.

Table 3. Range of parameters to be optimized.

Parameter	Lower Limit	Upper Limit
Rated power of EM_S (kW)	30	60
Rated speed of EM_S (r/min)	2500	4000
Rated power of EM_R (kW)	30	60
Rated speed of EM_R (r/min)	2500	4000
Planetary gear ratio	1.5	4
Final drive ratio	4	6.5

2.3.3. Optimization Process

To achieve superior energy efficiency while ensuring dynamic performance, we propose a two-layer hybrid optimization method based on the Particle Swarm Optimization (PSO) algorithm and Dynamic Programming (DP) algorithm for the DMCDS. This method effectively addresses the coupled effect of component parameters and system control. The two optimization layers work in conjunction to optimize the critical component parameters of the DMCDS. In the outer layer, the PSO algorithm optimizes the component parameters using the optimal control parameters provided by the inner layer. In the inner layer, the DP algorithm is applied to determine the optimal control parameters based on the component parameters given by the outer layer. This comprehensive approach enables global optimization of both component parameters and system control. The optimization process is visually illustrated in Figure 6.

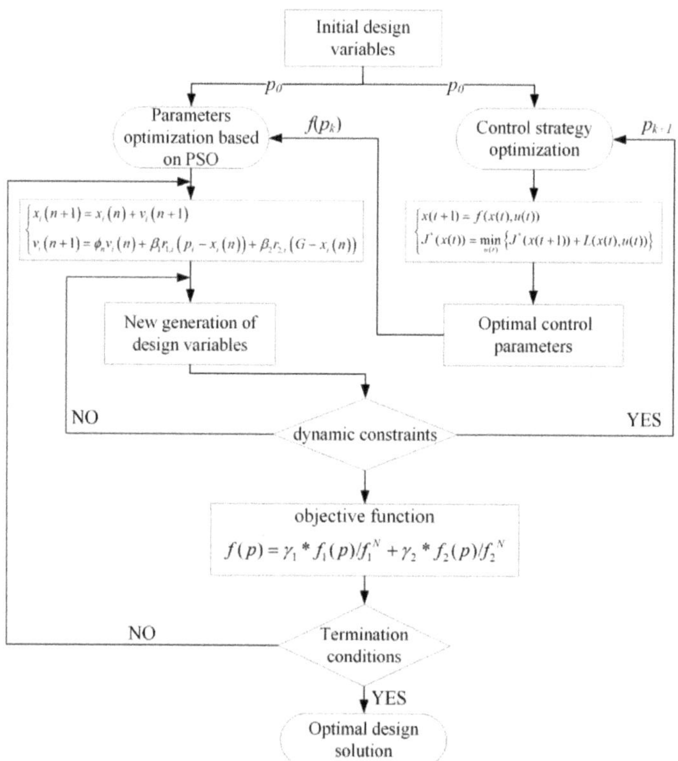

Figure 6. Optimization process of two-layer hybrid optimization method.

3. Results and Discussion

This section provides an overview of the Dual-Motor Coupled Drive System (DMCDS) parameter optimization results achieved through the two-layer hybrid optimization method, while highlighting the differences in energy-economic efficiency between pre-optimization and post-optimization. In this study, the DMCDS was specifically designed for urban SUVs, and the China light-duty vehicle test cycle and worldwide harmonized light-duty vehicles test cycles (CLTC and WLTC) were employed as the optimized driving cycles. Therefore, the vehicle speed over the driving cycles is shown in Figure 7.

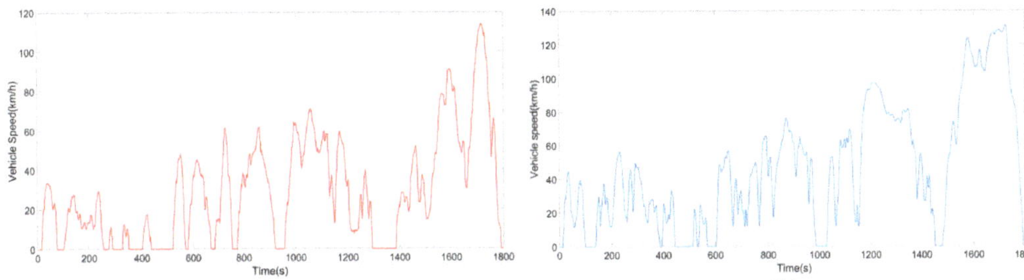

(a) (b)

Figure 7. Optimized driving cycles: (**a**) CLTC-P and (**b**) WLTC.

Under these two driving cycles, the component parameters of the DMCDS were optimized using the two-layer hybrid optimization approach, and the optimized parameters are listed in Table 4. A comparative analysis was conducted, which resulted in comparing the optimized configuration with the prototype configuration that utilized two identical baseline motors.

Table 4. Comparison of component parameters.

Optimization Parameter	Optimized Parameter Value	Prototype Parameter Value
Rated power of EM_S (kW)	33.5	32
Rated speed of EM_S (r/min)	3500	2250
Rated power of EM_R (kW)	31.5	32
Rated speed of EM_R (r/min)	4000	2250
Planetary gear ratio	2.26	1.86
Final drive ratio	5.15	4.93

Compared to the prototype scheme, the optimized motors (EM_S and EM_R) exhibit increased rated speeds and reduced peak torques. Consequently, the motor's constant power region shifts to higher speeds, accompanied by an expanded constant torque region. These enhancements significantly contribute to elevated operational load rates for both motors, improving the utilization of the high-efficiency operational range. Furthermore, the increase in the planetary gear ratio (k) leads to a higher proportion of output power from EM_R and amplifies the output torque of EM_S, thus elevating the operational efficiency during high-speed, high-power and low-speed, high-torque conditions, fully leveraging the advantages of the DMCDS for efficient operation across various modes.

In contrast, the prototype scheme merely aligns with extreme load conditions of the vehicle, neglecting the critical match between the motor's high-efficiency range and the driving cycles. This oversight increases the probability of the motor operating in low-load and low-efficiency regions, affecting the driving range of the electric vehicle.

To explore the effects of the two-layer hybrid optimization method on the energy-economic efficiency of the DMCDS, this study conducted a comparative analysis of three drive system schemes under the same vehicle parameters and driving cycle conditions. The three schemes include the Single Motor Drive System (SMDS), the DMCDS prototype scheme (DMCDS-pro), and the optimized DMCDS scheme (DMCDS-opt). The primary parameters of the SMDS can be found in Table 5. It is worth noting that the efficiency map of the motor in SMDS has the same shape as the baseline motor. To attain the optimal economic efficiency for the SMDS, we adopted the enumeration method for the gear ratio optimization.

Table 5. Parameters of SMDS.

Parameter	Value
Rated power of motor (kW)	64
Peak power of motor (kW)	106
Rated speed of motor (r/min)	2250
First gear ratio	3.27
Second gear ratio	1.98
Final drive ratio	2.826

Figure 8 illustrates the distribution of operating points for the motors in the three drive system configurations. It is evident that, compared to SMDS, both DMCDS-pro and DMCDS-opt have a higher number of operational points located in the high-efficiency region for motors EM_S and EM_R. This indicates the advantage of the dual-motor coupled drive system in achieving efficient operation through coordinated control in various modes. When comparing the operational point distribution of DMCDS-pro and DMCDS-opt, the optimized DMCDS demonstrates a greater number of operational points in the high-efficiency region. This can be attributed to the increase in rated speed and decrease in peak torque of the two motors, resulting in higher load rates for both motors. Consequently, the utilization efficiency of the high-efficiency region for the motors is enhanced, so the DMCDS-opt should have higher system efficiency in this case.

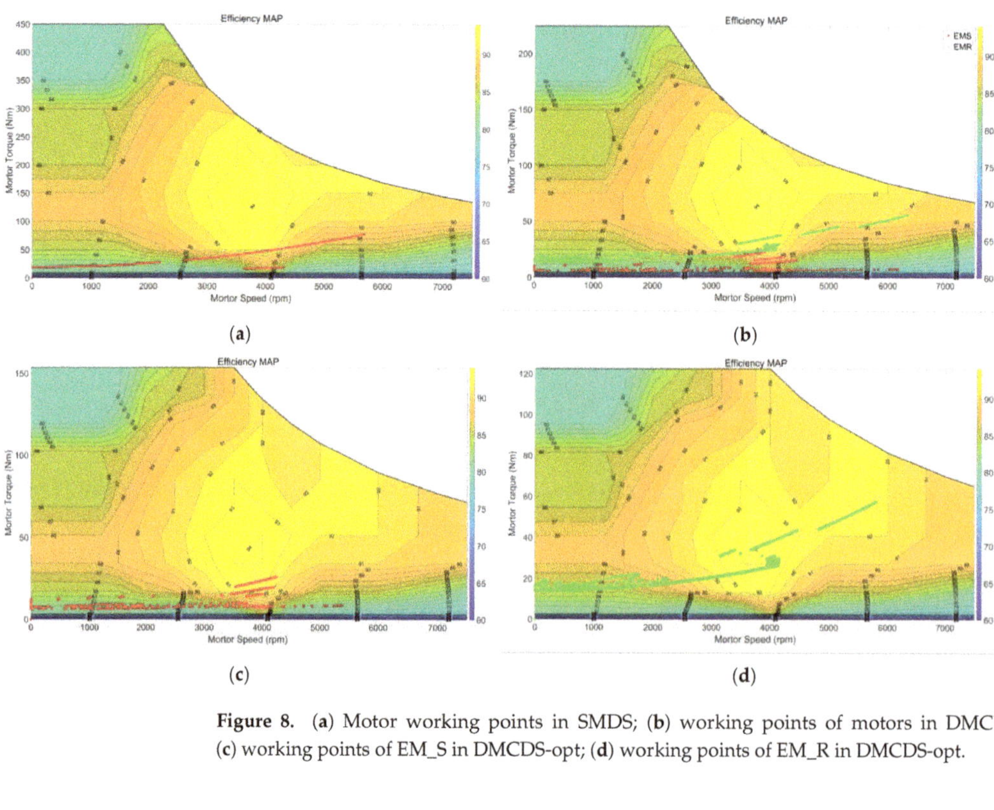

Figure 8. (**a**) Motor working points in SMDS; (**b**) working points of motors in DMCDS-pro; (**c**) working points of EM_S in DMCDS-opt; (**d**) working points of EM_R in DMCDS-opt.

The average drive efficiency (ADE) was adopted as one of the evaluation indicators for assessing the system economic performance in this work. For a given driving cycle, the system drive efficiency is represented as

$$\eta_{ADE} = \int_0^t \frac{P_w}{P_{ele}} dt \tag{21}$$

where η_{ADE} is the average drive efficiency and P_w is the vehicle demand power.

Figure 9 illustrates the time-varying drive efficiency of the three propulsion systems. The findings reveal that the SMDS has the lowest efficiency, followed by DMCDS-pro, which exhibits an average efficiency 1.75% higher than that of SMDS. Significantly, DMCDS-opt displays the highest average efficiency, exceeding DMCDS-pro by 2.5%.

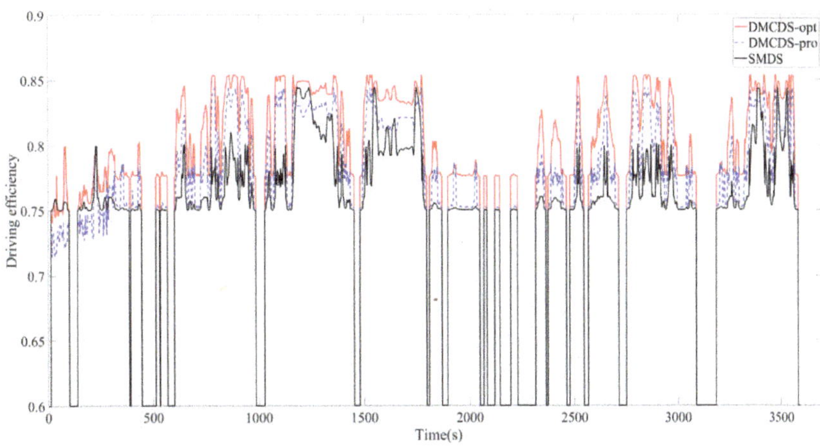

Figure 9. Efficiency of driving system in CLTC and WLTC.

Compared to SMDS, DMCDS-pro, which employs two identical motors, facilitates adaptation to varying vehicle speeds and loads through three driving modes. Moreover, the DMC mode enables continuous speed regulation, thereby enhancing motor operating efficiency. Consequently, DMCDS-pro exhibits a higher average drive efficiency compared to SMDS, which relies on a single motor and a two-gear transmission. However, the parameter matching of DMCDS-pro is based on vehicle performance indicators, without considering the alignment of the motor's high-efficiency range with the vehicle operating conditions. As a result, the utilization efficiency of the high-efficiency region for the two motors is relatively low.

Conversely, DMCDS-opt, featuring two motors with different efficiency characteristics, establishes three high-efficiency regions within the drive system. This design enhances the utilization efficiency of the motor high-efficiency regions and contributes to reducing energy consumption.

The high-efficiency region utilization and electricity consumption were adopted as another two economic evaluation indicators in this study. The indicator values for the three drive system schemes were obtained through simulations and are presented in Table 6.

Table 6. Economic evaluation indicators.

Indicator \ Schemes	SMDS	DMCDS-pro	DMCDS-opt
High-efficiency region utilization (efficiency > 90%)	6.6%	EM_R 30.2% EM_S 8.6%	EM_R 43.8% EM_S 12.8%
Electricity consumption (kWh/100 km)	18.18	17.58	16.95

The results indicate that the utilization efficiency of the high-efficiency region in the dual-motor coupled drive configuration was significantly superior to SMDS, resulting in SMDS exhibiting the highest electricity consumption (18.18 kWh/100 km). In contrast, DMCDS-pro and DMCDS-opt demonstrated reductions in electricity consumption of 3.3% and 6.8%, respectively. It is evident that DMCDS-opt exhibited a remarkable increase in high-efficiency region utilization after optimization, resulting in reduced electricity consumption. In conclusion, the DMCDS achieves a higher efficiency through the synergistic optimization of component parameters and system control, effectively enhancing the driving range.

4. Conclusions

In order to enhance the energy utilization efficiency and extend the driving range of electric vehicles, a dual-motor coupled drive system with a simple structure is proposed, and a dynamic mathematical model is established to analyze the power flow characteristics under different driving modes. Considering the interdependence between the optimization of component sizes and the system coordinated control in the dual-motor coupled drive system, this paper proposes a two-layer hybrid optimization approach. In the outer layer, a multi-objective particle swarm algorithm is employed to optimize the component sizes. Meanwhile, in the inner layer optimization, the given feasible component parameters from the outer layer are subjected to the DP algorithm to identify the optimal control parameters. The primary findings of this paper are summarized as follows:

(1) The selection of motor parameters and gear ratios exerts a substantial influence on the power losses and drive efficiency of the system. While keeping the system maximum output power unchanged, adjustments to the rated power, rated speed, and gear ratios can enhance the utilization efficiency of the high-efficiency region and effectively reduce electrical energy consumption.

(2) The optimized motors exhibit an increase in rated speed and a decrease in peak torque, resulting in a substantial improvement in the utilization efficiency of the high-efficiency region. Compared to the prototype scheme, motors EM_R and EM_S experience an increase of 45% and 48%, respectively. Moreover, the optimized DMCDS achieves an average drive efficiency 2.5% and 4.2% higher than that of DMCDS-pro and SMDS, respectively, leading to DMCDS-opt possessing the lowest energy consumption of 16.95 kWh/100 km.

Author Contributions: Conceptualization, T.L.; methodology, T.L. and N.Z.; software, T.L.; validation, T.L. and X.G.; formal analysis, T.L.; investigation, D.P.; data curation, T.L.; writing—original draft preparation, T.L.; writing—review and editing, T.L. and N.Z.; supervision, D.P.; project administration, T.L. All authors have read and agreed to the published version of the manuscript.

Funding: This research was supported by China research project for fundamental product innovation (K23F024).

Data Availability Statement: Not applicable.

Conflicts of Interest: The authors declare no conflict of interest.

References

1. Smith, W.J. Can EV (electric vehicles) address Ireland's CO_2 emissions from transport? *Energy* **2010**, *35*, 4514–4521. [CrossRef]
2. Nunes, P.; Farias, T.; Brito, M.C. Enabling solar electricity with electric vehicles smart charging. *Energy* **2015**, *87*, 10–20. [CrossRef]
3. Chopra, S.; Bauer, P. Driving range extension of EV with on-road contactless power transfer—A case study. *IEEE Trans. Indust. Electron.* **2013**, *60*, 329–338. [CrossRef]
4. Li, Z.; Khajepour, A.; Song, J. A comprehensive review of the key technologies for pure electric vehicles. *Energy* **2019**, *182*, 824–839. [CrossRef]
5. Walker, P.D. Modelling, Simulations, and Optimisation of Electric Vehicles for Analysis of Transmission Ratio Selection. *Adv. Mech. Eng.* **2013**, *5*, 1–13. [CrossRef]
6. Kamachi, M.; Miyamoto, H.; Sano, Y. Development of power management system for electric vehicle 'i-MiEV'. In Proceedings of the 2010 International Power Electronics Conference, Sapporo, Japan, 21–24 June 2010; pp. 2949–2955. [CrossRef]

7. de Santiago, J. Electrical Motor Drivelines in Commercial All-Electric Vehicles: A Review. *IEEE Trans. Veh. Technol.* **2012**, *61*, 475–484. [CrossRef]
8. Wu, J.; Liang, J.; Ruan, J.; Zhang, N.; Walker, P.D. Efficiency comparison of electric vehicles powertrains with dual motor and single motor input. *Mech. Mach. Theory* **2018**, *128*, 569–585. [CrossRef]
9. Zhang, C.; Zhang, S.; Han, G.; Liu, H. Power Management Comparison for a Dual-Motor-Propulsion System Used in a Battery Electric Bus. *IEEE Trans. Ind. Electron.* **2017**, *64*, 3873–3882. [CrossRef]
10. Hu, J.; Zu, G.; Jia, M.; Niu, X. Parameter matching and optimal energy management for a novel dual-motor multi-modes powertrain system. *Mech. Syst. Signal Process.* **2019**, *116*, 113–128. [CrossRef]
11. Tseng, S.; Tseng, C.; Liu, T.; Chen, J. Wide-range adjustable speed control method for dual-motor drive system. *IET Electr. Power Appl.* **2015**, *9*, 107–116. [CrossRef]
12. Holdstock, T.; Sorniotti, A.; Everitt, M.; Bertolotto, S. Energy consumption analysis of a novel four-speed dual motor drivetrain for electric vehicles. In Proceedings of the 2012 IEEE Vehicle Power and Propulsion Conference 2012, Seoul, Republic of Korea, 9–12 October 2012; pp. 295–300. [CrossRef]
13. Xu, L.; Ouyang, M.; Li, J.; Yang, F.; Lu, L.; Hua, J. Optimal sizing of plug-in fuel cell electric vehicles using models of vehicle performance and system cost. *Appl. Energy* **2013**, *103*, 477–487. [CrossRef]
14. Vagg, C.; Akehurst, S.; Brace, C.J. Stochastic dynamic programming in the real-world control of hybrid electric vehicles. *IEEE Trans. Control Syst. Technol.* **2016**, *24*, 853–866. [CrossRef]
15. Kim, N.; Cha, S.W.; Peng, H. Optimal equivalent fuel consumption for hybrid electric vehicles. *IEEE Trans. Control Syst. Technol.* **2012**, *20*, 817–825. [CrossRef]
16. Scordia, J.; Desbois-Renaudin, M.; Trigui, R.; Jeanneret, B.; Badin, F. Global optimisation of energy management laws in hybrid vehicles using dynamic programming. *Int. J. Veh. Des.* **2005**, *39*, 349–367. [CrossRef]
17. Sundström, O.; Ambühl, D.; Guzzella, L. On implementation of dynamic programming for optimal control problems with final state constraints. *Oil Gas. Sci. Technol. Rev. Inst. Fr. Pétrole* **2010**, *65*, 91–102. [CrossRef]
18. Wu, J.; Zhang, C.H.; Cui, N.X. PSO algorithm-based parameter optimization for HEV powertrain and its control strategy. *Int. J. Automot. Technol.* **2008**, *9*, 53–59. [CrossRef]
19. Ebbesen, S.; Dönitz, C.; Guzzella, L. Particle swarm optimisation for hybrid electric drive-train sizing. *Int. J. Veh. Des.* **2012**, *58*, 181–199. [CrossRef]
20. Zeng, X.H.; Wang, Z.W.; Song, D.F. Parameter Optimization of Dual-mode Power-split Hybrid Electric Bus Based on MIGA Algorithm. *J. Mech. Eng.* **2020**, *56*, 98–105. [CrossRef]
21. Nuesch, T.; Ott, T.; Ebbesen, S.; Guzzella, L. Cost and fuel-optimal selection of HEV topologies using particle swarm optimization and dynamic programming. In Proceedings of the 2012 American Control Conference, Montréal, QC, Canada, 27–29 June 2012. [CrossRef]
22. Bonnans, J.F.; Guilbaud, T.; Ketfi-Cherif, A. Parametric Optimization of Hybrid Car Engines. *Optim. Eng.* **2004**, *5*, 395–415. [CrossRef]
23. Wen, C.; Zhang, S.; Xie, B.; Song, Z.; Li, T.; Jia, F.; Han, J. Design and verification innovative approach of dual-motor power coupling drive systems for electric tractors. *Energy* **2022**, *247*, 1–22. [CrossRef]
24. Bayrak, A.E.; Kang, N.; Papalambros, P.Y. Decomposition-based design optimization of hybrid electric powertrain architectures: Simultaneous configuration and sizing design. *J. Mech. Des.* **2016**, *138*, 071405. [CrossRef]
25. Landolfi, E.; Minervini, F.J.; Minervini, N.; De Bellis, V.; Malfi, E.; Natale, C. Integration of a Model Predictive Control with a Fast Energy Management Strategy for a Hybrid Powertrain of a Connected and Automated Vehicle. *World Electr. Veh. J.* **2021**, *12*, 159. [CrossRef]
26. Coppola, A.; De Tommasi, G.; Motta, C.; Petrillo, A.; Santini, S. Double-layer control architecture for motion and torque optimisation of autonomous electric vehicles. *Transp. Res. Interdiscip. Perspect.* **2023**, *21*, 1–15. [CrossRef]
27. Fathy, H.K.; Reyer, J.A.; Papalambros, P.Y.; Ulsoy, A.G. On the coupling between the plant and controller optimization problems. In Proceedings of the American Control Conference 2001, Arlington, VA, USA, 25–27 June 2001; pp. 1864–1869. [CrossRef]
28. Fang, L.C.; Qin, S.Y.; Xu, G. Simultaneous optimization for hybrid electric vehicle parameters based on multi-objective Genetic Algorithms. *Energies* **2011**, *4*, 532–544. [CrossRef]
29. Zhuang, W. Optimal Design and Mode Shift Control of Multi-mode Hybrid Electric Vehicles. Ph.D. Thesis, Nanjing University of Science & Technology, Nanjing, China, 2017.
30. Nguyen, C.T.; Walker, P.D.; Zhou, S.; Zhang, N. Optimal sizing and energy management of an electric vehicle powertrain equipped with two motors and multi-gear ratios. *Mech. Mach. Theory* **2022**, *167*, 104513. [CrossRef]
31. Pennestrì, E.; Mariti, L.; Valentini, P.; Mucino, V. Efficiency evaluation of gearboxes for parallel hybrid vehicles: Theory and applications. *Mech. Mach. Theory* **2012**, *49*, 157–176. [CrossRef]

World Electric Vehicle Journal

MDPI

Article

A Quantitative Study on the Impact of China's Dual Credit Policy on the Development of New Energy Industry Based on Taylor Expansion Description and Cross-Entropy Theory

Jiantong Qiao [1,†], Shangru Yang [2,†], Jiaming Zhao [1,‡], Haoyuan Li [3,‡] and Yuezhen Fan [1,*]

1 School of Technology, Beijing Forestry University, No. 35, Qinghua East Road, Haidian District, Beijing 100083, China; qiaojiantong@bjfu.com (J.Q.); zhaojiaming@bjfu.edu.cn (J.Z.)
2 School of Mechanical Engineering and Automation, Beihang University, 37 Xueyuan Road, Haidian District, Beijing 100191, China; yangsr@buaa.edu.cn
3 School of Mechanical Engineering, Xi'an Jiaotong University, No. 28 Xianning West Road, Xi'an 710049, China; lihaoyuan@stu.xjtu.edu.cn
* Correspondence: fanyuezhen@bjfu.edu.cn
† These authors contributed equally to this work.
‡ These authors contributed equally to this work.

Abstract: The Dual Credit Policy is an important policy to promote the development of new energy vehicles unique to China. There is a lack of research that intuitively reflects the impact of the Dual Credit Policy on industrial development through an industry-based factual comparison of this policy. Based on the Taylor expansion and Cross-Entropy description, this article obtains the development regression function by the quantitative analysis of five indicators—the number of new energy vehicle-related patents, sales volume, production volume, the number of newly registered enterprises, infrastructure construction (the number of charging piles) before and after the implementation of the policy, and describes them quantitatively using the Taylor expansion to obtain the CPTI index. The CPCEI index is obtained by calculating the Cross-Entropy of the distribution of each indicator before and after policy implementation. The above two indices were compared for the growth trend and growth quantity, respectively. Finally, the following conclusions were obtained: 1. the Dual Credit Policy is more significantly promoted at the market level than the impact on the technical level; 2. although there is also incentive in infrastructure construction, it cannot fully react to the market demand; 3. the number of start-up's operating in the new energy field increases, but the overall growth trend gradually slows down and fails to significantly change the existing structure of the market. This study suggests that the government should launch a special incentive policy for charging piles, and new energy manufacturers should expand their production capacity to meet the market demand.

Keywords: Dual Credit Policy; policy impacts; Taylor series expansion; Cross-Entropy

Citation: Qiao, J.; Yang, S.; Zhao, J.; Li, H.; Fan, Y. A Quantitative Study on the Impact of China's Dual Credit Policy on the Development of New Energy Industry Based on Taylor Expansion Description and Cross-Entropy Theory. *World Electr. Veh. J.* **2023**, *14*, 295. https://doi.org/10.3390/wevj14100295

Academic Editors: Fachao Jiang, Yongyu Li and Weiwei Kong

Received: 22 August 2023
Revised: 13 September 2023
Accepted: 28 September 2023
Published: 16 October 2023

1. Introduction

China's Dual Credit Policy is a policy measure implemented to promote the development of New Energy Vehicles (NEVs). Its developmental timeline includes the introduction of the policy by the government in 2017, which required automobile manufacturers to ensure that a certain proportion of their total sales consisted of NEVs and set energy efficiency standards for vehicles. Subsequently, in 2019, the policy underwent modifications, further raising the requirements for manufacturers.

In 2021, the government released a new automotive industry development policy, continuing its support for the development of NEVs and announcing plans for carbon peaking and carbon neutrality aimed at further boosting the demand for NEVs. This policy plays a crucial role in the Chinese automotive market by encouraging manufacturers to increase the production and sales of NEVs. Additionally, it has significant implications for the global automotive market and the development of new energy technologies.

2. Literature Review

Research on China's Dual Credit Policy began to gain momentum around 2018, and to date, there are relatively few studies available for reference. Prior to 2020, research on the Dual Credit Policy primarily focused on whether the government subsidies and macro support measures under this policy were excessive. For instance, researchers like Zheng Jichuan [1] found that combining research and development (R&D) subsidies with the Dual Credit Policy could enhance market mechanisms, promote technological innovation, and facilitate the healthy development of automotive companies. Shiqi Ou and others [2] studied the policy's impact on the profits of plug-in electric vehicle companies, suggesting that pure electric cars with a range exceeding 250 km and plug-in hybrid SUVs might be popular. Some scholars also optimized the subsidies under the Dual Credit Policy [3].

In 2020, the subsidy intensity of this policy began to decrease, prompting scholars to shift their focus toward predicting the post-subsidy impact. As China's new energy vehicle industry developed and supporting infrastructure improved, a market-driven reshuffling of enterprises ensued. Companies that had profited by deceiving the government for subsidies were gradually phased out due to low product quality. After 2020, scholars began to concentrate on demonstrating whether the Dual Credit Policy truly had a driving effect on the industry and validating it through various models. For example, Yuchao Li conducted empirical tests on the "corner overtaking theory" using a difference-in-differences (DID) model with China's new energy vehicle industry as the research subject [4,5]. Ding Lian explored the impact of the Dual Credit Policy on production and cooperative research and development by constructing a game theory model [6]. Liu Chunling and others based their analysis on the Arrow-Karlin model, using the Hamilton function to analyze optimal control conditions for producing new energy vehicles in the context of cumulative points and carbon trading [6]. Their model was more comprehensive than others as it considered situations with no points, only points, or both points and carbon trading. Lu Chao and colleagues [7] argued that the Dual Credit Policy imposed higher requirements on the quality and energy efficiency of automotive products from the supply side. Their research considered coordination in the automotive supply chain regarding price, emissions reduction, and mileage, leading to three conclusions that provide guidance for supply chain coordination in automotive companies. Yu et al. [8] analyzed the impact of withdrawal and cumulative points on the optimal decisions of both automakers and dealers using the Stackelberg game.

To date, research on the aforementioned topics has become relatively abundant. The academic community largely agrees that the Dual Credit Policy effectively promoted the development of China's new energy vehicle industry and strongly recommends it as a policy that other countries can emulate. However, there has been limited discussion regarding the underlying mechanisms and reasons for the policy's effectiveness. Without a thorough study and analysis of the intrinsic logic and process by which the policy drives industry development, it may not be conducive to other countries attempting to formulate similar systems. Some scholars have conducted comparative analyses of the policy's implementation process and industry feedback paths [9,10]. Nevertheless, these studies tend to focus on fundamental industry development logic and lack comprehensive data support and mathematical logic. To gain a more direct and clear understanding of how China's Dual Credit Policy impacts industrial development, this study utilizes a large volume of data, condenses it into indices through mathematical modeling, and compares the indices before and after the policy's implementation.

There are various methods for policy research, and quantifiable approaches include text analysis, cost-benefit (utility) analysis, multi-criteria decision analysis, policy simulation, and dynamic modeling, among others. Li et al. [11] employed cost-benefit (utility) analysis using Chinese publicly listed new energy vehicle companies as research samples, applying a difference-in-differences model to analyze the Dual Credit Policy based on firm heterogeneity. They analyzed changes in investment scale, intensity, and structure of new energy enterprises under this policy and found that the Dual Credit Policy significantly promoted R&D investment by new energy vehicle companies, with the growth in the scale of R&D investment being more significant than the growth in intensity. Cheng et al. [12] analyzed the production decisions of automakers under the fuel consumption credit and new energy vehicle credit systems, proposing optimizations for production decisions under the credit system. Li and Xiong [13] analyzed the dynamic changes in the operational and environmental performance of new energy vehicle enterprises under the Dual Credit Policy from the dimensions of significance, agility, and stability. They concluded that the Dual Credit Policy had released positive effects during its incubation period, with more significant and stable growth in environmental performance and more agile responses in operational performance. Lu et al. [14] introduced externalities from the field of economics and the Pirlo theory to explain the intrinsic mechanism of the Dual Credit Policy. Their research provides theoretical support for the positioning and direction of the Dual Credit Policy in the development of the new energy vehicle industry.

In existing policy research, both domestic and international scholars have proposed numerous models. Zang Wei et al. [15] conducted quantitative AI policy research using policy tools and the PMC policy evaluation model, employing text mining and content analysis methods to quantitatively analyze current AI policy texts in China. They combined this with an analysis of China's AI research frontier trends to explore future policy development directions, providing specific and actionable recommendations for the formulation and revision of AI policies. To avoid the subjectivity of variable scoring, they used the results of policy text mining to assign values to variables based on the spatial vector model and used PMC indicator scores and PMC surface synthesis to reflect various dimensions of the policy. Although their policy quantification method is relatively traditional, it achieves multi-dimensional analysis and allows for a more comprehensive policy evaluation, making it very helpful for our research. Remal Abotah et al. [16] studied a comprehensive policy development evaluation model to assess the effectiveness of energy policy tools in increasing the adoption of renewable energy. They used a hierarchical decision model (HDM) to construct a comprehensive policy evaluation framework. Their policy evaluation model provides ideas from multiple perspectives for construction.

It can be seen that scholars have employed a wide range of models, all of which are reasonably explained. This paper proposes a new index to demonstrate the impact of policies on industrial development. The underlying logic is to approximate the coefficients of functions using the Taylor expansion method to analyze the trends in different variables. The Taylor expansion is widely used in fields such as physics, engineering, and the natural sciences [17]. It can be used to approximate the behavior of complex physical phenomena to better understand and predict experimental results. Similarly, this method has broad applicability in policy research. The Taylor expansion provides an effective tool for approximating complex industrial or economic models into simpler mathematical forms, making the analysis of policy impact more feasible. By analyzing the coefficients after Taylor expansion, researchers can understand the linear and nonlinear responses between different variables, revealing the potential impacts of policy changes. However, careful consideration of data quality and model applicability is needed, especially when policy changes are significant or nonlinear, to ensure the accuracy and reliability of the analysis.

3. Method and Proceeding

For a more concise description of the data and experiments, Table 1 presents the symbols used in this study to represent the relevant meanings.

Table 1. Symbol representation and related meaning.

Symbols	Meaning
b	before
a	after
t	time
Pa_b	Pre-policy patent growth
Pa_a	Patent growth after the policy
S_b	Pre-policy sales volume growth
S_a	Sales growth after the policy
Pr_b	Pre-policy production growth
Pr_a	Production growth after the policy
Co_b	Pre-policy new business growth
Co_a	New business growth after the policy
Ch_b	Pre-policy charging pile growth
Ch_a	Charging pile growth after the policy
Min	Minimum value
Max	Maximum value
T	Indicator data for the year
Pp_a	Patent probability distribution before policy implementation
Qp_a	Patent probability distribution after policy implementation
P_s	Probability distribution of sales volume (units) before policy implementation
Q_s	Probability distribution of sales volume (units) after policy implementation
Pp_r	Probability distribution of production (units) before policy implementation
Qp_r	Probability distribution of production (units) after policy implementation
Pc_O	Probability distribution of the number of newly established companies before the policy was implemented
Qc_O	Probability distribution of the number of newly established companies after the policy was implemented
Pc_h	Probability distribution of the number of charging posts before the policy was implemented
Qc_h	Probability distribution of the number of charging posts after the policy is implemented
H	Comparative data by year before and after the policy
CPTI	Dual Credit Policy Taylor expansion index
CPCEI	Dual Credit Policy Cross-Entropy index

Based on previous studies on policy quantification, this study will measure the impact of the Dual Credit Policy on the development of the new energy vehicle industry in terms of technology, market, capital, and infrastructure development. The specific indicators of these aspects are the number of patents granted for new energy, the production of new energy vehicles, the sales of new energy vehicles, the registration of companies providing new energy services, and the number of charging piles. According to the relevant documents from the Ministry of Industry and Information Technology of the People's Republic of China and combined with previous studies, 2017 is widely considered to be the first year when the Dual Credit Policy became widely known and played a role in the development of the industry, so this study looked for data from the five years before and five years after the policy was implemented, as shown in Table 2.

According to previous studies (some literature's serial numbers), the time series data can be fitted to the time series development function of a certain indicator by regression and make a forecast of the future trend. This function can better reflect the development pattern of this indicator at that stage and is a good quantitative tool. Therefore, in this study, regression analysis is conducted separately for the above indicators before and after the policy to obtain the statistically optimal fit function. Comparing the changes of the two functions of the same indicator before and after the policy, the impact of the Dual Credit Policy on different indicators in terms of development trends can be reflected more clearly, thus illustrating the impact of the Dual Credit Policy on the overall industry.

Table 2. Relevant data before and after 5 years.

	Patents	Sales Volume (Units)	Production (Volume)	Number of Newly Established Companies	Number of Charging Piles
2012	27	12,800	12,600	2491	18,000
2013	43	17,600	17,500	3102	22,528
2014	54	74,800	78,500	4766	31,000
2015	99	331,100	340,500	5758	49,000
2016	412	507,000	517,000	8755	141,000
2017	233	777,000	794,000	13,442	240,000
2018	643	1,256,000	127,0000	21,516	387,000
2019	677	1,206,000	1,242,000	23,311	516,000
2020	610	1,367,000	1,366,000	32,848	807,000
2021	993	3,521,000	3,545,000	72,707	1,147,000
Maximum value	993	3,521,000	3,545,000	72,707	1,147,000
Minimum value	27	12,800	12,600	2491	18,000
Average value	379.1	90,7030	918,310	18,869.6	335,852.8

Data source: CNKI, SAIC.

In the comparison between the two functions before and after, this study adopts two treatments, the Taylor expansion and Cross-Entropy, for continuous functions and discontinuous point sets, respectively. The Taylor formula can expand the functions of different shapes in the form of continuous sub-polynomials of the independent variable, which serves the purpose of standardizing and unifying the description of the functions. By summing up the complex weights of different term coefficients, the eigenvalues of the growth trend of the function—i.e., the acceleration of the function change—can be effectively defined and extracted, i.e., the Taylor index of the effect of the double integration policy. Cross-Entropy is now widely used in machine learning. It is used similarly to describe the mathematical treatment of the difference between the function obtained from machine learning and the target function; the value can effectively feedback on the difference between the two and feedback to the computer to correct its learning results and make it more accurate for continuous iteration. In this study, with the help of the concept of Cross-Entropy, the changes of each index before and after the double integration policy are analyzed, and the Cross-Entropy index of the policy impact is finally obtained. It should be noted that the Cross-Entropy index of double integration policy impact is used to describe the change of indicators, while the Taylor index of double integration policy impact is used to express the change of the indicator development trend.

To facilitate statistical analysis, some necessary processing of the raw data is required. In this study, the above data are normalized and standardized according to the indicators. Further, considering the requirements of regression analysis for the dependent variable, this study uses the standardization process of determining the range, and the data interval is set to [0.2,0.8]. The formula is

$$\text{Equivalent value of each item} = 0.2 + \frac{0.6}{\max - \min} * (\mathrm{T} - \min) \quad (1)$$

The processing results are shown in Table 3.

Based on the above processing results, this study conducted regression analysis using IBM SPSS Statistics 26 data processing software and obtained ten sets of parameter estimates and function images. Based on the statistical residual sum-of-squares test, F-test, and significance test, this study selected the most appropriate function type possible to fit the ten functions of the five variables before and after the policy, and the process was as follows.

Table 3. Processed data.

	Patents	Sales Volume (Units)	Production (Volume)	Number of Newly Established Companies	Number of Charging Piles
2012	0.2	0.2	0.2	0.2	0.2
2013	0.210	0.201	0.201	0.205	0.202
2014	0.217	0.211	0.211	0.219	0.207
2015	0.245	0.254	0.256	0.228	0.216
2016	0.439	0.285	0.286	0.254	0.265
2017	0.328	0.331	0.333	0.294	0.318
2018	0.583	0.413	0.414	0.363	0.396
2019	0.604	0.404	0.409	0.378	0.465
2020	0.5621	0.432	0.430	0.459	0.619
2021	0.8	0.8	0.8	0.8	0.8

Table 4 represents the selection of the growth equation of the number of patents before the implementation of the policy.

Table 4. Growth in the number of patents before the implementation of the policy.

	Square of R	F	Degree of Freedom 1	Degree of Freedom 2	Significance	Constants	b1	b2
Linear	0.654	5.659	1	3	0.098	0.108	0.051	
Logarithmic	0.470	2.657	1	3	0.202	0.158	0.108	
Secondary	0.923	12.056	2	2	0.077	0.303	−0.116	0.028
Index	0.712	7.423	1	3	0.072	0.149	0.173	
Logistic	0.712	7.423	1	3	0.072	6.712	0.841	

Figure 1 shows a fitting function image of the changes in patent index over the five years prior to policy implementation.

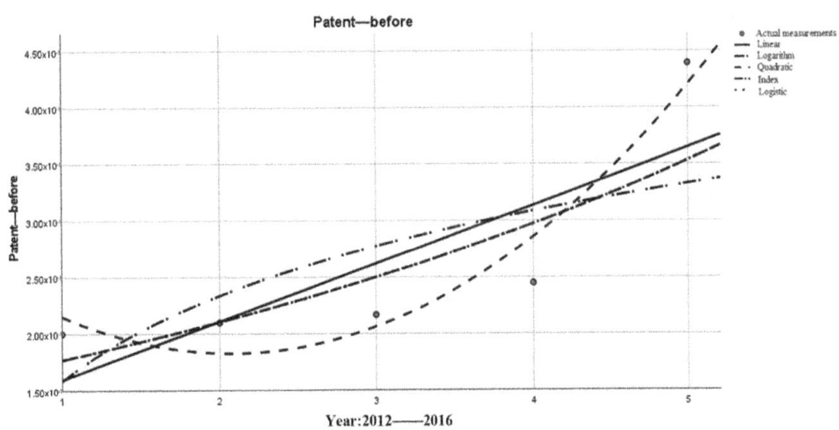

Figure 1. Growth in the number of patents before the implementation of the policy.

According to the statistical requirements, the model with a smaller sum-of-squared residual and a larger and less significant F-test was chosen as the most suitable expression. Considered together, the exponential function was chosen as the mathematical expression in this study.

$$Pa_b(t) = 0.149^{0.173t} \quad (2)$$

Table 5 represents the selection of the equation for the growth of the number of patents after the implementation of the policy. Figure 2 shows a fitting function image of the changes in patent index over the five years after policy implementation

Table 5. The number of patents increased after the implementation of the policy.

	Square of R	F	Degree of Freedom 1	Degree of Freedom 2	Significance	Constants	b1	b2
Linear	0.757	9.339	1	3	0.055	0.298	0.092	
Logarithmic	0.792	11.425	1	3	0.043	0.350	0.235	
Secondary	0.763	3.215	2	2	0.237	0.250	0.134	−0.007
Index	0.727	7.996	1	3	0.066	0.328	0.175	
Logistic	0.727	7.996	1	3	0.066	3.053	0.840	

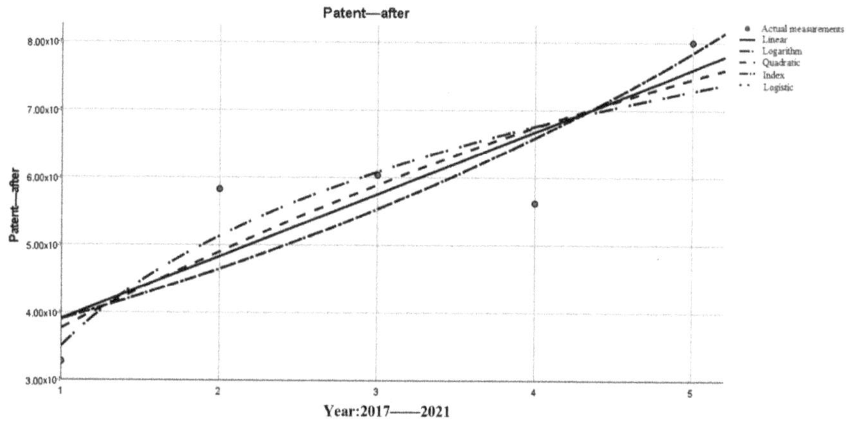

Figure 2. The number of patents increased after the implementation of the policy.

The model with a small sum-of-squared residual and a larger and less significant F-test was chosen as the most suitable expression according to statistical requirements. Considering the influence of chance factors such as the new crown epidemic in the fourth year after the implementation of the policy, data with a large deviation from the overall trend in the fourth year are not considered. Based on this study, the exponential function was chosen as the mathematical expression.

$$Pa_a(t) = 0.328^{0.328t} \tag{3}$$

Based on the same research approach and the way the expressions are chosen, the growth in the number of new energy vehicles sold, the growth in the number of productions, the growth in the number of newly registered enterprises, and the growth in infrastructure construction (taking the growth in the number of charging piles as an example) before and after the implementation of the policy are analyzed as follows.

The data were processed to obtain the model summary and parameter estimates, respectively, to obtain the following Tables 6–13. Figures 3–10 show the exponential fitting functions of sales volume, production, the number of newly registered enterprises in the field, and the number of charging stations in the five years before and after the policy.

$$S_b(t) = 0.172^{0.94t} \tag{4}$$

$$S_a(t) = 0.263^{0.181t} \tag{5}$$

$$Pr_b(t) = 0.171^{0.95t} \tag{6}$$

$$Pr_a(t) = 0.171^{0.95t} \tag{7}$$

$$Co_b(t) = 0.185^{0.185t} \quad (8)$$

$$Co_a(t) = 0.179^{0.224t} \quad (9)$$

$$Ch_b(t) = 0.179^{0.63t} \quad (10)$$

$$Ch_a(t) = 0.248^{0.229t} \quad (11)$$

Table 6. Pre-policy sales volume increase.

	Square of R	F	Degree of Freedom 1	Degree of Freedom 2	Significance	Constants	b1	b2
Linear	0.870	20.108	1	3	0.021	0.163	0.022	
Logarithmic	0.691	6.695	1	3	0.081	0.183	0.049	
Secondary	0.978	43.690	2	2	0.022	0.210	−0.017	0.007
Index	0.881	22.188	1	3	0.018	0.172	0.094	
Logistic	0.881	22.188	1	3	0.018	5.825	0.910	

Table 7. Increase in sales after the policy.

	Square of R	F	Degree of Freedom 1	Degree of Freedom 2	Significance	Constants	b1	b2
Linear	0.668	6.039	1	3	0.091	0.189	0.096	
Logarithmic	0.515	3.186	1	3	0.172	0.276	0.209	
Secondary	0.861	6.202	2	2	0.139	0.493	−0.165	0.044
Index	0.735	8.322	1	3	0.063	0.263	0.181	
Logistic	0.735	8.322	1	3	0.063	3.803	0.834	

Table 8. Pre-policy production growths.

	Square of R	F	Degree of Freedom 1	Degree of Freedom 2	Significance	Constants	b1	b2
Linear	0.873	20.610	1	3	0.020	0.163	0.023	
Logarithmic	0.694	6.819	1	3	0.080	0.183	0.050	
Secondary	0.977	42.568	2	2	0.023	0.209	−0.017	0.007
Index	0.884	22.791	1	3	0.017	0.171	0.095	
Logistic	0.884	22.791	1	3	0.017	5.834	0.909	

Table 9. Production growth after the policy.

	Square of R	F	Degree of Freedom 1	Degree of Freedom 2	Significance	Constants	b1	b2
Linear	0.665	5.945	1	3	0.093	0.192	0.095	
Logarithmic	0.513	3.156	1	3	0.174	0.278	0.208	
Secondary	0.856	5.964	2	2	0.144	0.494	−0.164	0.043
Index	0.731	8.134	1	3	0.065	0.265	0.179	
Logistic	0.731	8.134	1	3	0.065	3.770	0.836	

Table 10. Growth in new registrations before the policy.

	Square of R	F	Degree of Freedom 1	Degree of Freedom 2	Significance	Constants	b1	b2
Linear	0.936	44.106	1	3	0.007	0.182	0.013	
Logarithmic	0.801	12.063	1	3	0.040	0.193	0.030	
Secondary	0.985	66.011	2	2	0.015	0.200	−0.002	0.003
Index	0.951	58.204	1	3	0.005	0.185	0.058	
Logistic	0.951	58.204	1	3	0.005	5.397	0.944	

Table 11. Growth of newly registered businesses after the policy.

	Square of R	F	Degree of Freedom 1	Degree of Freedom 2	Significance	Constants	b1	b2
Linear	0.772	10.154	1	3	0.050	0.126	0.111	
Logarithmic	0.605	4.599	1	3	0.121	0.225	0.244	
Secondary	0.938	15.182	2	2	0.062	0.430	−0.150	0.044
Index	0.865	19.151	1	3	0.022	0.220	0.224	
Logistic	0.865	19.151	1	3	0.022	4.551	0.799	

Table 12. Pre-policy charging post growth.

	Square of R	F	Degree of Freedom 1	Degree of Freedom 2	Significance	Constants	b1	b2
Linear	0.714	7.496	1	3	0.071	0.175	0.014	
Logarithmic	0.527	3.349	1	3	0.165	0.189	0.031	
Secondary	0.948	18.239	2	2	0.052	0.224	−0.028	0.007
Index	0.738	8.447	1	3	0.062	0.179	0.063	
Logistic	0.738	8.447	1	3	0.062	5.572	0.939	

Table 13. Charger growth after the policy.

	Square of R	F	Degree of Freedom 1	Degree of Freedom 2	Significance	Constants	b1	b2
Linear	0.956	64.743	1	3	0.004	0.163	0.119	
Logarithmic	0.826	14.196	1	3	0.033	0.257	0.275	
Secondary	0.997	310.703	2	2	0.003	0.309	−0.006	0.021
Index	0.991	337.811	1	3	0.000	0.248	0.229	
Logistic	0.991	337.811	1	3	0.000	4.038	0.795	

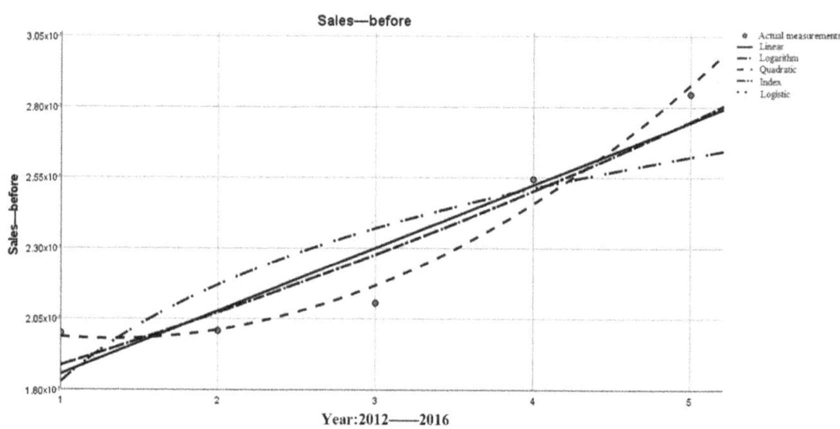

Figure 3. Changes in sales volume in the 5 years prior to policy implementation.

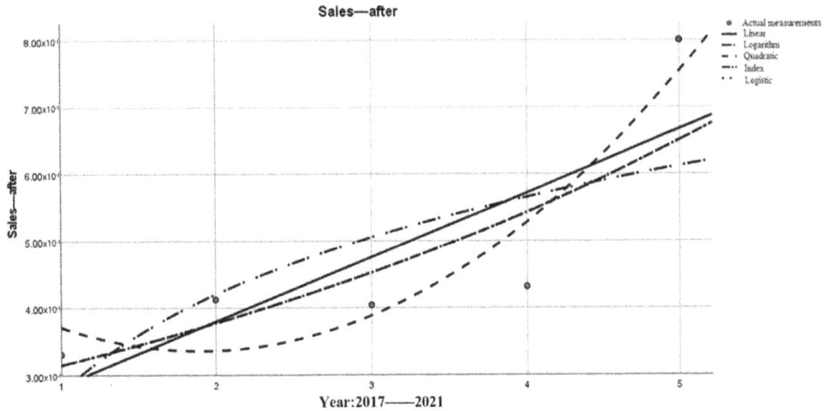

Figure 4. Changes in sales volume within 5 years after policy implementation.

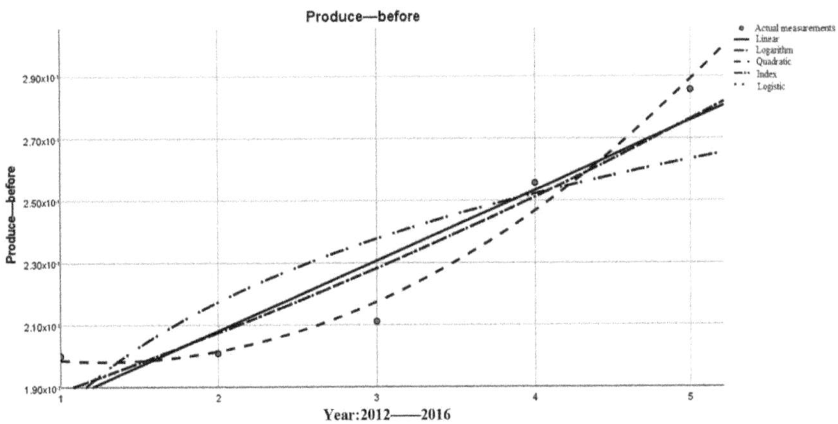

Figure 5. Changes in production in the 5 years prior to policy implementation.

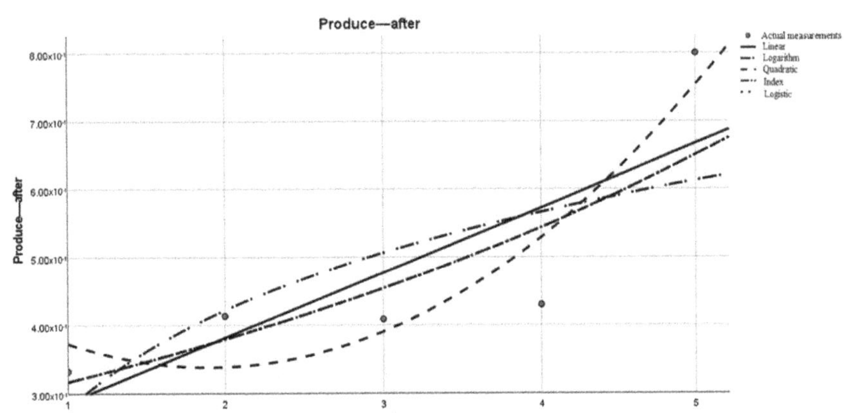

Figure 6. Changes in production over the past 5 years after policy implementation.

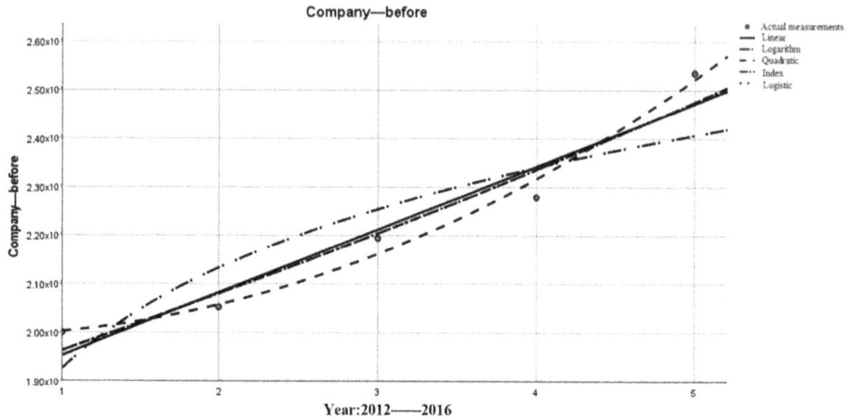

Figure 7. Changes in the number of newly registered enterprises in relevant fields in the 5 years before policy implementation.

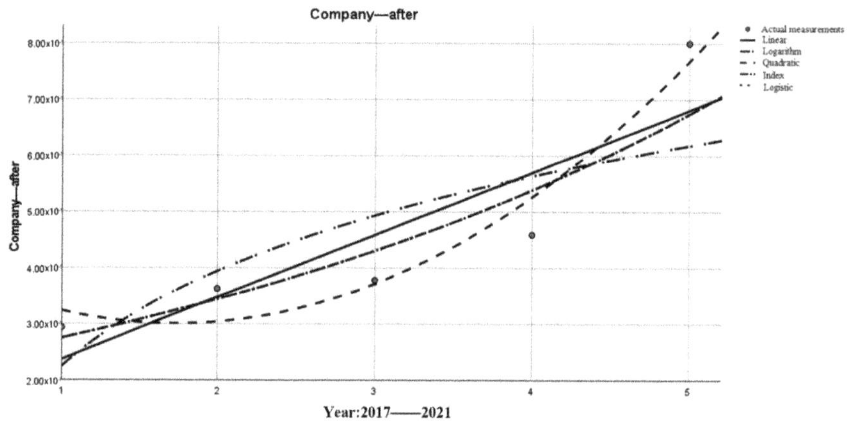

Figure 8. Changes in the number of newly registered enterprises in relevant fields within 5 years of the implementation of policies.

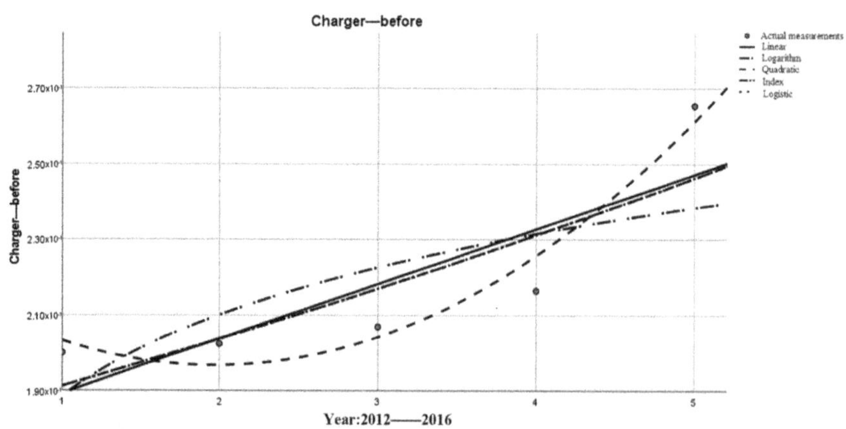

Figure 9. Changes in the number of charging power stations in the 5 years before the implementation of the policy.

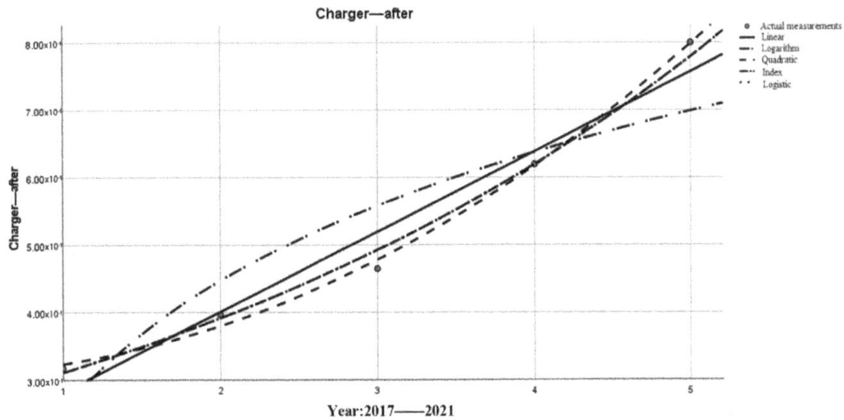

Figure 10. Changes in the number of charging power stations in the next 5 years after the implementation of the policy.

The above data are plotted against the corresponding images and the expressions are derived in the same manner as above, and the expressions and images are shown below.

The Taylor expansions of the above ten functions were studied and calculated, the parameters of the first five orders were listed, and the following table was obtained.

Through Table 14, it can be seen that the fourth- and fifth-order coefficients are close to zero, which is also consistent with the mathematical mechanism of the Taylor expansion. The higher the order of expansion, the smaller the overall impact on the function, that is, the higher order infinitesimal in higher mathematics. In order to facilitate the calculation and to consider the different degrees of influence of different orders on the overall expression of the function, this study selects the coefficients of the first three orders of the Taylor expansion and performs the summation of the complex weights to finally obtain the relative growth indices describing the different types of influencing factors before and after the policy. The first-order compound weight is the original value, the second-order compound weight is 1/2, and the third-order compound weight is 1/3. From this data, Table 15 is obtained.

Table 14. Taylor's Expanded Form.

Coefficient Matrix before Compound Weighting	Fifth Order Pole Number	Forth Order Pole Number	Third Order Pole Number	Secord Order Pole Number	First Order Pole Number	Constant Term
Patents-before	0.000	0.000	0.000	0.010	−0.143	1.000
Patents-after	0.000	0.000	−0.001	0.013	−0.159	1.000
Sales-before	−0.002	0.011	−0.062	0.258	−0.719	1.000
Sales-after	0.000	0.000	0.000	0.006	−0.105	1.000
Production-before	−0.002	0.012	−0.064	0.265	−0.729	1.000
Production-after	0.000	0.000	0.000	0.005	−0.103	1.000
Companies-before	0.000	0.000	0.000	0.009	−0.136	1.000
Companies-after	0.000	0.000	−0.001	0.014	−0.167	1.000
Charging-before	0.000	0.002	−0.017	0.111	−0.471	1.000
Charging-after	0.000	0.000	0.000	0.010	−0.139	1.000

Table 15. Empowering Taylor Unfolds.

Coefficient Matrix before Compound Weighting	Fifth Order Pole Number	Forth Order Pole Number	Third Order Pole Number	Secord Order Pole Number	First Order Pole Number	Relative Growth Index
Patents-before	0.000	0.000	0.000	0.005	−0.143	−0.138
Patents-after	0.000	0.000	0.000	0.006	−0.159	−0.153
Sales-before	0.000	0.003	−0.021	0.129	−0.719	−0.610
Sales-after	0.000	0.000	0.000	0.003	−0.105	−0.102
Production-before	0.000	0.003	−0.021	0.133	−0.729	−0.617
Production-after	0.000	0.000	0.000	0.003	−0.103	−0.101
Companies-before	0.000	0.000	0.000	0.005	−0.136	−0.131
Companies-after	0.000	0.000	0.000	0.007	−0.167	−0.161
Charging-before	0.000	0.001	−0.006	0.055	−0.471	−0.421
Charging-after	0.000	0.000	0.000	0.005	−0.139	−0.134

It should be noted that the positive, negative, and magnitude of the relative growth index do not represent the actual growth because the above data are obtained through independent normalization and standardization of different factors under the influence of policies, and it can only be compared with the relative growth index of similar factors to illustrate the changes in different indicators before and after the policy change.

The relative growth indices of similar factors are compared to obtain the policy impact coefficient for a single factor, as shown in Table 16.

Table 16. Policy Impact Factor.

Policy Impact Indicators	Patents	Sales Volume (Units)	Production (Volume)	Number of Newly Established Companies	Number of Charging Piles
Policy Impact Factor	0.904	5.964	6.135	0.816	3.142

Cross-Entropy is using the same set of events to measure the information about the variability between two probability distributions. In information theory, Cross-Entropy is denoted as two probability distributions, P and Q, where P denotes the true distribution and Q denotes the untrue distribution in the same set of events, where the untrue distribution Q is used to denote the average number of bits required for an event to occur. Let the two probability distributions of different indicators before and after the policy study in this study be P and Q, where P is the pre-policy distribution and Q is the post-policy distribution.

In this study, the variability of the pre-policy indicator P is expressed in terms of the post-policy indicator Q, which can be given by the following equation.

$$H(P,Q) = \sum_{t} P(t) \cdot \ln\left(\frac{1}{Q(t)}\right) \quad (12)$$

The base of the logarithmic function is chosen to be e, which is in line with the general convention of Cross-Entropy calculation, and the value reflects only the relative influence without strict requirements of taking values, as shown in Table 17.

Table 17. Data comparison before and after the policy.

Indicator Category	Patents	Sales Volume (Units)	Production (Volume)	Number of Newly Established Companies	Number of Charging Piles
Comparative data by year before and after the policy	0.223	0.221	0.220	0.245	0.229
	0.113	0.1778	0.177	0.208	0.187
	0.109	0.191	0.189	0.214	0.159
	0.141	0.214	0.216	0.177	0.104
	0.0980	0.0635	0.0637	0.057	0.059

By summing the above data according to the indicator categories, the Cross-Entropy index of the impact of the Dual Credit Policy for a particular indicator is obtained, and the results are shown in Table 18.

Table 18. Policy Impact Cross-Entropy Index.

	Patents	Sales Volume (Units)	Production (Volume)	Number of Newly Established Companies	Number of Charging Piles
Dual Credit Policy affects Cross-Entropy index	0.685	0.867	0.866	0.901	0.738

4. Data Analysis and Interpretation

According to the formula of the policy impact Taylor index, when the value is less than 1, it means that the growth rate of the indicator has slowed down after the policy, and when the value is greater than 1, it means that the growth rate of the indicator has significantly increased after the policy implementation, and the larger the value, the faster the growth rate, which means the higher the sensitivity of the policy introduction to the indicator. According to the formula and definition of the Cross-Entropy index of policy impact, the larger the value, the more obvious the change of the indicator before and after the policy.

Observing the table, it is obvious that the Taylor index of policy influence on the sales, production, and the number of charging piles of new energy vehicles is greater than one, which indicates that the introduction of the policy has a greater incentive effect on the production of enterprises, the purchase intention of consumers, and the construction of infrastructure. According to Qiao Jiantong et al. [18], it is the above three factors that play a dominant role in industry evaluation. Therefore, this study concludes that China's Dual Credit Policy has a greater driving effect on the development of the new energy vehicle industry.

Among the three indicators greater than one, the policy impact coefficients of sales volume and production volume are close, indicating that the policy introduction has a relatively similar driving effect on these two indicators. The number of charging piles, however, is much lower than the incentive effect on production and sales, although it also shows a significant incentive effect after the policy is introduced. With the accumulation of time, the number of new energy vehicles grows year by year, while the growth of the number of charging piles has difficulty matching the considerable number of new energy vehicles. As a result, China has experienced a significant shortage of charging piles in recent years during peak travel periods such as holidays. Further, the data results of this study are in line with what is really happening in the current usage scenario.

The results of the study also show that there was a slowdown in the growth of patents and the number of new companies formed after the policy was introduced. This does not mean that the policy has had a negative impact on industry development. First, according to Qiao Jiantong et al. [16], the weight of IPR and willingness to capital in the industry development index is small in measuring the current industry development. These two can be screened in terms of intuitive factors such as the number of patents granted and the number of new companies formed. Of course, this study does not suggest that these two are not important for the industry. According to Li Xueqing [19], intellectual property rights and willingness to invest are the basis for the development of the industry, and both are prerequisites for the formation of good products and market feedback. Therefore, there has been a lot of technology research and investment in this study even before the national Dual Credit Policy was introduced.

Comparing the difference between the two above, the policy impact coefficient of IP is closer to 1, indicating that the growth rate of patent applications has slowed down since the policy came out, but basically followed its original development trajectory. This also indicates that technology research is relatively less sensitive to policy. Scientific research is something that requires technology accumulation over a long time, meaning that generations of scientists and engineers keep innovating a little bit at a time, until its properties are stable and its development law is inherent. The number of newly established companies, on the other hand, has slowed down more significantly, with a numerical decrease of nearly 20%.

Through the comparison of the Cross-Entropy index, we can see that all indicators continue to grow after the policy, and the Cross-Entropy index of sales and production is basically the same, with sales slightly larger than production. Among them, new start-ups have, relatively, the most growth in terms of number. This data indicates that there are more entrepreneurs joining the new energy track in an attempt to gain a foothold in the market.

5. Conclusions

Overall, the Dual Credit Policy has strongly promoted the development of China's new energy vehicle industry and can be a policy widely studied and referred to around the world. Synthesizing the results of the above research and data analysis, the following specific conclusions and recommendations can be obtained from this study.

China's Dual Credit Policy has played an important role in promoting the development of the new energy vehicle industry, which is mainly reflected in the production capacity of enterprises, consumers' willingness to purchase, and infrastructure construction. The output of scientific research is hardly affected by the policy in the short term. The policy offers far less incentive for infrastructure than the market needs and expects, and there will continue to be a large gap in the coming years.

The incentive of the Dual Credit Policy makes investors more willing to invest in original enterprises rather than in new enterprises, which also reduces the possibility of cheating the national policy preferences and rubbing the hot spot to rub the wind from the dimension of the capital market. The willingness of capital to invest in new energy companies is decreasing, and they are more willing to allocate resources to companies with accumulated technology and long production experience. The number of newly established companies exceeds the market needs. Further, a large part of the new energy enterprises cannot meet the market requirements, and this lack of technical production capacity will be shut down. At present, market supply and demand have basically reached equilibrium, and consumers' purchasing power will continue to be maintained for a period of time. With the replacement of traditional fuel and vehicles, the market dividend will continue for some time.

We believe that, with the passage of time, especially after 2030, pure electric models will occupy the absolute market dominance and become the mainstream of the automotive market.

Author Contributions: Conceptualization, J.Q. and S.Y.; methodology, J.Q. and S.Y.; software, J.Q.; validation, S.Y., J.Z., H.L. and Y.F.; formal analysis, J.Q.; investigation, S.Y.; resources, S.Y.; data curation, H.L. and Y.F.; writing—original draft preparation, J.Q. and J.Z.; writing—review and editing, J.Q.; visualization, S.Y.; supervision, J.Z., H.L. and Y.F.; project administration, J.Q., H.L. and Y.F. All authors have read and agreed to the published version of the manuscript.

Funding: This research received no external funding.

Data Availability Statement: The data can be found here: https://kdocs.cn/l/ccapF7e9wtfF (accessed on: 7 July 2023).

Conflicts of Interest: The authors declare no conflict of interest.

References

1. Zheng, J.; Zhao, H.; Li, Z. Research on R&D subsidies for new energy vehicle industry under Dual Credit Policy. *Sci. Res. Manag.* **2019**, *40*, 126–133. [CrossRef]
2. Ou, S.; Lin, Z.; Qi, L.; Przesmitzki, S. The dual-credit policy: Quantifying the policy impact on plug-in electric vehicle sales and industry profits in China. *Energy Policy* **2018**, *121*, 597–610. [CrossRef]

3. Zhu, Y. Optimization of Subsidy Strategies for the New Energy Vehicle Supply Chain under the Dual Credit Policy. Master's Thesis, Qingdao University, Qingdao, China, 2020. Available online: https://kns.cnki.net/KCMS/detail/detail.aspx?dbname=CMFD202101&filename=1020384579.nh (accessed on 16 June 2020).
4. Li, W.; Dai, L.; Guo, B.; Wu, S. Cooperative Innovation Game Analysis of Upstream and Downstream Enterprises in the New Energy Vehicle Industry under the Composite Traction Mechanism in the Post-Subsidy Era. *Soft Sci.* **2021**, *35*, 81–88. [CrossRef]
5. Li, Y.; Zhang, L.; Liu, J.; Qiao, X. Can the Dual-Credit Policy Help China's New Energy Vehicle Industry Achieve Corner Overtaking? *Sustainability* **2023**, *15*, 2406. [CrossRef]
6. Liu, C.-L.; Zhao, L.; Li, J.; Gao, M.-Y.; Deng, X.-D. Analysis of New Energy Vehicle Production Decision Based on Optimal Control in the Context of Considering Cumulative Points and Carbon Trading. *Comput. Integr. Manuf. Syst.* 1–18. Available online: http://kns.cnki.net/kcms/detail/11.5946.TP.20220317.2003.005.html (accessed on 8 April 2022).
7. Lu, C.; Wang, Q.; Chen, Q. Supply chain coordination of vehicles considering price, emission reduction, and range under the Dual Credit Policy. *Syst. Eng. Theory Pract.* **2021**, *41*, 2595–2608.
8. Yu, X.H.; Ye, Z.X.; Li, M. Optimization analysis of two-level supply chain production decision under subsidy withdrawal-Dual Credit Policy. *Oper. Res. Manag.* **2021**, *30*, 42–49.
9. Zhao, D. A Comparison and Implications of Development Policies for the New Energy Vehicle Industry in China and the United States. *Knowl. Action* **2021**, *1*, 45–48. [CrossRef]
10. Rao, Y.; Shu, T. Analysis of Technological Innovation Behavior of New Energy Vehicle Enterprises Based on China's Dual-Credit Policy. *Manag. Rev.* **2023**, *35*, 74–85+111. [CrossRef]
11. Li, X.; Xiong, Y.Q. Analysis of the Impact of Dual Credit Policy on R&D investment of new energy vehicle enterprises. *Sci. Res.* **2021**, *39*, 1770–1780. [CrossRef]
12. Cheng, Y.W.; Mu, D. Production decision optimization of automobile manufacturers under cumulative points system. *Syst. Eng. Theory Pract.* **2018**, *38*, 2817–2830.
13. Li, X.; Xiong, Y.Q. Stages of the impact of the Dual Credit Policy on new energy vehicles from the perspective of both operational and environmental performance. *Resour. Sci.* **2021**, *43*, 1–11.
14. Lu, Y.; Chen, H.; Liu, J. Analysis of the economic impact of the Dual Credit Policy on the development of China's new energy vehicle industry. *Chin. Foreign Entrep.* **2019**, *9*, 121–122.
15. Zang, W.; Zhang, Y.; Xu, L. A quantitative study of China's artificial intelligence policy text--the current state of policy and frontier trends. *Sci. Technol. Prog. Countermeas.* **2021**, *38*, 125–134.
16. Abotah, R.; Daim, T.U. Towards building a multi-perspective policy development framework for transition into renewable energy. *Sustain. Energy Technol. Assess.* **2017**, *21*, 67–88. [CrossRef]
17. Blaszczyk, M.; Jantos, D.R.; Junker, P. Application of Taylor series combined with the weighted least square method to thermodynamic topology optimization. *Comput. Methods Appl. Mech. Eng.* **2022**, *393*, 114698. [CrossRef]
18. Qiao, J.; Yang, S.; Chen, Z.; Zhuang, Z.; Chen, L. A quantitative study of policy-driven changes and forecasts in the development of the hydrogen transportation industry. *Energy Rep.* **2022**, *8* (Suppl. 4), 1218–1225. [CrossRef]
19. Li, X. *Scientific and Technological Progress, Economic Growth and Hilbert Space*; People's Publishing House: Beijing, China, 2014.

World Electric Vehicle Journal

MDPI

Article

Research on the Multimode Switching Control of Intelligent Suspension Based on Binocular Distance Recognition

Chen Huang [1,2], Kunyan Lv [1,*], Qing Xu [2] and Yifan Dai [2]

1 Institute of Automotive Engineering, Jiangsu University, Zhenjiang 212013, China; huangchen@ujs.edu.cn
2 State Key Laboratory of Automotive Safety and Energy, Tsinghua University, Beijing 100084, China; qingxu@tsinghua.edu.cn (Q.X.); df08@mails.tsinghua.edu.cn (Y.D.)
* Correspondence: 17712603097@163.com

Abstract: As the upgrade of people's requirements for automotive driving comfort, conventional passive suspensions for cars have fallen short of existing demands due to their nonadjustable damping and stiffness, so semiactive suspensions and active suspensions have gained growing acceptance. Compared with active suspensions, semiactive suspensions offer the advantages of a low manufacturing cost and reliable structure, and thus have become the preferred choice for most vehicles. To optimize the control effect of semiactive suspensions under different working conditions, this paper completed the modeling of magnetorheological semiactive suspension system dynamics and road inputs; then, the design of binocular camera sensing algorithms was performed to obtain the real-time distance of the target using the point cloud ranging function, and the parameters required for suspension control were also obtained. This was followed by the completion of the control-mode-switching rules and the design of the suspension controller. According to the different control objectives, the mode could be divided into the obstacle-road mode, straight-road mode, and curved-road mode. The suspension controller included the BP-PID (neural network PID controller) controller and the force distributor. Finally, the effectiveness of the mode-switching rules and the control method was verified through system simulation and the hardware-in-the-loop test.

Keywords: semiactive suspension; vehicle control strategy; target recognition; BP neural network

Citation: Huang, C.; Lv, K.; Xu, Q.; Dai, Y. Research on the Multimode Switching Control of Intelligent Suspension Based on Binocular Distance Recognition. *World Electr. Veh. J.* **2023**, *14*, 340. https://doi.org/10.3390/wevj14120340

Academic Editors: Fachao Jiang, Yongyu Li and Weiwei Kong

Received: 19 October 2023
Revised: 22 November 2023
Accepted: 2 December 2023
Published: 7 December 2023

1. Introduction

As major component of the automotive chassis system, the suspension directly affects the ride comfort of a vehicle [1]. Conventional passive suspensions have not been able meet the existing demand because they cannot change the damping and stiffness. Active suspensions have excellent performance but cannot be put into large-scale application. In contrast, semiactive suspensions are less costly and have an adjustable damping force. By applying control algorithms to semiactive suspensions, they can be adaptively adjusted according to the current and future road conditions, which is a better choice for suspension systems currently [2–5].

As a widely used senor in vehicles, the vehicle camera has a significant contribution in vehicle control and assisted driving. As a sensor built on the principle of binocular stereo vision, the binocular camera has both image-acquisition and range-measurement functions. This camera captures images from different angles simultaneously using a binocular stereo vision system consisting of two cameras, which is then processed by a matching algorithm to obtain a parallax map. Finally, it calculates the distance information of the object to be measured by combining the camera calibration parameters. A simple binocular system can be completed using two on-board monocular cameras, which has great application prospects [6–9].

For semiactive suspensions, many studies have been conducted at home and abroad. In terms of the suspension structure, Hu et al. proposed a magnetorheological damper with an improved structure. They increased the magnetic field range by connecting the radial gap and axial gap, which improved the response speed of the damper and increased the adjustment range [10]. Brooks in the UK designed a current–variable liquid damper using the current–variable effect to control the flexible diaphragm, resulting in a smoother and more homogeneous process of the damping force change [11]. In terms of suspension control, there are currently the PID, LQG, MPC, sliding film control methods, and so on [12–16]. Neural networks for the online tuning of PID parameters have been widely used because of their flexibility and relatively mature technology [17–20]. For example, Li Mei et al. proposed a fuzzy neural network parameter optimization algorithm combining the particle swarm optimization algorithm and the gradient descent method. They proposed to adjust the parameters of the active suspension PID controller in real time with body acceleration as the main optimization objective [1]. However, prescanning control, which can be adjusted in real time according to the road conditions ahead, has gradually gained popularity among researchers in recent years. Bender first proposed the concept of prescanning control and applied it to the control of a single-degree-of-freedom suspension system [21]. R.S. Sharp et al. applied wheelbase prescanning to the design of an active suspension. This system relied on the prescanning information provided by the front wheels to apply control to the rear wheels, and the results indicated that wheelbase prescanning significantly improved the performance of the rear suspension [22]. C. Gohrle proposed two model-predictive approaches for the preview of active suspension controllers and compared them with the well-known optimal-preview control approach. The first model-predictive controller optimized the actuator displacements on a nonequidistant grid over the preview horizon. The second controller optimized the trajectories for the heave, pitch, and roll of the vehicle body over the preview horizon using a quadratic program [23]. Panshuo Li et al. from the University of Hong Kong proposed a multiobjective control method for active suspension based on wheelbase prescanning. They compared the active suspension performance under different operating conditions and found that the body droop and angular accelerations were significantly improved as compared to the control method without wheelbase prescanning [24].

However, most of the current semiactive suspension control methods do not consider the main factors affecting ride comfort under different working conditions. This results in the fact that a single control mode cannot be adapted to all applications. Therefore, in this paper, a binocular camera was adopted to complete target recognition and prescanning information extraction. That is, the speed bumps and the start points of the lane-line circle curve were used as the judgment rule for the suspension mode so as to optimize the body posture under different scenarios and improve the ride comfort and handling stability.

The rest of this paper is organized as follows. System modeling is described in Section 2, including the seven-degrees-of-freedom (DoF) full-vehicle model, the road-surface input model, and the mathematical model of magnetorheological dampers. A target-distance-recognition method based on the binocular camera is described in Section 3. In Section 4, an intelligent-suspension control system is presented. In Section 5, an HIL experiment is performed to illustrate the effectiveness and advantages of the proposed scheme. Finally, the work of the full paper is summarized and the validity of the proposed program is clarified in Section 6.

2. System Modeling

2.1. Vehicle Dynamics Modeling

The seven-degrees-of-freedom whole-vehicle dynamics model is shown in Figure 1. The model is more consistent with the actual vehicle dynamics, so this paper chooses to use this model to develop the control algorithm [2]. It includes the body roll, body pitch, and body droop, as well as the droop movement of all four wheels. Where the subscripts n represent the left front wheel, right front wheel, left rear wheel, and right rear wheel, respectively; f_n is the controlled damping force; z refers to the vertical displacement of the body centroid; θ represents the body pitch angle; ϕ is the body roll angle; z_{sn} is the vertical displacement of the body at the corresponding position; z_{un} signifies the vertical displacement of the corresponding wheel; m is the upper spring mass; m_{un} is the lower spring mass at each position; a, b, c, and d, respectively, denote the distance from the center of vehicle to the front, rear, left, and right ends; k_{tn} is the stiffness of each tire; k_1 and k_2 are the stiffness of the front suspension elastic element; k_3 and k_4 refer to the stiffness of the rear suspension elastic element; c_n is the passive suspension damping force; I_y is the pitch inertia, while I_x denotes the lateral tilt inertia; q_n signifies the road input.

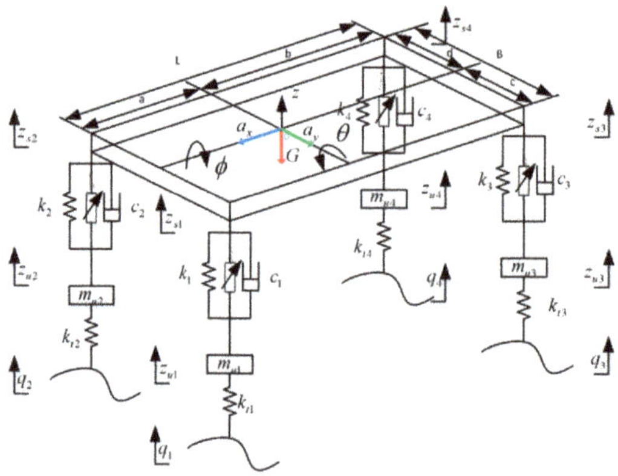

Figure 1. Vehicle dynamics model.

The vehicle dynamics equations include:

$$\begin{gathered} m\ddot{z} - [k_1(z_{u1} - z_{s1}) + c_1(\dot{z}_{u1} - \dot{z}_{s1}) + k_2(z_{u2} - z_{s2}) + c_2(\dot{z}_{u2} - \dot{z}_{s2}) + k_3(z_{u3} - z_{s3}) \\ + c_3(\dot{z}_{u3} - \dot{z}_{s3}) + k_4(z_{u4} - z_{s4}) + c_4(\dot{z}_{u4} - \dot{z}_{s4})] + \\ f_1 + f_2 + f_3 + f_4 = 0 \end{gathered} \tag{1}$$

$$\begin{gathered} I_y\ddot{\theta} - \left[k_1(z_{u1} - z_{s1}) + c_1\left(\dot{z}_{u1} - \dot{z}_{s1}\right) + k_2(z_{u2} - z_{s2}) + c_2\left(\dot{z}_{u2} - \dot{z}_{s2}\right) - f_1 - f_2\right]a + \\ \left[k_3(z_{u3} - z_{s3}) + c_3\left(\dot{z}_{u3} - \dot{z}_{s3}\right) + k_4(z_{u4} - z_{s4}) + c_4\left(\dot{z}_{u4} - \dot{z}_{s4}\right) - f_3 - f_4\right]b = 0 \end{gathered} \tag{2}$$

$$\begin{gathered} I_x\ddot{\phi} - \left[k_2(z_{u2} - z_{s2}) + c_2\left(\dot{z}_{u2} - \dot{z}_{s2}\right) + k_4(z_{u4} - z_{s4}) + c_4\left(\dot{z}_{u4} - \dot{z}_{s4}\right) - f_2 - f_4\right]c + \\ \left[k_1(z_{u1} - z_{s1}) + c_1\left(\dot{z}_{u1} - \dot{z}_{s1}\right) + k_3(z_{u3} - z_{s3}) + c_3\left(\dot{z}_{u3} - \dot{z}_{s3}\right) - f_1 - f_3\right]d = 0 \end{gathered} \tag{3}$$

$$m_{u1}\ddot{z}_1 - k_{t1}(q_1 - z_{u1}) + k_1(z_{u1} - z_{s1}) + c_1\left(\dot{z}_{u1} - \dot{z}_{s1}\right) - f_1 = 0 \tag{4}$$

$$m_{u2}\ddot{z}_2 - k_{t2}(q_2 - z_{u2}) + k_2(z_{u2} - z_{s2}) + c_2\left(\dot{z}_{u2} - \dot{z}_{s2}\right) - f_2 = 0 \tag{5}$$

$$m_{u3}\ddot{z}_3 - k_{t3}(q_3 - z_{u3}) + k_3(z_{u3} - z_{s3}) + c_3\left(\dot{z}_{u3} - \dot{z}_{s3}\right) - f_3 = 0 \tag{6}$$

$$m_{u4}\ddot{z}_4 - k_{t4}(q_4 - z_{u4}) + k_4(z_{u4} - z_{s4}) + c_4\left(\dot{z}_{u4} - \dot{z}_{s4}\right) - f_4 = 0 \tag{7}$$

2.2. Mathematical Model of the Magnetorheological Dampers

The magnetorheological damper mathematical model is of great importance for the calculation of the damping force of the semiactive suspension, and a good mathematical model can significantly improve the effectiveness of the control algorithm. For the consideration of computational efficiency, as well as the accuracy, a nonparametric model established by the motion parameters of the damper is widely used at present. The polynomial model proposed by Choi [25] is used to model the damping force based on the speed of the motion of the damper, with the damping force being:

$$F_d = \sum_{i=0}^{n} a_i v^i \tag{8}$$

where i is the polynomial order $i = 1, 2, 3, \ldots, n$, which is used to simulate the magnetorheological damper hysteresis characteristics and improve the model accuracy, with $n > 5$ in general; v is the speed of the damper motion; a_i represents the polynomial model coefficient, which is obtained by fitting to the experimental data, and its relationship with the current is:

$$msa_i = B_i + C_i I \tag{9}$$

So, the damper damping force is:

$$F_d = \sum_{i=0}^{n} (B_i + C_i I)v^i \tag{10}$$

This provides the magnitude of the control current:

$$I = \frac{F_d - \sum_{i=0}^{n} B_i v^i}{\sum_{i=0}^{n} C_i v^i} \tag{11}$$

Characterization tests were conducted on a prototype magnetorheological damper. The arrangement of the test bench was referred to as the "Test Method of Cartridge Damper Bench QC/T395-1999". The test rig included the upper computer, controller, excitation head, sensors, and other equipment. During the test, the magnetorheological damper was excited in the vertical direction to produce a simple harmonic motion with a fixed stroke in the middle of the test bench, and the test bench can be found in Figure 2. The supply current used in the test was varied in the range of 0–2 A at an interval of 0.2 A. After each current value was stabilized, cyclic loading was performed at different excitation frequencies. The average value of the recorded data was then taken as the test result to obtain the damper characteristics data under different currents. The results can be observed in Figure 3. It could be seen that, when the current was certain, the damping force was proportional to the speed of the magnetorheological damper. And, as the speed increased, the separation of the upper and lower branch curves of the magnetorheological damper demonstrated the characteristics of nonlinearity.

Figure 2. Bench test layout.

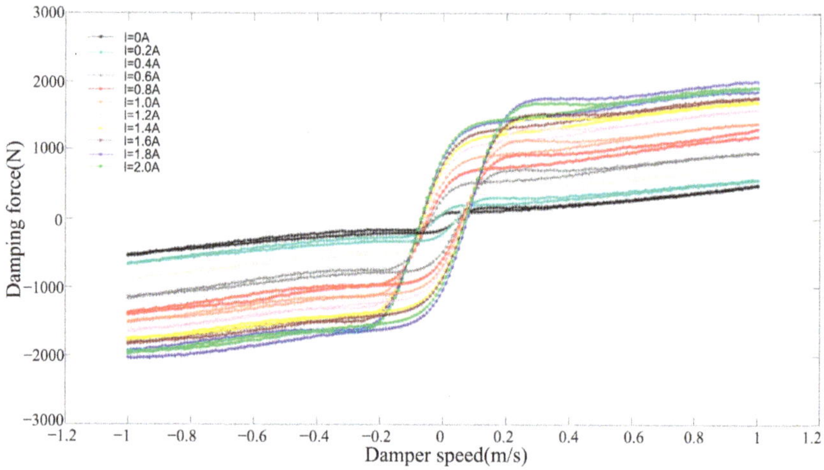

Figure 3. Damp characteristic curve.

The maximum current 2 A has been specified in the text. From the figure, it can be seen that the magnitude of the damping force is proportional to the speed of the magnetorheological damper when the current is certain. And, with the increase of speed, the separation of the upper and lower branch curves of the magnetorheological damper is more obvious. The simulation and experimental comparison diagrams are shown in Figure 4.

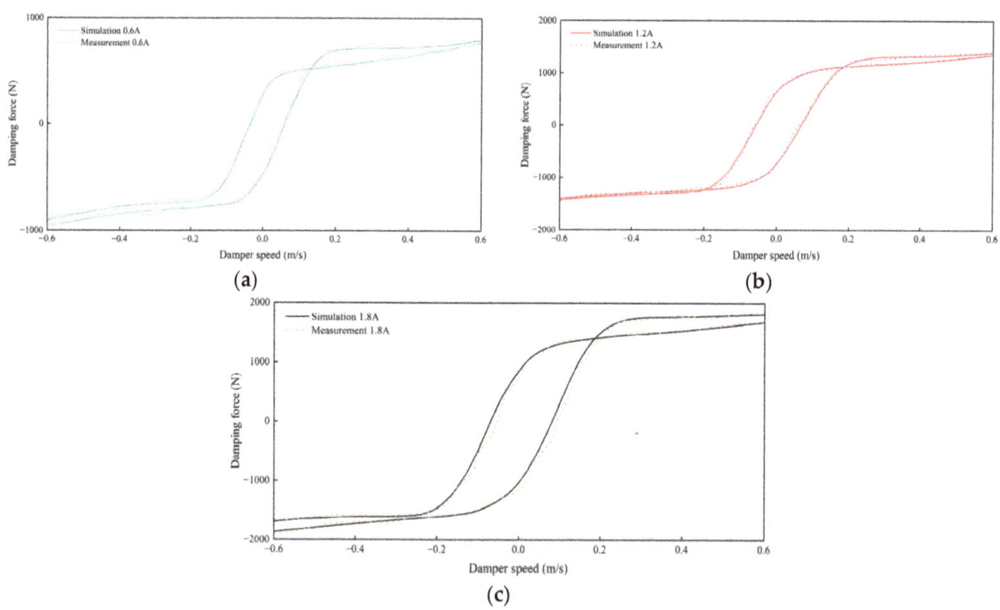

Figure 4. Simulation and measurement comparison diagrams: (**a**) I = 0.6 A; (**b**) I = 1.2 A; (**c**) I = 1.8 A.

2.3. Road-Surface Input Model

The power spectral density of the road can be expressed according to the Draft Method for Representation of Road Unevenness proposed by the International Organization for Standardization in ISO/TC108/SC2N67 [26]. Then, the spatial frequency power spectral density was converted to the temporal frequency power spectral density to obtain the filtered white noise pavement model, and the equation is shown below:

$$\dot{q}(t) = -2\pi f_0 q(t) + 2\pi n_0 w(t)\sqrt{G_q(n_0)v} \tag{12}$$

where $w(t)$ is the Gaussian white noise and v is the vehicle speed; f_0 refers to the lower limit as of frequency, which is generally taken as 0.01.

Since this paper is based on the whole-vehicle model for control-method development, it is necessary to build a four-wheel input model with a single-wheel pavement-unevenness input. Assuming that the vehicle travels at a uniform speed, there is a time delay between the road excitation of the front and rear wheels, which can be calculated by the following equation:

$$t = \frac{L}{v} \tag{13}$$

where L is the front and rear wheelbase, and v denotes the vehicle speed.

Considering the actual suspension structure and the connection of various parts of the vehicle body, there is a mutual interference relationship between the left and right wheel road-surface-unevenness input. Normally, the coherence is stronger in the low-frequency region and weaker in the high-frequency region, so a low-pass filter was applied to save the low-frequency strong-coherence region of the left wheel road-surface-unevenness input, and the high-pass filter was used to save the high-frequency low-coherence region of the random road surface unevenness; the two regions could be synthesized to generate the right wheel road-surface-unevenness input [27].

When the vehicle speed reached 10 m/s and the road surface was B-grade, the four-wheel pavement-unevenness input obtained from the simulation could be shown, as in Figure 5:

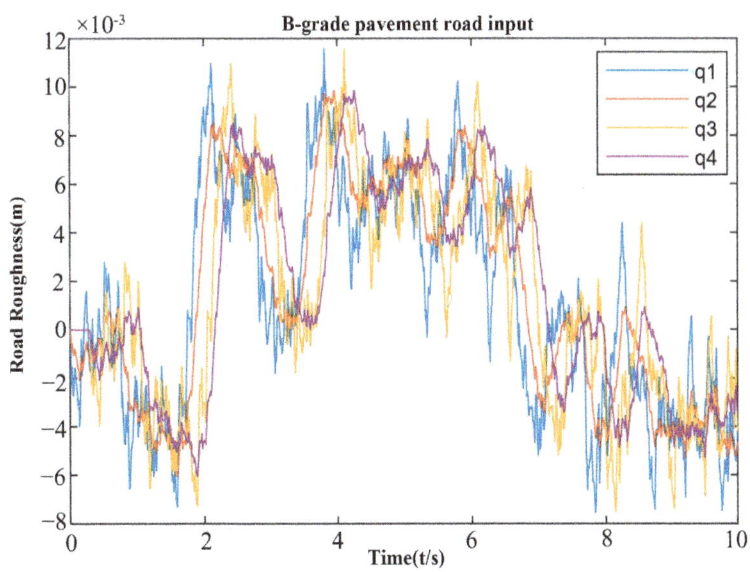

Figure 5. Four-wheel random-road input.

The value of the road input oscillates repeatedly around 0 and there is a time difference between front and rear axes, but their value is roughly the same.

3. Design of the Target-Distance-Recognition Method Based on the Binocular Camera

3.1. Speed-Bump-Detection Model Based on Deep Learning

In recent years, the emergence of convolutional neural networks (CNNs) has made great contributions to the development of target-detection technology, in which the one-stage target-detection algorithm YOLO (You Only Look Once) has been widely used because of its fast detection speed and strong model-generalization capability. Also, it is in the process of continuous updating and iteration [28]. In this paper, we adopted YOLOv4-tiny, a lightweight algorithm based on the improvement of the YOLOv4, that could meet the real-time accuracy requirements for the target detection of speed bumps during vehicle driving due to its simplified feature-extraction process and improved operation rate.

Then, the speed bump images were collected by real vehicles, and the photos of speed bumps with different shooting perspectives and different exposure degrees were selected to obtain effective data. To increase the sample size and improve the richness of the data, the dataset was expanded by the method of data augmentation. The images obtained are shown in Figure 6, in which a is the brightness enhancement, b signifies the contrast enhancement, c is the horizontal flip, and d is the random direction rotation. Finally, 4810 valid data were obtained, and the obtained image data were then target-labeled by the neural network dataset creation toolbox LabelImg to generate the VOC dataset, which produces a neural network readable dataset format.

Figure 6. Data augmentation.

In the dataset, the training set was set to 2886 sheets, and both the test and validation sets were 962 sheets. To improve the training efficiency and training effect, migration training was performed using the existing pretraining weights. In the freeze training phase, the network parameters were fine-tuned without changing the parameters of the backbone network. By then, the Batch_size = 8 and epoch = 50. In the thaw training phase, all the network parameters were adjusted, with the Batch_size = 2 and epoch = 2950. After 3000 training cycles, the model converged and the training effect reached the real-time accuracy requirements. The main performance indicators of the model can be seen in Table 1.

Table 1. Model training performance index.

	Precision	**F1 Score**	**Recall**	**mAP**	**AP**
IoU = 0.5	98.57%	0.99	98.57%	98.53%	98.53%

The IOU is the intersection and concurrency ratio of the bounding box predicted by the model and the true detection box. Precision is the accuracy of model. Recall represents the proportion of positive classes predicted by the model that are indeed positive classes.

The F1 score is the reconciled average of precision and recall. The mAP is the mean average precision, and AP is the average precision.

3.2. Algorithm for Detecting the Start Point of a Circular Curve at the Lane Line

3.2.1. Lane Line Detection

The recognition of the lane-line circle curve was based on lane line detection, which was achieved by preprocessing, lane line positioning, and lane line fitting.

The purpose of preprocessing was to remove noise and information that was not related to the detection target. After the original image optimization, ROI extraction, inverse perspective transformation, and binary image edge detection, the processed lane line image can be obtained, and the results are found in Figure 7.

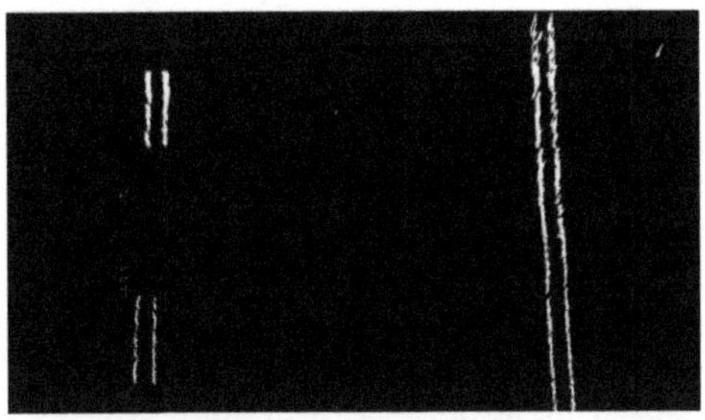

Figure 7. Lane line edge extraction.

3.2.2. Lane Line Positioning

The white pixels in the bird's-eye view represent the lane outline and the black pixels denote the background information; therefore, it is possible to locate the lane line by obtaining the position of the white pixels. A sliding window is typically used to calculate the number of white pixels in each column of the pixel coordinate system, and the result is shown in Figure 8.

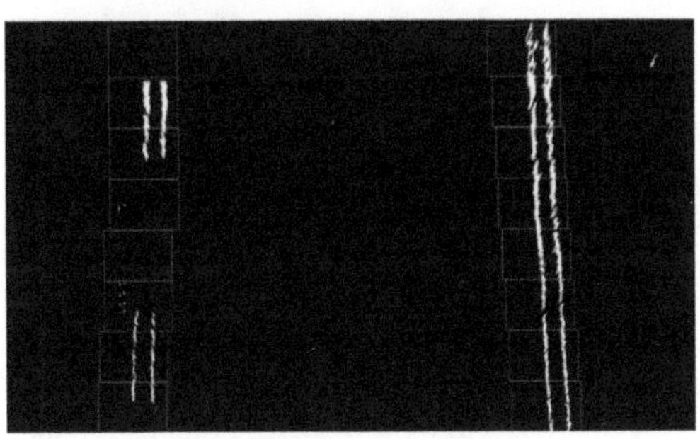

Figure 8. Sliding window positioning method.

To complete the sliding window lane line positioning, we need to determine the starting point of the lane line. The bird's-eye view from the bottom to the top represents a gradual distance from the camera. So, the sliding window is used to count the number of white pixel blocks in the window starting from the bottom, and the starting point of the lane line is determined based on the peaks of the number. Then, a new sliding window is drawn centered on the starting point towards the top, and the number of white pixels in the window is counted. The peak value is used as the new starting point to make a new sliding window. Then, the process is repeated to determine the direction of the lane line. The algorithm flow is shown in Figure 9.

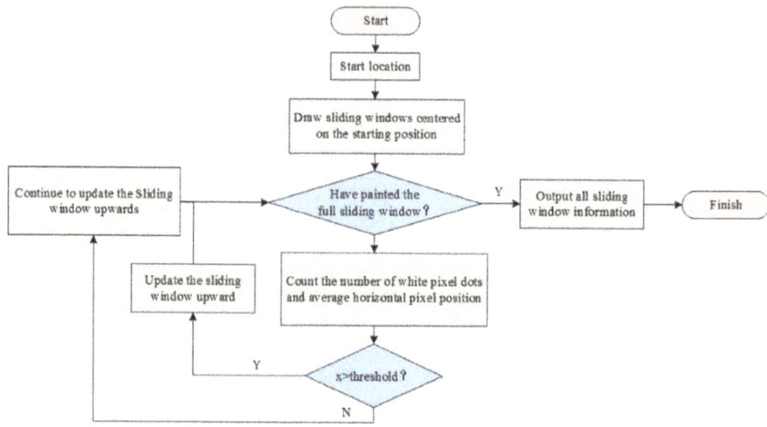

Figure 9. Sliding window positioning algorithm process.

3.2.3. Lane Line Fitting

The fitting algorithms applicable to the curved lane were mainly random sampling, the consistent RANSAC method, the Bessel curve fitting method, and the polynomial fitting method based on least squares. Given that the curvature of the lane lines in the curves within the urban roads would not change quickly, the lane line could be approximated as parabolic. The least-squares method was used in this paper to fit the lane lines [29,30].

The results obtained by fitting the lane lines based on this method are shown in Figure 10:

Figure 10. Lane line fitting.

3.2.4. Kalman Filtering

Since the least-squares method fitted the lane lines by traversing all the data only once, this method was less computationally intensive, but led to poor interference immunity. This method extracted lane lines based on fixed colors, and could be easily affected by the environment; as a consequence, errors and fitting jumps would be induced. The application of the Kalman filtering algorithm reduced this effect and fitted the lane lines more accurately [31]. Figure 11 is a comparison chart of lane-line-detection effect, and it can be seen that the application of Kalman filtering could improve the robustness of lane-line detection.

3.2.5. Lane-Line Curve Start-Point Location Detection

The minimum radius of the curvature of a circular curve at the design speed of an urban road section was found to be 100–200 m by checking the "Highway Route Design Specification" (JTG D20-2006). The recommended value was five to eight times the minimum value. Therefore, whether the curvature of the circular curve was greater than 1000 m was chosen as the basis for judging whether to enter the curve. According to the radius of curvature calculation formula, it can be obtained:

$$R = \frac{1}{k} \tag{14}$$

Then, the curvature needs to satisfy:

$$k < \frac{1}{1000} \tag{15}$$

where R is the radius of curvature and k is the curvature:

$$k = \frac{y''}{\left[1 + (y')^2\right]^{\frac{3}{2}}} \tag{16}$$

Assuming that the position of the starting point of the circular curve under the image coordinate system is (x_s,y_s), the expression of the fitted curve of the lane line obtained according to the Kalman filtering is:

$$y = a_f x^2 + b_f x + c_f \tag{17}$$

Equations (15) and (16) are joined to provide:

$$\frac{y''}{\left[1 + (y')^2\right]^{\frac{3}{2}}} < \frac{1}{1000} \tag{18}$$

The derivation of Equation (17) gives:

$$y'' = 2a_f \tag{19}$$

$$y' = 2a_f x + b_f \tag{20}$$

Bringing $x = x_s$ into Equations (19) and (20), and taking the result into Equation (18), gives:

$$\frac{2a_f}{\left[1 + \left(2a_f x_s + b_f\right)^2\right]^{\frac{3}{2}}} < \frac{1}{1000} \tag{21}$$

Figure 11. Comparison of lane-line-detection effects.

From the above equation, we could identify the range of x_s. The greatest value was selected and bought into the fitting curve to complete the positioning of (x_s,y_s). The result is shown in Figure 12.

Figure 12. Lane-line circle-curve starting-point detection.

3.3. Target-Distance-Recognition Algorithm

In extracting the speed bump distance, the center position of the target-detection frame was selected as the distance-recognition extraction point because the target-detection frame could contain the speed bump, which was approximately rectangular in shape. If the speed bump was detected, the camera could get the upper-left vertex position (u_{lu},v_{lu}) and lower-right vertex position (u_{rl},v_{rl}) of the detection frame, and the principle is shown in Figure 13. The three coordinate systems from left to right are the camera coordinate system, the pixel coordinate system and the world coordinate system. The center position of the detection frame (u_c,v_c) is derived from Equations (22) and (23):

$$u_c = \frac{(u_{rl} - u_{lu})}{2} + u_{lu} \tag{22}$$

$$v_c = \frac{(v_{rl} - v_{lu})}{2} + v_{lu} \tag{23}$$

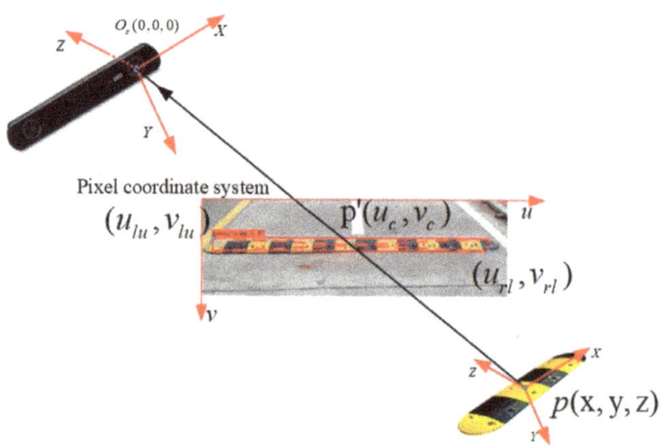

Figure 13. Coordinate relationship of the center point of the detection frame.

After determining the position of the center of the detection frame, the distance between the target and the camera could be calculated according to the coordinate conversion relationship of the camera. Then, the distance d between the speed bump and the center of the front wheel could be calculated according to the parameters of the camera. The calculation formula is shown in Equation (24).

$$d = \sqrt{{d_t}^2 - h^2} - l_{cw} \tag{24}$$

where h refers to the height of the camera from the horizontal road surface and l_{cw} denotes the horizontal distance from the camera to the center of the front axis.

The camera used in this paper was a binocular camera with depth-perception function, which could provide 3D point cloud information of the subject. In extracting the distance of the starting point of the circle curve, since the pixel-coordinate-location identification method of the starting point of the road circle curve has been given in the previous paper, the distance of the starting point of the circle curve could be extracted from the 3D point cloud data of the camera under the condition of the known pixel coordinates. The workflow of this extraction method is shown in Figure 14. If the lane-line circular curve start point is detected, the pixel coordinates are obtained and the point cloud information is calculated. Then, the distance from the target is calculated based on the distance conversion. If the start point is not detected, the lane line is continued to be detected.

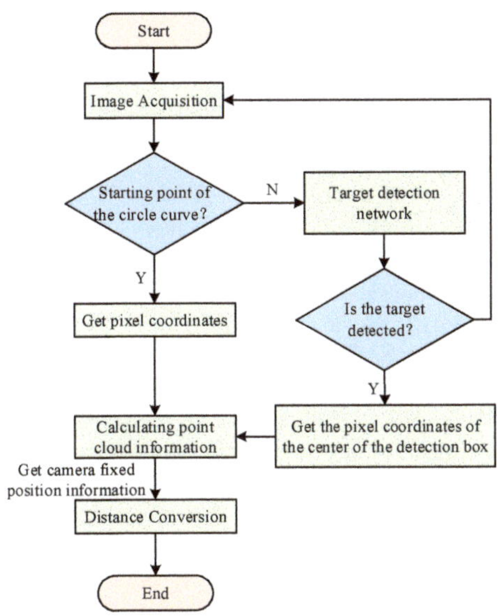

Figure 14. Target-distance extraction algorithm flow chart.

4. Design of the Intelligent Suspension Control System

The whole control system consisted of the road input, perception algorithm, control-mode-switching strategy, BP-PID controller, and force distributor. Firstly, the target was detected by the visual-perception system, and the time to reach the target was calculated according to the distance and driving speed. This was used to determine which control strategy the system switched to. Then, the difference between the desired output and the actual output of the control target was adopted as the input of the BP-PID controller in different control modes. The output of the control force was then obtained by self-tuning the PID parameters through the BP neural network. Subsequently, the output of control force was input to the force distributor to obtain the damping force on the four suspensions. Finally, the control current was calculated by the magnetorheological inverse model identification result, and the result could be input to the magnetorheological dampers to obtain the actual control force. The framework of the whole control system is shown in Figure 15:

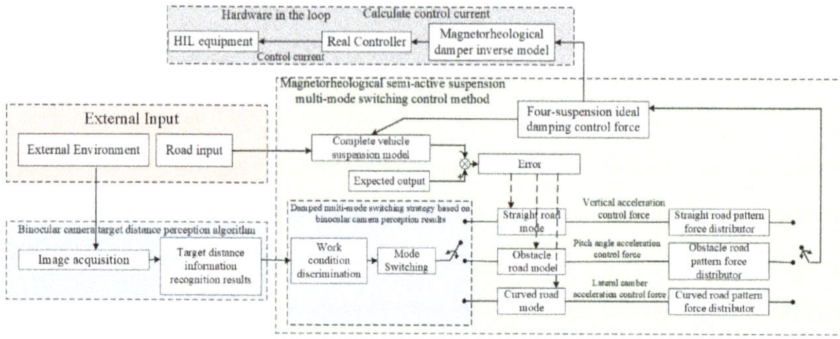

Figure 15. Semiactive suspension control system framework.

4.1. Mode Switching

The suspension control mode could be divided into the obstacle-road mode, straight-road mode, and curved-road mode according to different working conditions. The straight-road mode mainly controls the body dip acceleration to reduce the bumps caused by the road undulation, the obstacle-road mode mainly controls the body pitch acceleration to reduce the vehicle pitch-angle change due to the speed bump, and the curved-road mode mainly controls the body roll acceleration to reduce the rapid body tilt caused by the inertia of the spring mass when the vehicle enters the curve. At the same time, the priority of the curved-road mode was set to the highest to prevent the vehicle from being in a dangerous situation and causing injury to the passengers. The default control mode was the straight-road mode.

The working-condition identification method is shown in Figure 16. The vehicle started with $m = 0$ and entered the straight-road mode. Then, the vehicle determined whether to enter the curved-road mode. When $c = 1$, the first recognized time t and the first distance L_{t_s} will be recorded. When $t \geq \frac{L_{t\_s}}{v}$, the vehicle enters the curved-road mode, otherwise continues to stay in the straight-road mode. Here, c equals 1 when traveling in a curve. When the vehicle is about to exit the curve, it will recognize the lane-line round curve start point for the last time. At this time, t_e as well as L_{t_e} will be recorded. If $t_e \geq \frac{L_{t\_s}}{v}$, it means that the vehicle enters into the straight road and returns to the straight-road mode, otherwise it continues to stay in the curved-road mode.

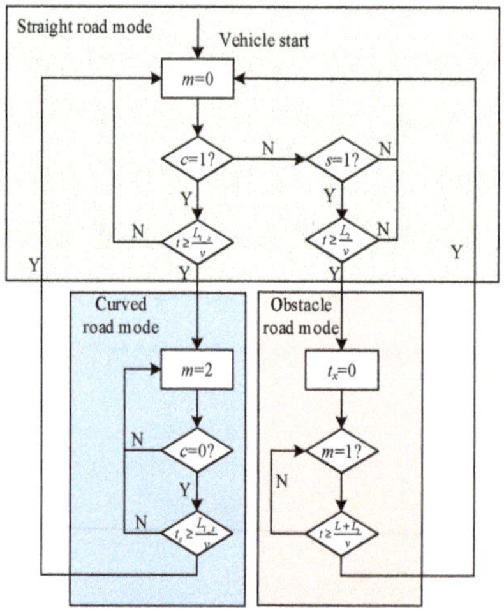

Figure 16. Condition identification method.

When the vehicle does not satisfy the judgement condition of the curve mode, a judgement will be made to determine whether the vehicle satisfies the switching condition of the obstacle-road mode. If $s = 1$, the first recognized time t and the target distance L_t will be recorded. If $t \geq \frac{L_t}{v}$, it means that the vehicle is passing the speed bump and goes into the obstacle-road mode. At this time, $m = 1$. The time required for vehicles to completely pass the speed bump is used as the judgment basis for the duration of the obstacle-road mode. If $t \geq \frac{L_t+L}{v}$, it means that the vehicle completely passes the speed bump and it returns to the straight-road mode. At this time, $m = 0$.

4.2. BP-PID Controller Design

The controller consists of the BP neural network and the PID controller. In 1986, Rumelhart et al. first introduced the concept of the BP neural network, which is a multilayer feedforward network based on the error back-propagation algorithm for network training. It continuously adjusted the network weights through sample data training, so that the error value decreased and the actual output results were close to the desired value. The error back-propagation algorithm ensures that the BP neural network has good generalization and self-learning ability, so it is widely used in the design of nonlinear system controllers. The input of the BP neural network includes the desired input, actual output, and the error of the system. The output consists of the adjustment parameters K_p, K_i, and K_d to realize the online adjustment of the PID parameters. As such, it can realize a fast response under the uncertain input conditions of the time-varying system. The framework of the controller is shown in Figure 17.

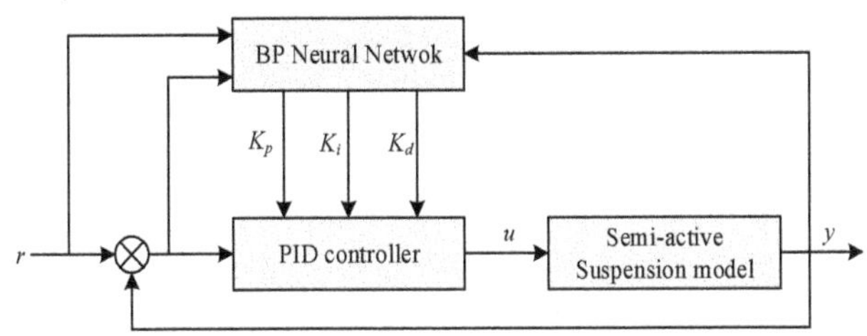

Figure 17. BP-PID controller.

The number of neurons m in the neural network was three. We selected the single-hidden-layer network, and the number of output neurons l was three. The sampling period $T_{bp} = 0.001$ was designed as per the Simulink simulation step. The weights of each layer were initialized using random numbers from 0 to 1, and the Softmax function was used as the activation function.

In the case of the controller for the straight-ahead mode, the control objective of this mode was to keep the body smooth and reduce the body droop acceleration. Therefore, the body droop acceleration was adopted as the actual input of the controller, and the desired output was set as 0. The suspension control force was calculated by the neural network.

4.3. Parameter Optimization

Since the control forces obtained by the BP-PID controller were distributed to the four suspensions in the same direction and in equal amounts, the road inputs of the left and right wheels were not equal, and the effect of suppressing the body roll and pitch was not significantly improved. The next step was to design a force distributor to achieve a coordinated distribution of the damping forces.

4.3.1. Improvement of the Salp Swarm Algorithm

The salp swarm algorithm, proposed by Mirjalili et al. in 2017, was inspired by the idea of a "leader-follower chain" moving together to forage for food. The leader searches globally over a large area, while the followers follow the previous leader to explore locally and eventually lead the group to food. The method improves global exploration and local exploitation, and reduces the number of local optima [32].

The flow of salp swarm algorithm is shown in Figure 18:

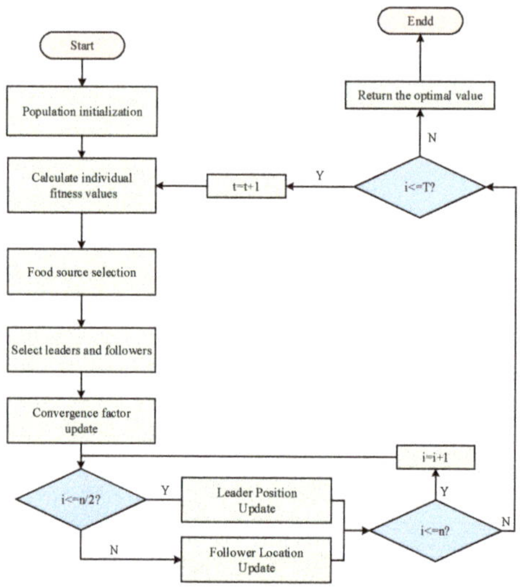

Figure 18. Flow of the salp swarm algorithm.

The leader position is updated according to Equation (26) and the follower position is updated according to Equation (27):

$$X_d^1 = \begin{cases} F_d + c_1(ub - lb)c_2 + lb, c_3 \geq 0.5 \\ F_d - c_1(ub - lb)c_2 + lb, c_3 < 0.5 \end{cases} \tag{25}$$

$$X_d^{i'} = \frac{X_d^i + X_d^{i-1}}{2} \tag{26}$$

$$c_1 = 2e^{-\left(\frac{4l}{L}\right)^2} \tag{27}$$

where X_d and F_d represent the dth dimensional leader position and food position; u_b and l_b denote the upper and lower bounds; c refers to the convergence factor, which is updated by Equation (28); l is the number of current iterations and L is the maximum number of iterations; c_2 and c_3 are random parameters in the range [0, 1].

The initialization of the bottle sheath group affects the computational speed as well as the accuracy of the algorithm. The use of the Tent Map to randomly generate a uniformly distributed initialization of the bottle sheath groups in the search space can be beneficial in improving the algorithm to find the optimal bottle sheath group, which can be expressed by Equation (29):

$$z_{j+1}^i = \begin{cases} \frac{z_j^i}{u}, 0 \leq z_j^i \leq u \\ \frac{1-z_j^i}{1-u}, 0 \leq z_j^i \leq 1 \end{cases} \tag{28}$$

where i = 1, 2, 3, … N is the number of populations, j signifies the number of current iterations, and u is the chaos control parameter. Thus, the Tent is applied to initialize the salp swarm as:

$$X_{D*N} = z_j^i(ub_{D\times N} - lb_{D\times N}) + lb_{D\times N} \tag{29}$$

In addition, the position of the food source moves all the time when the salp swarm is actually searching for food. So, the "crazy concept" is introduced to model this phenomenon, and by introducing a madness operator in the leader position update equation to avoid the population from falling into local optimal solutions. The leader update equation is shown in Equation (31):

$$X_d^1 = \begin{cases} F_d + P(c_4)sign(c_4)x_{cr} + c_1(ub - lb)c_2 + lb, c_3 \geq 0.5 \\ F_d + P(c_4)\text{sign}(c_4)x_{cr} - c_1(ub - lb)c_2 + lb, c_3 < 0.5 \end{cases} \tag{30}$$

$$P(c_4) = \begin{cases} 1, c_4 \leq P_{cr} \\ 0, c_4 > P_{cr} \end{cases} \tag{31}$$

$$sign(c_4) = \begin{cases} -1, c_4 \geq 0.5 \\ 1, c_4 < 0.5 \end{cases} \tag{32}$$

where c_4 denotes a random number of [0, 1] conforming to a uniform distribution and x_{cr} refers to a smaller constant; $P(c_4)$ is taken according to Equation (32), where P_{cr} is the crazy probability; sign(c_4) is taken according to Equation (33).

The dependence of followers on the leader depends on the inertia weight. A larger weight enhances the global search ability of the salp swarm and a smaller weight helps to achieve local exploitation. To balance the global search and local exploitation ability, linearly decreasing weights $w(t)$ were introduced, and the new follower update formula can be seen as follows:

$$X_d^{i'} = \frac{X_d^i + w(t)X_d^{i-1}}{2} \tag{33}$$

$$w(t) = \tfrac{w_s(w_s - w_e)(L-l)}{L} \tag{34}$$

where w_s is the initial weight, w_e represents the maximum number of iterations weight, L is the maximum number of iterations, and l refers to the current number of iterations.

4.3.2. Force Distributor Design

The objective functions to be optimized in the three modes can be constructed by the improved salp swarm algorithm, and the coordination weights of the damping forces in the different modes can be obtained. Taking the straight-road mode as an example, the forces assigned to the four dampers are:

$$f_i = d_{i_{\ddot{z}}} f_{bp_{pid}} \tag{35}$$

where $d_{i_{\ddot{z}}}$ is the damping force coordination weight.

Due to the different units and orders of magnitude of the vehicle acceleration and dynamic tire load, it is necessary to perform normalization. Taking the root mean square value of the corresponding performance index of the passive suspension as a reference, the subperformance function could be defined as follows:

$$p_\delta = \frac{RMS(\delta_{sa})}{RMS(\delta_{pas})} \tag{36}$$

where $RMS(\delta_{sa})$ is the root mean square value of the semiactive suspension performance index; $RMS(\delta_{pas})$ is the root mean square value of passive suspension; δ represents the body droop acceleration, body pitch-angle speed, body roll-angle acceleration, and tire dynamic load, respectively, while its subscript pas represents the passive suspension and *sa* represents the semiactive suspension.

The integrated optimization objective function is the sum of the subobjective functions, and the integrated optimization objective functions for the straight-road, obstacle-road, and curved-road mode are as follows:

$$P_{straight} = p_{\ddot{z}} + p_{F_{d1sa}} + p_{F_{d2sa}} + p_{F_{d3sa}} + p_{F_{d4sa}} \tag{37}$$

$$P_{obstacle} = p_{\ddot{\theta}} + p_{F_{d1sa}} + p_{F_{d2sa}} + p_{F_{d3sa}} + p_{F_{d4sa}} \tag{38}$$

$$P_{curved} = p_{\ddot{\phi}} + p_{F_{d1sa}} + p_{F_{d2sa}} + p_{F_{d3sa}} + p_{F_{d4sa}} \tag{39}$$

In addition, the suspension motion restraint and tire dynamic load restraint should be included to prevent the effects of prolonged abnormal operation on the suspension performance, as well as the vehicle. Suspension motion constraints and the dynamic tire loads can be expressed by Equations (40) and (41):

$$\frac{RMS(z_{us_n})}{[z_{us_n}]} \leq \frac{1}{3} \tag{40}$$

$$\frac{RMS(F_{dn})}{\left(m + \sum_{n=1}^{4} m_{un}\right)g} \leq \frac{1}{3} \tag{41}$$

where z_{us_n} represents the suspension dynamic deflection and $[z_{us_n}]$ signifies the suspension dynamic travel.

Due to the addition of the constrained set of equations, it is necessary to remove the solutions outside the constraints by adding penalty terms to the integrated optimization objective function. The following equations can be used as the penalty terms:

$$p_{motion} = RMS(z_{us_n}) - \frac{1}{3}[z_{us_n}] \leq 0 \tag{42}$$

$$p_{force} = RMS(F_{dn}) - \frac{1}{3}\left(\mathrm{m} + \sum_{n=1}^{4} m_{un}\right)g \leq 0 \tag{43}$$

Therefore, the integrated optimization objective function for the straight-road model is changed as follows:

$$P_{daily} = p_{\ddot{z}} + p_{F_{d1sa}} + p_{F_{d2sa}} + p_{F_{d3sa}} + p_{F_{d4sa}} + w_p p \tag{44}$$

where w_p is the penalty weight and the value is 0 when the penalty term holds, otherwise it is 1; p refers to the penalty factor and usually takes a larger value to exclude the unconditional solution.

The length of a salp was 1–10 cm and the median length was 5 cm, so the population number chain of the salp swarm did not exceed 300, and the number of populations was set to (50,4) considering the operation speed. The maximum number of iterations was 200, the inertia weight w_s = 0.9, w_c = 0.4, the madness probability C_{cr} = 0.3, and x_{cr} = 0.0001. When the value was less than three, or the maximum number of iterations was reached, the calculation was ended and the results were output.

The search range of the weight coefficients of the force distributor was set at [−1, 1], and the obtained weight coefficients can be seen in Table 2. In the straight-road mode, the four forces are in the same direction. Under the obstacle-road mode, the rear axle forces are in the opposite direction to the front axle. In the curved-road mode, the left and right suspension forces are in opposite directions.

Table 2. Force controller weighting factor.

Control Mode	Left Front	Right Front	Left Rear	Right Rear
Straight Road Mode	0.951	0.986	0.933	0.915
Obstacle Road Mode	0.986	0.978	−0.948	−0.954
Curved Road Mode	−0.972	0.961	−0.981	0.986

4.4. Simulation Results and Analysis

In this paper, the joint simulation was realized by Python and MATLAB/Simulink. The perception algorithm was implemented based on Python3.8, which was the control algorithm. The whole vehicle and road-input model was implemented by Simulink2021a. Python passed to MATLAB2021a the lane-line circle-curve start-point result c, the speed-bump target-recognition result s, the time t, the target-object distance L_t, and other parameters. MATLAB obtained the control-mode-switching parameters m through Simulink with Python's parameters. The flow is shown in Figure 19.

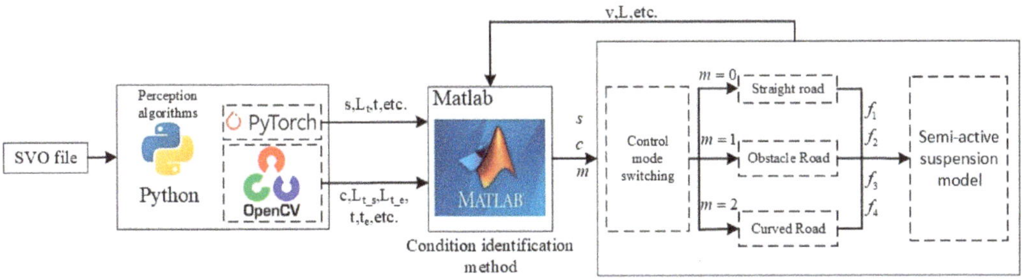

Figure 19. Flow of joint simulation.

The vehicle driving speed was stabilized at 10 m/s, and the total length of the test section was 250 m. After the measurement, the test section entered the curve at the position of 50 m, drove out of the curve after 110 m in the curve, and then had a speed bump at 70 m in the straight road. Figure 20 shows the satellite map of the collected road section and the result of the target-feature labeling.

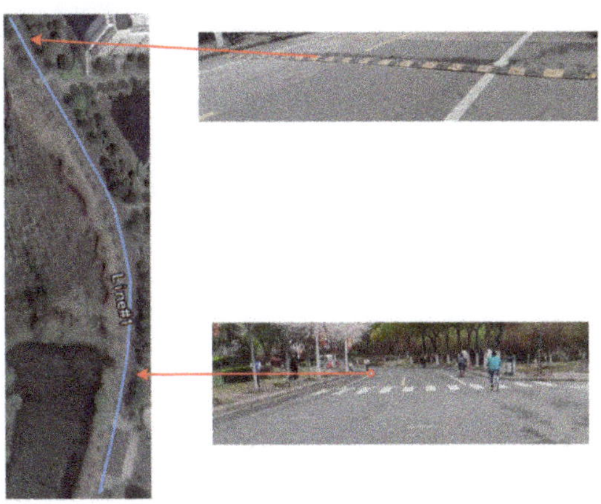

Figure 20. Image-acquisition section satellite map and feature mark.

The results of the joint simulation control mode switching are shown in Figure 21. The simulation switching curve was 0.009 s earlier than the theoretical switching curve into the straight-road mode, when the vehicle got out of the curve. The simulation test under the semiactive suspension control mode switched to the obstacle-road mode and was 0.008 s earlier than the theoretical switching curve when the vehicle was about to pass the speed bump target. After the vehicle passed the target of the speed bump, the simulated switching curve of the control method back to the straight-road mode was ahead of the theoretical switching curve by 0.007 s. The reason may be that the vehicle driving speed was slightly higher than 10 m/s, and the theoretical switching curve was calculated as 10 m/s. However, the difference between the simulation and theoretical model was only on the order of milliseconds, which would not affect the switching strategy efficiency.

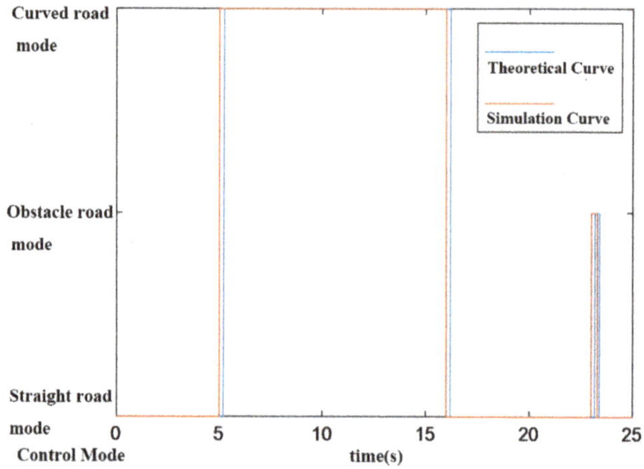

Figure 21. Control-mode-switching process.

The joint simulation of the suspension control effect is shown in Figure 22, and the results obtained are shown in Table 3:

Table 3. Decreased values of the semiactive suspension performance indicators.

	Droop Acceleration	Roll-Angle Acceleration	Pitch-Angle Acceleration	Dynamic Deflection of Left Front	Dynamic Deflection of Right Rear	Dynamic Tire Load of Left Front	Dynamic Tire Load of Right Rear
Decrease percentage	20.18%	10.34%	14.36%	2.3%	1.2%	0.13%	1.2%

As can be seen, compared with the passive suspension, the semiactive suspension significantly improved the droop acceleration, and there was an over 10 percent improvement in the pitch and roll acceleration compared to the passive suspension. Meanwhile, the suspension dynamic deflection and the wheel dynamic load performance achieved a small improvement.

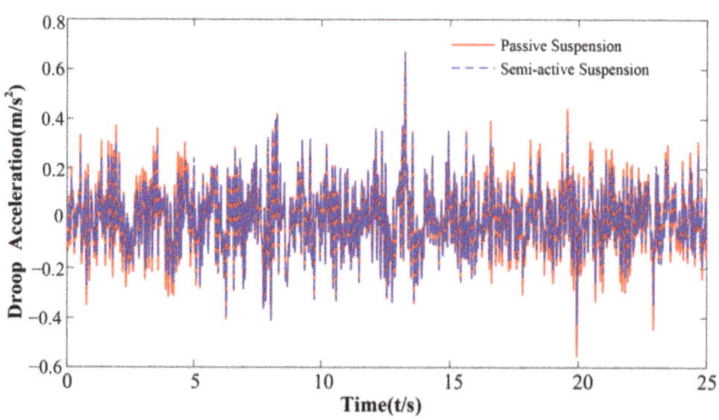

(a)

Figure 22. *Cont.*

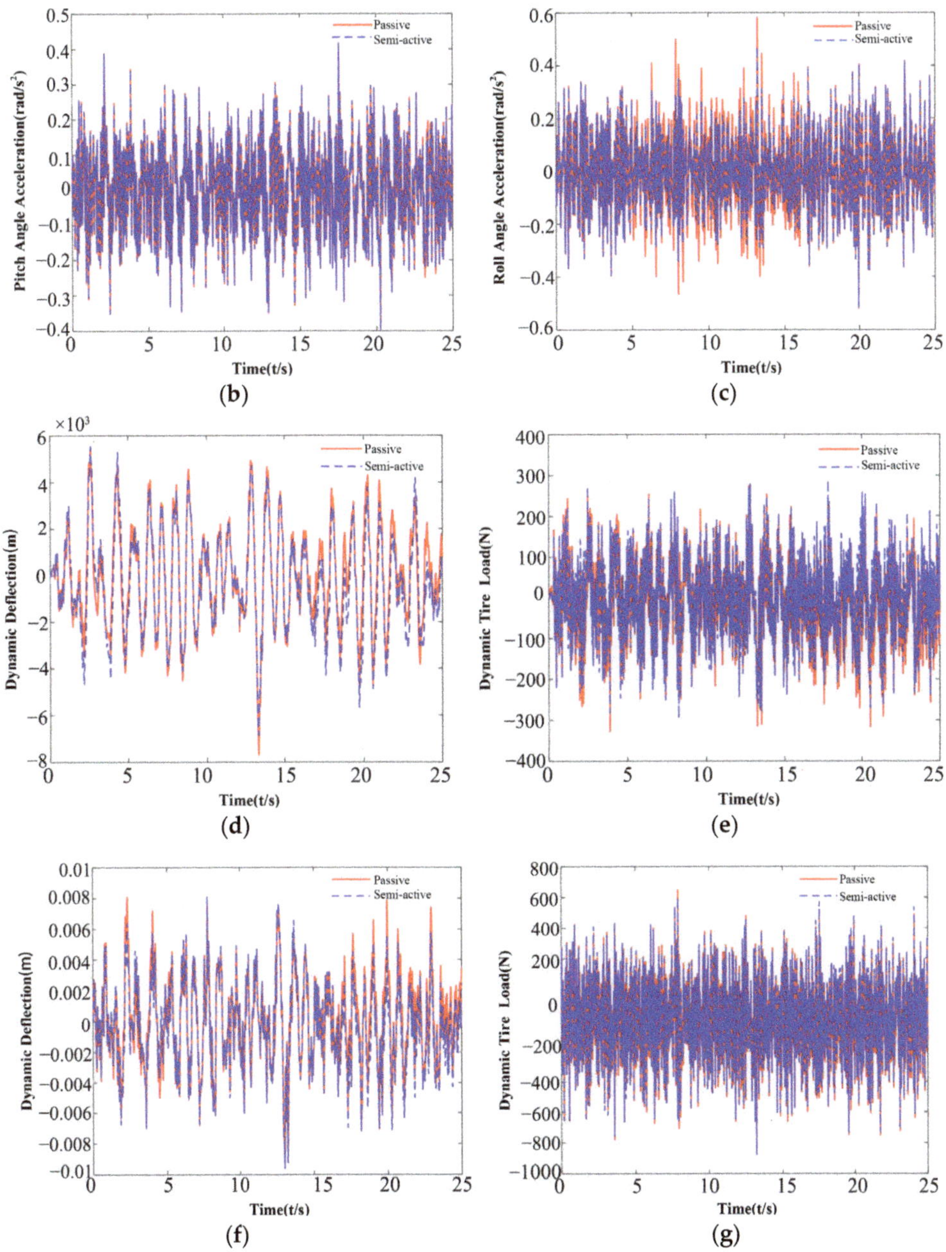

Figure 22. Simulation effect: (**a**) Comparison of the droop acceleration; (**b**) Comparison of the body pitch-angle acceleration; (**c**) Comparison of the body roll-angle acceleration; (**d**) Dynamic deflection of the right rear suspension; (**e**) Dynamic tire load of the right rear suspension; (**f**) Dynamic deflection of the left front suspension; (**g**) Dynamic tire load of the left front suspension.

5. HIL Experiment

The hardware-in-the-loop experimental test environment constructed in this paper consisted of the D2P (development to prototype), the host computer software (LABCAR-

OPERATOR V5.0) and the LABCAR test cabinet(from Vehinfo, Shanghai, China) [33]. The D2P connects and interacts with the host computer through the CAN bus and provides abundant I/O interfaces. The host computer was equipped with MATLAB, LCO, and other software, and the test interface for the host computer could be constructed to detect variables. LABCAR provided the I/O boards and CAN communication boards, which could test the operation effect, transmit the required signals to the host computer, and transmit the simulated signals during the simulation operation. The flow of the test is shown in Figure 23.

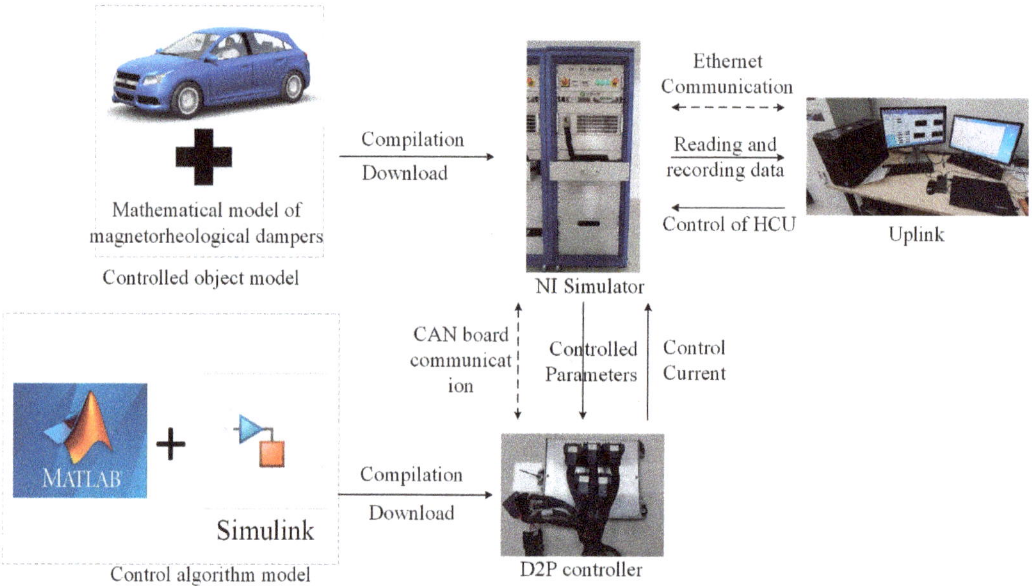

Figure 23. Flow of the HIL.

The hardware-in-the-loop test results are shown in Figure 24, the results obtained are shown in Table 4. And the analysis indicated that the root mean square value of body roll acceleration are more significantly improved compared to the pitch and droop acceleration. And the multi-mode control method improves the performance of suspension and tire compared to the passive suspension and PID control methods.

Table 4. Decreased values of the semiactive suspension performance indicators in the HIL.

Decrease Percentage	Droop Acceleration	Roll-Angle Acceleration	Pitch-Angle Acceleration	Dynamic Deflection of Left Front	Dynamic Deflection of Right Rear	Dynamic Tire Load of Left Front	Dynamic Tire Load of Right Rear
Compared with PID	1.2%	2.4%	10.12%	3.22%	2.75%	0.56%	0.92%
Compared with Passive	3.79%	4.45%	22.27%	5.62%	6.62%	1.85%	2.86%

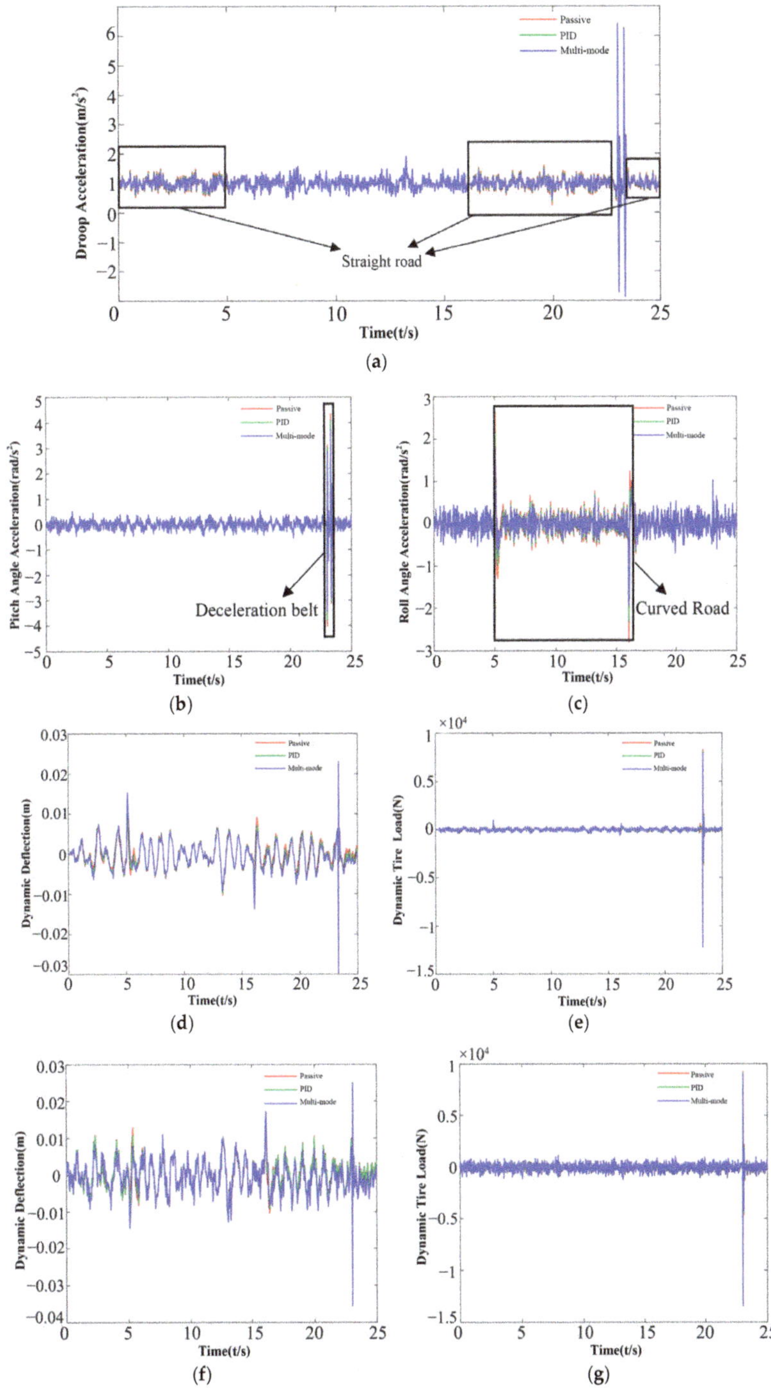

Figure 24. HIL test results: (**a**) Comparison of the droop acceleration; (**b**) Comparison of the body pitch-angle acceleration; (**c**) Comparison of the body roll-angle acceleration; (**d**) Dynamic deflection of the right rear suspension; (**e**) Dynamic tire load of the right rear suspension; (**f**) Dynamic deflection of the left front suspension; (**g**) Dynamic tire load of the left front suspension.

6. Conclusions

In this paper, a multimode control strategy for a semiactive suspension with a magnetorheological damper based on binocular camera target-distance recognition was designed. The aim was to realize the adaptive switching of semiactive suspension control methods under different road conditions so as to improve the smoothness, comfort, and handling stability of the vehicles. Although we made a little progress, there are still many of problems that need to be solved in the future due to the limitations of the experimental equipment, experimental conditions, and time:

(1) The effectiveness of the control method needs to be further verified from the perspective of a real vehicle. Due to the lack of a real vehicle equipped with a magnetorheological semiactive suspension, and the time relationship, this paper only carries out MIL and HIL experiments through Simulink and Carsim, and does not carry out real-vehicle tests.

(2) This paper makes less use of the function of the binocular camera. As a sensor, it can measure the 3D point cloud so that it has a powerful function that is not inferior to Laser Radar. By using the binocular camera to realize the real-time scanning of the front terrain, it can provide richer road-surface information for the suspension control algorithm, and realize the suspension control algorithm with a more excellent effect.

Author Contributions: Conceptualization, C.H. and K.L.; methodology, C.H.; software, K.L.; validation, C.H. and K.L.; formal analysis, C.H., K.L. and Q.X.; investigation, K.L.; resources, Y.D.; data curation, K.L.; writ-ing—original draft preparation, K.L.; writing—review and editing, C.H.; visualization, Q.X.; su-pervision, Q.X.; project administration, C.H.; funding acquisition, Y.D. All authors have read and agreed to the published version of the manuscript.

Funding: This work is supported by the State Key Laboratory of Vehicle Safety and Energy Saving of Tsinghua University, grant number [KFY2207], International Cooperation Fund of Jiangsu, grant number [BZ2022050], and General Program of China Postdoctoral Science Foundation, grant number [2021M691847].

Data Availability Statement: Data are contained within the article.

Acknowledgments: Thanks to the State Key Laboratory of Vehicle Safety and Energy Saving of Tsinghua University, and International Cooperation Fund of Jiangsu for the support of this paper.

Conflicts of Interest: The authors declare no conflict of interest.

References

1. Li, M.; Li, J.; Li, G.; Xu, J. Analysis of Active Suspension Control Based on Improved Fuzzy Neural Network PID. *World Electr. Veh. J.* **2022**, *13*, 226. [CrossRef]
2. Zheng, X.; Zhang, H.; Yan, H.; Yang, F.; Wang, Z.; Vlacic, L. Active Full-Vehicle Suspension Control via Cloud-Aided Adaptive Backstepping Approach. *IEEE Trans. Cybern.* **2020**, *50*, 3113–3124. [CrossRef]
3. Haemers, M.; Derammelaere, S.; Ionescu, C.M.; Stockman, K.; Viaene, J.D.; Verbelen, F. Proportional-Integral State-Feedback Controller Optimization for a Full-Car Active Suspension Setup using a Genetic Algorithm. *IFAC-PapersOnLine* **2018**, *51*, 1–6. [CrossRef]
4. Scharstein, D.; Szeliski, R. A Taxonomy and Evaluation of Dense Two-Frame Stereo Correspondence Algorithms. *Int. J. Comput. Vis.* **2002**, *47*, 7–42. [CrossRef]
5. Lu, Y.; Zou, L.; Chen, Y.; Mao, Y.; Zhu, J.; Lin, W.; Wu, D.; Chen, R.; Qu, J.; Zhou, J. Rapid Alternate Flicker Modulates Binocular Interaction in Adults With Abnormal Binocular Vision. *Investig. Ophthalmol. Vis. Sci.* **2023**, *64*, 15. [CrossRef] [PubMed]
6. Blake, R.; Wilson, H. Binocular vision. *Vis. Res.* **2011**, *51*, 754–770. [CrossRef] [PubMed]
7. Yuan, J.; Zhang, G.; Li, F.; Liu, J.; Xu, L.; Wu, S.; Jiang, T.; Guo, D.; Xie, Y. Independent Moving Object Detection Based on a Vehicle Mounted Binocular Camera. *IEEE Sens. J.* **2021**, *21*, 11522–11531. [CrossRef]
8. Rovira-Más, F.; Wang, Q.; Zhang, Q. Design parameters for adjusting the visual field of binocular stereo cameras. *Biosyst. Eng.* **2010**, *105*, 59–70. [CrossRef]
9. Ding, J.; Klein, S.A.; Levi, D.M. Binocular combination in abnormal binocular vision. *J. Vis.* **2013**, *13*, 14. [CrossRef]
10. Hu, G.; Liao, M.; Li, W. Analysis of a compact annular-radial-orifice flow magnetorheological valve and evaluation of its performance. *J. Intell. Mater. Syst. Struct.* **2017**, *28*, 1322–1333. [CrossRef]
11. Brooks, D.A. High Performance Electro-Rheological Dampers. *Int. J. Mod. Phys. B* **1999**, *13*, 2127–2134. [CrossRef]

12. Tang, J.; Dai, W.; Archer, C.; Yi, J.; Zhu, G. Borderline knock adaptation based on online updated surrogate models. *Int. J. Engine Res.* **2022**, *24*, 2958–2972. [CrossRef]
13. Ye, H.; Zheng, L. Comparative study of semi-active suspension based on LQR control and H_2/H_∞ multi-objective control. In Proceedings of the 2019 Chinese Automation Congress (CAC), Hangzhou, China, 22–24 November 2019; pp. 3901–3906.
14. Tang, J.; Dai, W.; Archer, C.; Yi, J.; Zhu, G. Cycle-based LQG knock control using identified exhaust temperature model. *Int. J. Engine Res.* **2022**, *24*, 3047–3060. [CrossRef]
15. Li, H.; Yu, J.; Hilton, C.; Liu, H. Adaptive Sliding-Mode Control for Nonlinear Active Suspension Vehicle Systems Using T–S Fuzzy Approach. *IEEE Trans. Ind. Electron.* **2013**, *60*, 3328–3338. [CrossRef]
16. Li, H.; Jing, X.; Lam, H.K.; Shi, P. Fuzzy Sampled-Data Control for Uncertain Vehicle Suspension Systems. *IEEE Trans. Cybern.* **2014**, *44*, 1111–1126. [CrossRef] [PubMed]
17. Xu, C.; Xie, F.; Zhou, R.; Huang, X.; Cheng, W.; Tian, Z.; Li, Z. Vibration analysis and control of semi-active suspension system based on continuous damping control shock absorber. *J. Braz. Soc. Mech. Sci. Eng.* **2023**, *45*, 341. [CrossRef]
18. Wang, M.; Pang, H.; Wang, P.; Luo, J. BP neural network and PID combined control applied to vehicle active suspension system. In Proceedings of the 2021 40th Chinese Control Conference (CCC), Shanghai, China, 26–28 July 2021; pp. 8187–8192.
19. Qiu, R. Adaptive Control of Vehicle Active Suspension Based on Neural Network Optimization. In Proceedings of the 2021 7th International Conference on Energy Materials and Environment Engineering (ICEMEE 2021), Zhangjiajie, China, 23–25 April 2020; Volume 261.
20. Khan, L.; Qamar, S.; Khan, U. Adaptive PID control scheme for full car suspension control. *J. Chin. Inst. Eng.* **2016**, *39*, 169–185. [CrossRef]
21. Bender, E.K. Optimum Linear Preview Control With Application to Vehicle Suspension. *J. Basic Eng.* **1968**, *90*, 213–221. [CrossRef]
22. Louam, N.; Wilson, D.A.; Sharp, R.S. Optimization and Performance Enhancement of Active Suspensions for Automobiles under Preview of the Road. *Veh. Syst. Dyn.* **2007**, *21*, 39–63. [CrossRef]
23. Gohrle, C.; Schindler, A.; Wagner, A.; Sawodny, O. Design and Vehicle Implementation of Preview Active Suspension Controllers. *IEEE Trans. Control Syst. Technol.* **2014**, *22*, 1135–1142. [CrossRef]
24. Li, P.; Lam, J.; Cheung, K.C. Multi-objective control for active vehicle suspension with wheelbase preview. *J. Sound Vib.* **2014**, *333*, 5269–5282. [CrossRef]
25. Choi, S.-B.; Lee, S.-K.; Park, Y.-P. A Hysteresis Model for the Field-Dependent Damping Force of a Magnetorheological Damper. *J. Sound Vib.* **2001**, *245*, 375–383. [CrossRef]
26. Zhao, Q.; Zhu, B. Multi-objective optimization of active suspension predictive control based on improved PSO algorithm. *J. Vibroeng.* **2019**, *21*, 1388–1404. [CrossRef]
27. Zhang, L.; Yin, X.; Shen, J.; Yu, H. Cloud-aided moving horizon state estimation of a full-car semi-active suspension system. In Proceedings of the 2016 IEEE International Conference on Systems, Man, and Cybernetics (SMC), Budapest, Hungary, 9–12 October 2016; pp. 527–532.
28. Mahendru, M.; Dubey, S.K. Real Time Object Detection with Audio Feedback using Yolo vs. Yolo_v3. In Proceedings of the 2021 11th International Conference on Cloud Computing, Data Science & Engineering (Confluence), Noida, India, 28–29 January 2021; pp. 734–740.
29. Troelstra, M.A.; Van Dijk, A.-M.; Witjes, J.J.; Mak, A.L.; Zwirs, D.; Runge, J.H.; Verheij, J.; Beuers, U.H.; Nieuwdorp, M.; Holleboom, A.G.; et al. Self-supervised neural network improves tri-exponential intravoxel incoherent motion model fitting compared to least-squares fitting in non-alcoholic fatty liver disease. *Front. Physiol.* **2022**, *13*, 942495. [CrossRef] [PubMed]
30. Mohammed-Azizi, B.; Mouloudj, H. Least-squares fitting applied to nuclear mass formulas. Solution by the Gauss–Seidel method. *Int. J. Mod. Phys. C* **2022**, *33*, 2250076. [CrossRef]
31. Wang, T.; Tao, X.; Zhang, J.; Li, Y. Predicting Traffic Congestion Time Based on Kalman Filter Algorithm. *Adv. Res. Rev.* **2020**, *1*. [CrossRef]
32. Mirjalili, S.; Gandomi, A.H.; Mirjalili, S.Z.; Saremi, S.; Faris, H.; Mirjalili, S.M. Salp Swarm Algorithm: A bio-inspired optimizer for engineering design problems. *Adv. Eng. Softw.* **2017**, *114*, 163–191. [CrossRef]
33. Soeiro, L.G.G.; Filho, B.J.C. Vehicle Power System Modeling and Integration in Hardware-in-the-Loop (HIL) Simulations. *Machines* **2023**, *11*, 605. [CrossRef]

World Electric Vehicle Journal

MDPI

Article

Carbon Footprint Enhancement of an Agricultural Telehandler through the Application of a Fuel Cell Powertrain

Valerio Martini †, Francesco Mocera *,† and Aurelio Somà †

Department of Mechanical and Aerospace Engineering (DIMEAS), Politecnico di Torino, Corso Duca degli Abruzzi 24, 10129 Torino, Italy; valerio.martini@polito.it (V.M.); aurelio.soma@polito.it (A.S.)
* Correspondence: francesco.mocera@polito.it
† These authors contributed equally to this work.

Abstract: The growing awareness about climate change and environmental pollution is pushing the industrial and academic world to investigate more sustainable solutions to reduce the impact of anthropic activities. As a consequence, a process of electrification is involving all kind of vehicles with a view to gradually substitute traditional powertrains that emit several pollutants in the exhaust due to the combustion process. In this context, fuel cell powertrains are a more promising strategy, with respect to battery electric alternatives where productivity and endurance are crucial. It is important to replace internal combustion engines in those vehicles, such as the those in the sector of Non-Road Mobile Machinery. In the present paper, a preliminary analysis of a fuel cell powertrain for a telehandler is proposed. The analysis focused on performance, fuel economy, durability, applicability and environmental impact of the vehicle. Numerical models were built in MATLAB/Simulink and a simple power follower strategy was developed with the aim of reducing components degradation and to guarantee a charge sustaining operation. Simulations were carried out regarding both peak power conditions and a typical real work scenario. The simulations' results showed that the fuel cell powertrain was able to achieve almost the same performances without excessive stress on its components. Indeed, a degradation analysis was conducted, showing that the fuel cell system can achieve satisfactory durability. Moreover, a Well-to-Wheel approach was adopted to evaluate the benefits, in terms of greenhouse gases, of adopting the fuel cell system. The results of the analysis demonstrated that, even if considering grey hydrogen to feed the fuel cell system, the proposed powertrain can reduce the equivalent CO_2 emissions of 69%. This reduction can be further enhanced using hydrogen from cleaner production processes. The proposed preliminary analysis demonstrated that fuel cell powertrains can be a feasible solution to substitute traditional systems on off-road vehicles, even if a higher investment cost might be required.

Keywords: fuel cell; hydrogen; GHG emissions reduction; hybrid electric vehicle; telehandler; innovative powertrain; Non-Road Mobile Machineries

Citation: Martini, V.; Mocera, F.; Somà, A. Carbon Footprint Enhancement of an Agricultural Telehandler through the Application of a Fuel Cell Powertrain. *World Electr. Veh. J.* **2024**, *15*, 91. https://doi.org/10.3390/wevj15030091

Academic Editors: Fachao Jiang, Yongyu Li and Weiwei Kong

Received: 1 February 2024
Revised: 27 February 2024
Accepted: 28 February 2024
Published: 1 March 2024

1. Introduction

In recent years, the scientific community has deeply investigated the effects of anthropic activities in terms of environmental pollution, as well as the consequences on human health, climate change and economics [1–6]. Indeed, almost all human acitivities involve systems that are sources of emissions. The emissions produced can differ in quantity, depending on the specific pollutant or greenhouse gas under consideration, for the different sectors of anthropic activity. Industry, agriculture and transport sectors are characterized by high emission levels due to the adoption of internal combustion engines, which are required to accomplish several tasks. Indeed, internal combustion engines (ICEs) are one of the major contributors to air pollution, mainly due to fuel extraction processes and by-products of combustion [7,8]. In this context, several efforts are made, both from academic and industrial worlds, to study and develop innovative powertrains with lower emission levels, in order to reduce the impact related to the transport sector [9]. These efforts are supported by policies that will force the introduction of electric and alternatives powertrains as substitutions for their traditional diesel and gasoline counterparts [10]. If this trend already has a clearly visible effect on passenger cars, with several countries experiencing a quick introduction of electric vehicles on the market [11], the sector of Non-Road Mobile Machinery (NRMM) will still be at an earlier stage of electrification, even if studies have demonstrated that these vehicles have a high impact in terms of life cycle emissions [12,13]. The reason for that is related to the operative requirements that these vehicles must fulfill, with high productivity and endurance that represent a barrier to the development of pure battery electric powertrains [14]. As a consequences, several studies focused on hybrid powertrains [15–19]. Indeed, the hybridization of the powertrain allows for a downsizing of the thermal unit, which can lead to a higher efficiency and better fuel economy. Moreover, the adoption of a smaller engine can allow for simplier aftertreatment systems since the emission limits are usually defined according to the rated power of the ICE. As a consequence, different manufacturers have presented prototypes of hybrid off-road vehicles [20,21]. However, hybrid powertrains featuring the presence of an internal combustion engine still produce several harmful pollutants in the exhaust. To overcome this limit, an alternative solution for the electrification of the sector of NRMM is represented by fuel cell powertrains [22–28]. Fuel cell hybrid electric powertrains have gained attention due to their characteristics that combine the advantage of having approximately zero local emissions with high endurance and low refuelling time [29]. These properties, along with the high energy density of hydrogen, are of particular interest for the sector of NRMM. As a consequence, fuel cells, powered using hydrogen or other fuels, such as ammonia, are a promising solution to decarbonize the so-called hard-to-abate sectors, such as the marimite one [30]. Indeed, fuel cell powertrains can operate for several hours straight, which is a severe operational requirement for off-road vehicles, without having the issue of range anxiety. Moreover, fuel cell systems have higher efficiency with respect to thermal engines; thus, a better fuel economy is expected. From an economical point of view, using hydrogen as fuel can allow for energy independence and self-sufficiency, along with the possibility of defining circular economy scenarios [31]. The most promising type of fuel cell for vehicular applications is the proton-exchange membrane fuel cell (PEMFC), due to its high efficiency, low working temperature, compactness and long operational life [32,33]. However, the benefits of introducing fuel cell systems in terms of greenhouse gases emission reduction strongly depends on the hydrogen production method [34–36], with production through steam methane reforming that, at present, is the most adopted one and contributes to more than 60% of the global hydrogen production [37]. Other issues related to fuel cell systems are represented by their high purchasing costs and the inadequate state of the hydrogen refuelling network, which are two of the major challenges that must be addressed in the near future to promote their diffusion [38,39]. From an applicative point of view, fuel cell powertrains can have different topologies. Indeed, to avoid the fast degradation of fuel cells, one or more auxiliary units, generally batteries or supercapacitors, should be introduced to the powertrain to help manage sudden changes in the external load [40]. Indeed, fuel cell degradation is related to start and stop cycles, idling, high power conditions and load changes [41]. With the introduction of other power sources comes the mandatory development of an energy management strategy (EMS) that must determine how the electrical power requested by the electric motor is split among the different units [42]. Given these premises, in the present paper a fuel cell hybrid electric powertrain for a off-road heavy duty vehicle, namely a telehandler, is presented. The specifications of the traditional vehicle under investigation are defined according to existing and commercially available models. In detail, the Merlo Turbofarmer 42.7 vehicle was taken as a reference for the analysis [43]. This vehicle was designed specifically for agricultural applications. These vehicles are characterized by the presence of an hydraulic system for the actuation of the mechanical arm, thus the total load is determined by the sum of the power requested by the driveline and the power requested by the hydraulic system [?]. The powertrain architecture is composed of a PEMFC and a battery pack, with DC-DC power converters for the connection of both the fuel cell and the battery pack with the DC bus. As for the EMS, a simple power follower strategy was developed. Numerical models of both the fuel cell powertrain and the traditional counterpart were built in MATLAB/Simulink. Simulations were carried out to evaluate performances and fuel consumptions, in order to compare the two powertrains. Moreover, the environmental impacts, considering the global warming potential, of the two powertrains were compared using Well-to-Wheel (WtW) coefficients for both Diesel and Hydrogen. This paper is structured as follows: Section 2 introduces the case study and the proposed fuel cell powertrain, Section 3 presents the numerical models used for the simulations, Section 4 describes the EMS and the simulated work scenario, Section 5 shows and discusses the results obtained from the simulations, and finally Section 6 summarizes the conclusions.

2. Case Study

2.1. Traditional Vehicle

The traditional vehicle under investigation is a 105 kW diesel-powered telehandler, with a unladen mass of 7800 kg, a maximum load capacity of 4200 kg and a maximum lifting height of 7 m [43]. A telehandler is a machine with a telescopic boom that can be extended to lift, handle and place loads. The boom can be equipped with different implements to complete different tasks depending on the specific application. Due to their properties, telehandlers are widely adopted in the industry sector and in agriculture. Indeed, their particular design configuration allows for moving loads from and to places that are unreachable for the other vehicles. A schematic representation of the powertrain is reported in Figure 1, while the main properties of the vehicle are reported in Table 1. The transmission is a hydrostatic transmission with a variable displacement pump and fixed displacement motor. This type of transmission is adopted since it allows for a continously variable transmission, thus reducing the complexity of the whole driveline, and, moreover, it features high power transmission capabilities without requiring expensive and bulky components. However, the higher power losses with respect to a pure mechanical driveline have a non-neglibile impact on the overall vehicle efficiency. Downstream from the hydrostatic transmission, there is a two-speed gearbox, which allows for operating in low-speed and high-speed conditions. This gearbox is useful for having optimized gear ratios for the two most common work conditions. Indeed, during work scenarios involving the use of the telescopic boom, the vehicle generally operates in a speed range below 15 km/h, while during road transportation it operates at speeds up to 40 km/h. The low-speed regime is also useful for overcoming slopes. As for the hydraulic system for work operations, it comprises a lifting arm and an extension boom.

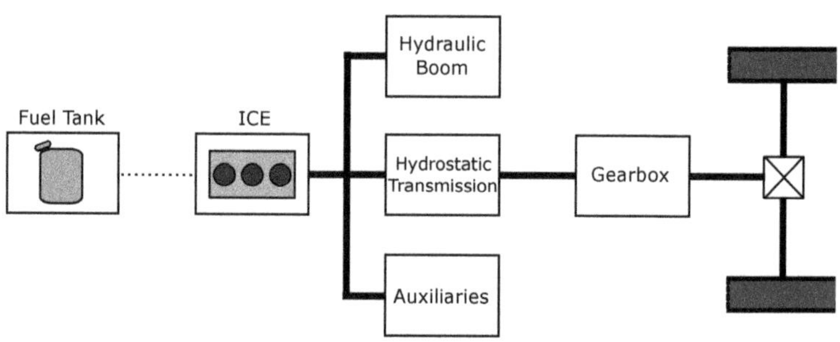

Figure 1. Schematic representation of the traditional reference powertrain.

Table 1. Traditional telehandler's main properties.

Parameter	Value
Unladen Mass	7800 kg
Max Load capacity	4200 kg
Max lift height	7 m
Max vehicle speed	40 km/h
Transmission	Hydrostatic transmission
Diesel Engine	4-cyl 3.6 L 105 kW

2.2. Fuel Cell Hybrid Electric Vehicle

The proposed fuel cell configuration is shown in Figure 2. Apart from the powertrain system, comprising the fuel cell system, the batteries, the power converters and the electric motor, the other elements of the vehicle, namely the hydrostatic transmission, hydraulic system and gearbox, were considered the same as the traditional vehicle. The main properties of the electric motor, fuel cell stack and battery pack for the proposed configuration are reported in Table 2. The fuel cell system was dimensioned considering the average expected power required by the vehicle, to guarantee a high productivity comparable with that of the traditional counterpart. As for the battery pack, it was sized to have enough power capabilities to satisfy sudden and abrupt changes in the power request without excessive C-rates. Indeed, considering a discharge current of 5C, the battery pack is able to provide an electric power that is equal to 75% of the electric motor nominal power. Moreover, the two units were selected considering reasonable space availability constraints for the on-board integration. The vehicle mass was assumed to be the same, or at least not substantially different, as the traditional case.

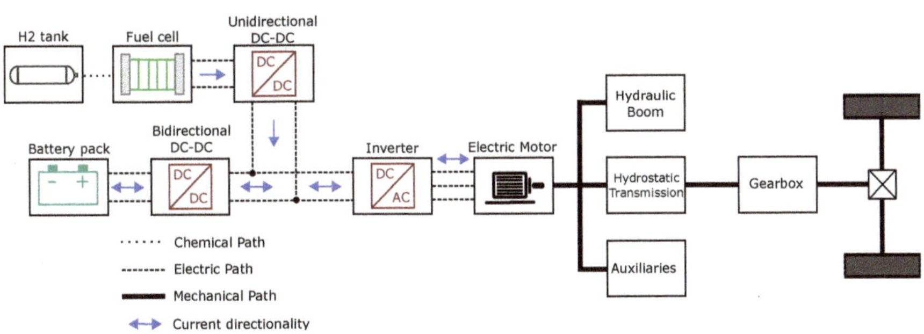

Figure 2. Schematic representation of the fuel cell hybrid electric powertrain.

Table 2. Proposed fuel cell powertrain main properties.

Element	Parameter	Value
Fuel cell system	Stack max power	53 kW
	Number of cells	300
	Max operating point	312.5 A @ 170 V
	Stack efficiency	47.5% @ 50 kW
Battery Pack	Rated Capacity	50 Ah
	Rated Voltage	320 V
Electric Motor	Rated Power	105 kW
	Rated Torque	502 Nm
	Maximum Efficiency	95%

3. Numerical Modelling

3.1. Traditional Vehicle Numerical Modelling

The numerical models were built in MATLAB/Simulink using the Simscape tool. As a consequence, a Physical Network modelling approach was used [45]; thus, each entity is considered as a physical entity capable of exchanging energy with all the other elements to which it is connected. The same modelling approach was used for previous studies from the author's research group [15,16,18,22]. During the development of the numerical model, the following aspects were covered:

- Vehicle dynamics;
- Transmission;
- Hydro-mechanical system;
- Lifting arm and extension boom;
- Engine power output and fuel consumption.

As for the vehicle dynamics, the approach adopted by the authors consisted of a 1D longitudinal model, represented in Figure 3, and was characterized by the following equations:

$$m\dot{V_x} = 2(F_{xf} + F_{xr}) - F_{aero} - mg \cdot sin(\beta) \tag{1}$$

$$F_{zf} = \frac{-h(F_{aero} + mg \cdot sin\beta) + b \cdot mg \cdot cos(\beta)}{2(a+b)} \tag{2}$$

$$F_{zr} = \frac{+h(F_{aero} + mg \cdot sin\beta) + a \cdot mg \cdot cos(\beta)}{2(a+b)} \tag{3}$$

where:

- a, b, and h represent the relative position of the center of gravity of the vehicle with respect to the front and rear axles.
- m is the tractor mass; g is the acceleration of gravity.
- β is the road slope angle.
- V_x is the vehicle longitudinal speed.
- F_{aero} is the aerodynamic drag force as $F_{aero} = 0.5\rho C_d A {V_x}^2 sign(V_x)$, with ρ being the air density, C_d being the drag coefficient and A being the frontal cross-sectional area of the vehicle.
- F_{xf} and F_{xr} are the contact forces between the wheels and the ground on the longitudinal direction (front and rear axel); these forces are determined by the tire–soil interaction.
- F_{zf} and F_{zr} are the contact forces between the wheels and the ground on the longitudinal direction (front and rear axel).

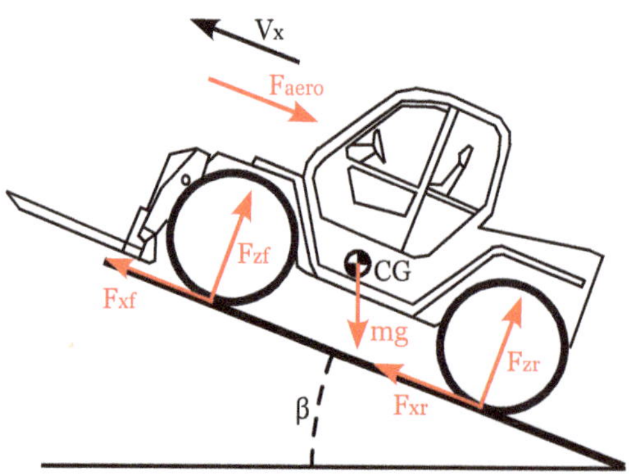

Figure 3. One-dimensional longitudinal model for the vehicle dynamics.

The contact between the tires and the soil was parameterized in terms of static and kinetic coefficients. The first determines the applied torque at which the tire loses traction and begins to slip, and the second determines the amount of torque the tire transmits to the pavement once it begins to slip. Thus, the traction force was evaluated according to the following equations:

$$F_{xi} = \begin{cases} \frac{T_{wheel}}{R_{wheel}} & \text{if } \frac{T_{wheel}}{R_{wheel}} \leq \mu_{static} * F_{zi} \\ \mu_{kinetic} * F_{zi} & \text{if } \frac{T_{wheel}}{R_{wheel}} > \mu_{static} * F_{zi} \end{cases} \quad (4)$$

where F_{xi} is the traction force on the i-axle, F_{zi} is the normal force on the i-axle, T_{wheel} is the torque at the wheel downstream from the driveline, R_{wheel} is the wheel radius, and μ_{static} and $\mu_{dynamic}$ are, respectively, the static and kinetic friction coefficients.

As for the transmission, the hydrostatic transmission was modelled considering three main elements: a variable displacement pump, a fixed displacement motor and a pressure relief valve. The characteristics considered for those elements are reported in Table 3. The no-load torque and friction torque vs. pressure gain/drop coefficients of the hydraulic units were modelled so that the hydrostatic transmission was characterized by the efficiency curve shown in Figure 4. That efficiency curve was obtained considering that the datasheets of hydraulic pumps and motors had similar applications and nominal specifications in the proposed case study.

Table 3. Hydrostatic transmission characteristics.

Parameter	Value
Pump max displacement	150 cm^3
Motor displacement	150 cm^3
Pump nominal pressure gain	250 bar
Motor nominal pressure drop	250 bar
Pump and motor nominal shaft speed	2000 rpm
Pump nominal volumetric efficiency	92%
Motor nominal volumetric efficiency	92%
Valve pressure setting	300 bar

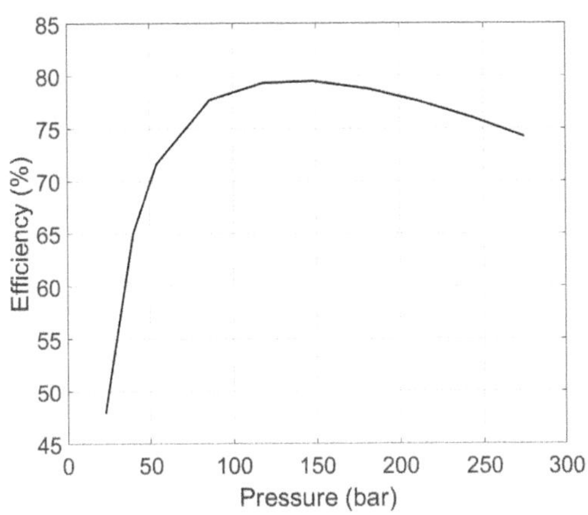

Figure 4. Hydrostatic transmission efficiency (pump displacement = 150 cm^3; pump shaft speed = 2000 rpm).

As for the hydraulic system for the lifting arm and the extension boom, the following approach was adopted: firstly, the hydro-mechanical system was modelled considering an hydraulic pump, two four-way directional valves and two double-acting hydraulic cylinders. Secondly, the lifting arm and extension boom were modelled using Simscape

Multibody. For the development of the Multibody, the kinematics of the system was defined in accordance with [46]. In detail, the Multibody model included two hydraulic cylinders, one to control the lifing angle of the telescopic arm, and the other to control the extension of the boom. The hydraulic pistons were modelled as cylindrical solids and were connected to the cylinder barrels, modelled as revolved solids, by means of prismatic joints. The head of the hydraulic piston for the regulation of the lift angle was connected to the body of the lifting arm by means of a bearing joint. The bottom of the hydraulic cylinders was connected to the chassis of the vehicle using revolute joints. The body, representing the bucket, was linked to the head of the hydraulic boom using a revolute joint that was controlled so that it maintained the same angle with respect to the pavament during lifting operations. To take into account the presence of a mass in the bucket, an external force, as a function of time, was added and placed in the middle of the bucket. After defining the Multibody system, it was interfaced with the hydro-mechanical network so that the translational force coming from the physical network model was used to evaluate the movement of the multibody joints, and, as a consequence, of the lifting arm and extension boom. As a feedback from the multibody network, the hydraulic cylinders received the information of position and speed. The two four-way directional valves were used as actuators for the hydraulic cylinders. A pressure relief valve, with a pressure setting of 300 bar, was introduced in the hydraulic system to prevent excessive peaks in the circuit.

The internal combustion engine was modelled using its speed-torque profile. As for the fuel consumption, the fuel consumption estimation model described in [47] was adopted. According to this model, the brake-specific fuel consumption (BSFC) is evaluated using a polynomial curve that is a function of the engine speed and torque:

$$Z = b_1 + b_2 \cdot X + b_3 \cdot Y + b_4 \cdot X^2 + b_5 \cdot X \cdot Y + b_6 \cdot Y^2 \tag{5}$$

where:

- X is the normalized engine speed: $X = \frac{n}{n_{nom}} \cdot 100$.
- Y is the normalized brake torque: $Y = \frac{T}{T_{nom}} \cdot 100$.
- Z is the normalized BSFC: $Z = \frac{BSFC}{BSFC_{min}} \cdot 100$.
- $b_{i=1,\dots,6}$ are the polynomial coefficients.

According to this model, the region of the minimum BSFC is usually located at about 73–77% of the nominal engine rotational speed and at a high load, namely 85–95% of the nominal torque. Finally, the power required by the vehicle auxiliaries was considered, assuming that they require around 8% of the rated engine power. The numerical model developed in MATLAB/Simulink is shown in Figure 5.

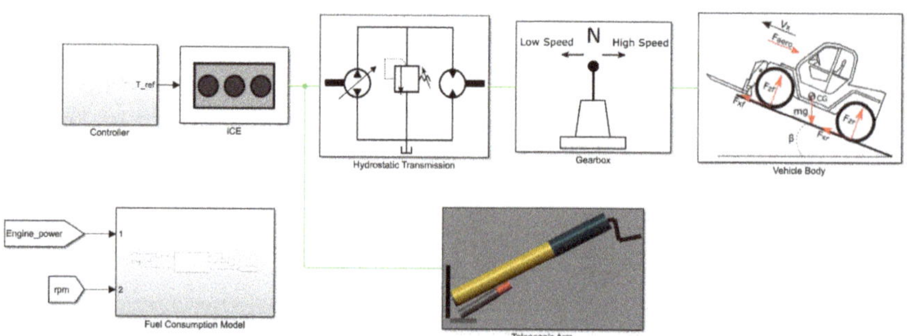

Figure 5. Numerical model for the traditional powertrain.

3.2. Fuel Cell Vehicle Numerical Modelling

As for the fuel cell vehicle, the authors assumed that, apart from the engine, all the other subparts are the same as the traditional counterpart. Thus, the vehicle chassis, transmission, hydraulic system, lifting arm and extension boom were not modified. As for the powertrain, the following elements were considered during the modelling:

- The fuel cell system.
- The battery pack.
- The power converters.
- The electric motor.

The fuel cell system was modelled as an equivalent circuit using the following equation [22,48]:

$$V_{stack} = N_{cell} * (E_{Nernst} - V_{act} - V_{ohm} - V_{conc}) \quad (6)$$

where V_{stack} is the overall voltage of the fuel cell stack, N_{cell} is the number of cells in the stack, E_{Nernst} is the Nernst voltage, V_{act} represents the voltage loss due to activation, V_{ohm} represents the voltage loss due to internal ohmic resistance and V_{conc} stands for the voltage loss due to concentration (mass transport processes). Given the operative conditions, the following equations can be used to determine the voltage of the fuel cell stack:

$$E_{Nernst} = 1.229 + (T - 298) * \frac{-44.43}{2F} + \frac{R_g T}{2F} * ln(\frac{p_{H_2} p_{O_2}^{1/2}}{p_{H_2O}}) \quad (7)$$

$$V_{act} = \frac{R_g T}{2F\alpha} * log(\frac{i_{dens}}{i_0}) \quad (8)$$

$$V_{ohm} = R_{ohm} * i_{dens} \quad (9)$$

$$V_{act} = \frac{R_g T}{2F} * log(1 - \frac{i_{dens}}{i_{lim}}) \quad (10)$$

where:

- T is the stack temperature.
- F is the Faraday constant, equal to 96,485.33 C/mol.
- R_g is the ideal gas constant.
- p_{H_2}, p_{O_2} and p_{H_2O} represent, respectively, the hydrogen, oxygen and water partial pressures.
- α is the charge transfer coefficient.
- i_{dens} is the current density.
- i_0 is the exchange current density.
- R_{ohm} is the ohmic resistance.
- i_{lim} is the maximum current density.

Using the aforementioned equations with operational parameters generally adopted for these systems, the voltage–current curve reported in Figure 6 was obtained. As for the hydrogen consumption, the following equation was used:

$$q_{H_2} = \frac{N_{cell} i_{FC} MM_{H_2}}{2F} \quad (11)$$

where q_{H_2} is the hydrogen mass flow that reacts at the anode, MM_{H_2} is the H_2 molar mass and i_{FC} is the current delivered by the fuel cell stack.

However, the sole fuel cell stack model is not enough to properly simulate the behaviour of the whole fuel cell system. Indeed, to effectively evaluate the system efficiency, the power absorbed by the Balance of Plant (BoP) system should be considered [22,26,33]. According to [49], the fuel cell BoP approximately absorbs around 13–19% of the power delivered by the stack. To address this issue, the power required by the BoP was added into the numerical model as an additional electrical load on the DC bus.

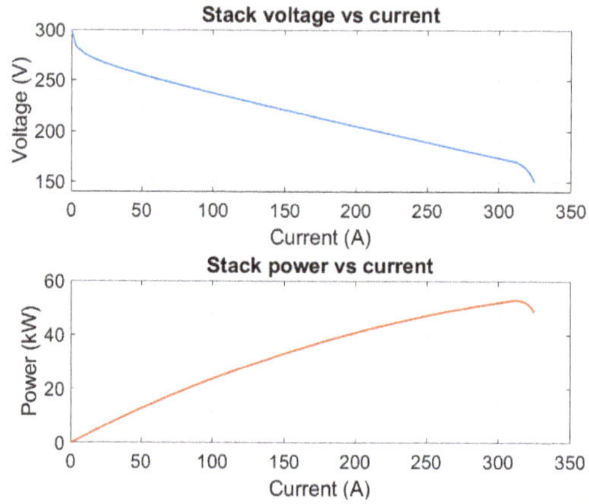

Figure 6. Fuel Cell stack curves.

As for the battery pack, it was modelled using a dynamic equivalent circuit model [50,51]. In detail, the dual polarization model shown in Figure 7 was used. In the adopted model, resistors and capacitors were considered constant, while the open circuit voltage was modelled as a function of the SOC. The SOC was evaluated using a simple Coulomb counting mode, neglecting more detailed models.

Concerning the power converters, their efficiency was considered constant and equal to 95%. Finally, the electric motor was modelled considering the torque-dependent electrical losses and the speed-dependent electrical losses. Moreover, a series resistance was considered between the DC bus and the electric motor, in order to take into account the ohmic losses along the wires. The numerical model of the fuel cell hybrid electric telehandler is shown in Figure 8.

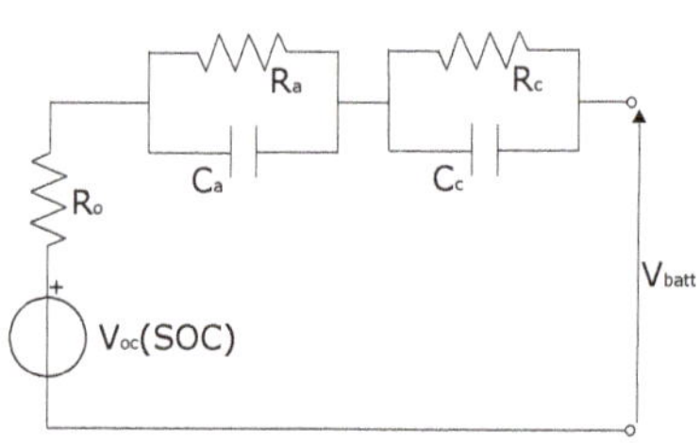

Figure 7. Dynamic equivalent circuit model for the battery pack.

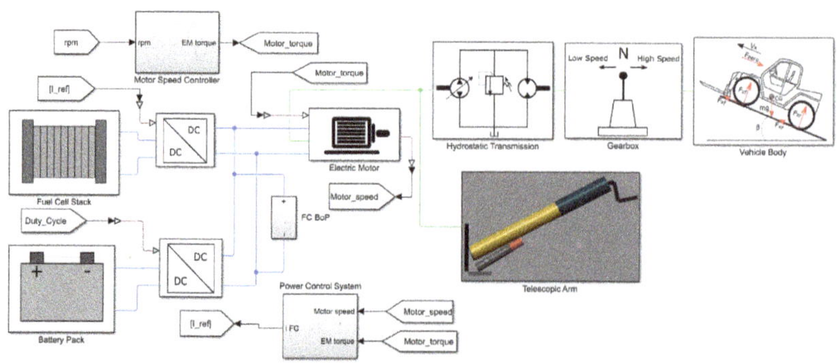

Figure 8. Fuel Cell hybrid electric telehandler numerical model.

4. Numerical Simulations

4.1. Powertrain Control Strategy

Regarding the traditional vehicle, the internal combustion engine is set at a fixed speed of 2000 rpm and is controlled using a proportional-integral (PI) controller. The vehicle speed is determined by changing the displacement of the pump in the hydrostatic transmission. As for the hydraulic system, it is controlled acting on the pump and on the valves. As for the fuel cell hybrid electric vehicle, the approach is the same. However, since there are two power sources, namely the fuel cell stack and the battery pack, an energy management strategy must be defined to determine how to split the power required by the motor between the two different sources. For this preliminary analysis, the authors opted for a simple power follower control strategy. Power follower strategies, like other rule-based strategies, have the advantage of easy implementation and integration in embedded controllers, and are able to provide good and stable control even if they could lead to not optimal fuel economy or component degradation with respect to optimization-based energy management strategies [52]. However, the authors deemed that a simple power follower strategy was sufficient for this preliminary analysis. The main goal of the strategy is to operate in a charge-sustaining mode without exceeding the power limits of the fuel cell. To prevent fast degradation, the fuel cell stack should operate following the low frequency component of the load. Indeed, sudden and intense changes in the fuel cell power output could reduce its lifetime due to reactant starvation or membrane flooding processes [41,53–55]. In fuel cell powertrains, the power conditioning units are represented by the DC-DC converters, which can be controlled both in the voltage or the current reference mode. Therefore, to perform the power split between the fuel cell and the battery pack, the power control system evaluates a reference current for the fuel cell, which is used to control the unidirectional DC-DC converter. In detail, the current reference is evaluated, using a predefined set of rules, according to the current required by the electric motor. To operate in a charge-sustaining mode, the control strategy adopts penalty factors depending on the batteries' State of Charge (SOC). The fuel cell minimum output was set to be equal to its idle power and assumed to be approximately 10% of the nominal stack power. A schematic representation of the control strategy for the fuel cell powertrain is represented in Figure 9. As for the bidirectional DC-DC converter on the batteries' side, it was used to control the voltage of the DC-DC bus, which was set to 640 V.

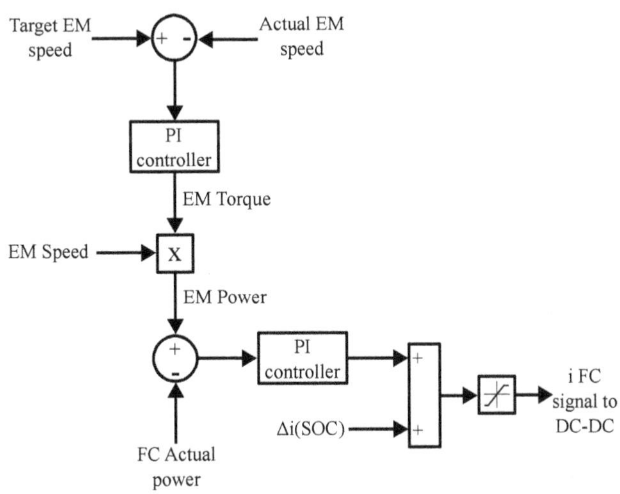

Figure 9. Fuel Cell hybrid electric powertrain control strategy.

4.2. Simulated Work Scenarios

To assess the powertrain performance and to perform a comparison between the traditional vehicle and the fuel cell hybrid electric counterpart, the simulations were defined in order to evaluate the behaviour of the two vehicles in terms of acceleration time, overcoming slopes, fuel economy and environmental impact. Moreover, the simulations were also used to assess the maximum change rate, expressed in kW/s, in the fuel cell power output. For the performance assessment, the following tests were performed:

- Acceleration from 0 to 40 km/h with no load.
- Acceleration from 0 to 40 km/h, handling a trailer with a total weight of 4200 kg.
- Maximum speed at 20% of slope.
- Maximum approachable slope, handling a load of 4200 kg.
- Maximum approachable slope at 15 km/h.

The acceleration tests were conducted using the high-speed regime gear. Thus, during the acceleration, no gearshifting is performed. As for the slope tests, they were conducted with a standing start at the prescribed slope. The friction coefficients for the wheel-ground contact were defined according to a pavement road. To assess the fuel economy, a work cycle based on a typical real operartive scenario was defined. The proposed real work scenario was based on the one defined in [56], which corresponds to the telehandler picking and handling a load, and is composed of the following phases:

- Approaching: the vehicle approaches to the load that must be moved.
- Loading: the vehicle picks up the load using the telescopic arm; in this phase, the telescopic arm lift angle starts to increase, lifting the load and reaching a maximum angle of approximately 50 degrees, and then decreases to 25 degrees, which is the angle at which the vehicle handles the load during the transfer phase.
- Release: the vehicle moves back from the point where the load was located; the telescopic arm maintains a constant lift angle of 25 degrees.
- Transfer with load: the vehicle handles the load up to 15 km/h and reaches the point where the load must be placed; the lifting angle remains constant at 25 degrees during the whole phase.
- Unloading: the vehicle deposits the load by means of the telescopic arm and the extension boom; in this phase, the lift angle increases to 50 degrees, then the extension of the telescopic boom starts to increase, reaching 1000 mmm, and then the load is placed; after that, the extension boom returns to 0 mm and the lift angle decreases to 0 degrees.
- Transfer without load: the vehicle moves back without the load.

The work cycle is reproduced twice with two different loads, one corresponding to approximately 3000 kg and the other to 1500 kg. As for the wheel–ground contact, in a real work scenario the friction coefficients were defined considering off-road conditions.

5. Results and Discussion

5.1. Simulations Results

5.1.1. Performance Tests

The results of the performance evaluation tests are reported in Table 4. As expected, the two powertrains showed approximately the same performances. The traditional vehicle performed slightly better due to the higher power of the internal combustion engine in the range 1400–1900 rpm with respect to the electric motor. As for the slope tests, it should be highlighted that the maximum performance was limited, not by the capacity of the wheel to transmit force to the ground, but by the available power at the driveline downstream from the hydrostatic transmission.

Table 4. Performance test results.

Performed Test	Traditional Vehicle	Fuel Cell Hybrid Vehicle
Time for 0–40 km/h no load	12 s	12.5 s
Time for 0–40 km/h with 4200 kg	20 s	21 s
Max speed with 20% slope	9 km/h	8 km/h
Max slope with 4200 kg	15% @ 5 km/h	15% @ 5 km/h
Max slope at 15 km/h	12%	10%

However, the aim of the performance tests was also to evaluate the behaviour of the fuel cell powertrain under peak power conditions. Therefore, the maximum instantaneous hydrogen consumption, the maximum C-rates in charge and discharge for the battery pack and the mean change rate in the fuel cell power output were evaluated. The results are shown in Table 5. As stated, the acceleration tests were the most impactful in terms of stress on both the FC system and the battery pack.

Table 5. Fuel cell powertrain behaviour during the performance tests.

	FC System		Battery Pack	
	Max Change Rate (kW/s)	Max H_2 Flow (g/s)	Max Charge C-Rate	Max Discharge C-Rate
0–40 km/h no load	8.7	0.69	0.6	6.3
0–40 km/h 4200 kg	5.1	0.66	0.35	6.1
Top speed 20% slope	4.2	0.47	0.35	5.6
Max slope with 4200 kg	3.4	0.52	0.35	5.2
Max slope at 15 km/h	4.0	0.55	0.35	4.9

5.1.2. Real Work Scenario

The simulation results for the real work scenario are reported in Figures 10 and 11. As can be observed from the figure, the 0–15 km/h acceleration phase is the most demanding one due to the high vehicle mass, with a peak power of approximately 55 kW when carrying a 3000 kg load. Also in this case, the simulation was used to evaluate the behaviour of the fuel cell powertrain, and the results are summarized in Table 6.

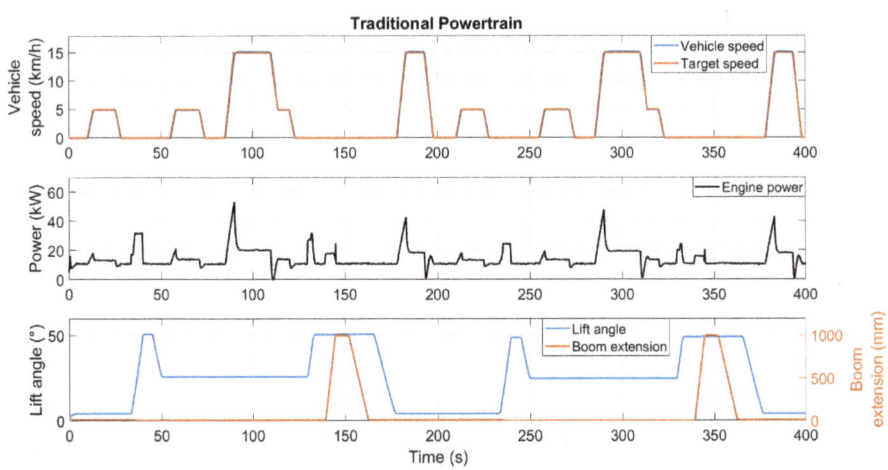

Figure 10. Real work scenario results for the traditional powertrain.

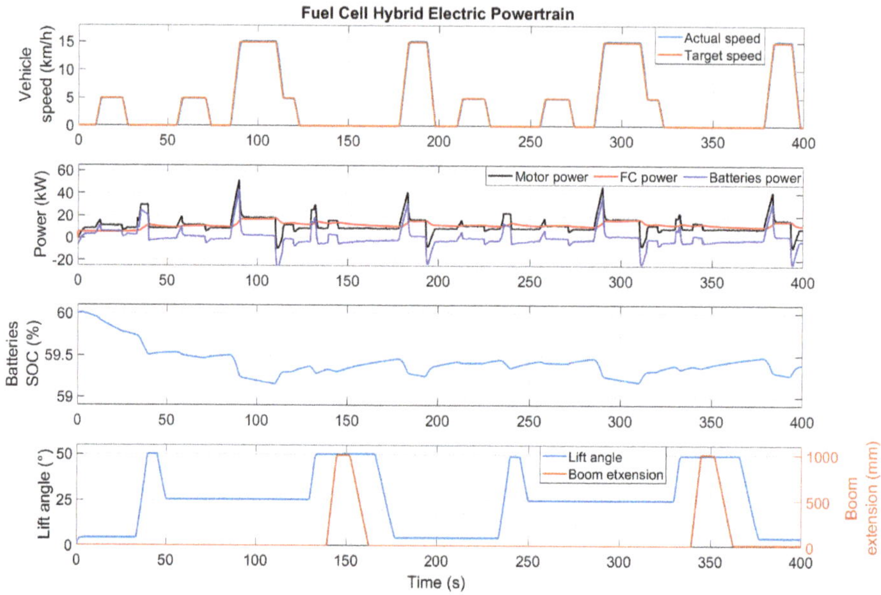

Figure 11. Real work scenario results for the fuel cell hybrid electric powertrain.

Table 6. Fuel cell powertrain behaviour during the real work scenario test.

Parameter	Value
Max FC change rate (kW/s)	4.04
Max H_2 consumption (g/s)	0.23
Max charging C-rate	1.7 C
Max discharging C-rate	2.7 C
Total H_2 consumption (g)	60.4

5.2. Fuel Cell System Degradation Analysis

To estimate the fuel cell system durability, in this section a degradation analysis, based on models available in the literature, is proposed. According to [41], the lifetime of a fuel cell system can be defined as follows:

$$T_{lifetime,FC} = \frac{\Delta P}{k_p(\beta_1 n_1 + \beta_2 n_2 + \beta_3 t_1 + \beta_4 t_2)} \tag{12}$$

where $T_{lifetime,FC}$ is the expected lifetime, ΔP is the maximum acceptable reduction in the fuel cell output performance, generally fixed at 10%; $\beta_{i=1,\ldots,4}$ are the performance degradation rates related, respectively, to load change cycling, start-stop cycling, idle condition and high power condition; n_1, n_2, t_1 and t_2 are the load changing cycles, start and stop cycles, idle condition time and high-power condition time; and finally, k_p is an accelerating coefficient. The main limitations of this model are related to the coefficients that may not be updated since the study was published in 2008. However, this model is considered a key reference for fuel cells' degradation evaluation and was used in more recent studies that proposed this kind of analysis [53,54,57,58]. Nevertheless, fuel cell systems' durability has experienced relevant improvements in the last few years. Indeed, according to [59], in 2007, the state of the art in terms of the durability for fuel cell systems in automotive applications corresponded to 1250 h. Considering the present DOE targets [60], the ultimate targets in terms of durability are set to 8000 h. However, to be conservative, the authors opted for considering the coefficients used in [41]. In [53], an additional coefficient, related to the natural degradation of the fuel cell due to material aging, was added. However, no details are provided about how that coefficient was determined. Therefore, in the present paper, the authors opted for not considering it. The adopted values for the degradation analysis of the fuel cell systems are reported in Table 7. The degradation analysis was conducted considering the real work scenario, which should be the most common operating condition, and the 0–40 km/h acceleration test with no load, which was the test that exhibited the highest change rate in the fuel cell power output. The results in terms of the expected life, considering a maximum decrease of 10% in the system performance for the two scenarios, are reported in Table 8. As it can be stated, severe and frequent accelerations can lead to a reduction in the lifetime expectancy. On the contrary, the lifetime estimation considering the real work scenario exceeded the DOE ultimate target of 8000 h. However, the authors wanted to highlight that the adopted powertrain control strategy was not optimized through an optimization process, but was a simplified control based on a power follower algorithm with a low dynamic response of the fuel cell system. The aim was to roughly reduce the load changes in the fuel cell power output, but no optimization algorithm was used. Future works might focus on a more developed aging-aware control strategy.

Table 7. Adopted coefficients for the degradation analysis of the fuel cell system. The coefficients were defined according to [41].

Factor	Condition	Value	Unit
β_1	Load cycling	5.93×10^{-5}	%/cycle
β_2	Start and Stop	1.96×10^{-3}	%/cycle
β_3	Idling ($P_{out,FC} \leq 10\%$)	1.26×10^{-3}	%/h
β_4	High power ($P_{out,FC} \geq 90\%$)	1.47×10^{-3}	%/h
k_p	-	1.72	-

Table 8. Degradation analysis results for the two considered scenarios: real work scenario and 0–40 km/h acceleration test with no load.

Test Scenario	FC System Expected Life
0–40 km/h acceleration test no load	1257 h
Real work scenario	9410 h

5.3. Environmental Analysis

To assess the environmental performances of the fuel cell powertrain with respect to the traditional model, the authors opted for using a WtW approach. This approach is useful for evaluating the environmental performances of a vehicle during its use phase, since it accounts for the emissions related to fuel extraction, treatment, distribution and conversion [61]. The authors opted for considering the use of hydrogen from the actual hydrogen production mix, which is mainly from fossil fuel resources; hydrogen from steam methane reforming but with a carbon capture system, namely blue hydrogen; and hydrogen from electrolysis, based on the grid, nuclear and renewables. The authors also decided to consider the global warming potential as an impact category, since the climate change issue is one of the major challenges that humankind has to face within the near future, and the efforts to mitigate it are among the main reasons behind the electrification of powertrains. For hydrogen from the actual production mix and Diesel consumption, the WtW factors for the equivalent CO_2 emissions were derived from [8], while the coefficients for the other considered hydrogen production methods were taken from [34]. The WtW emission factors adopted for the environmental impact analysis are summarized in Table 9. The factor for the hydrogen from electrolysis based on renewables was determined considering the average values for electrolysis based on biomass, wind and solar. As for the electrolysis based on the grid, the WtW emission factor may vary depending on the country, since it mainly depends on the electricity production mix. The analysis was conducted considering the real work scenario. The results are shown in Table 10. As can be stated, with the present hydrogen production mix, the fuel cell powertrain can reduce the CO_2 equivalent emissions of 68.6%. However, with cleaner and greener production methods, the emission reduction can be enhanced. In detail, the emission reduction can reach 97.5% using electrolysis with electricity from nuclear, and 92.6% using hydrogen from electrolysis powered with electricity from renewables. Furthermore, a noticeable reduction in GHG emissions can also be achieved using hydrogen produced from fossil fuel resources with carbon capture technologies. Instead, with the production of hydrogen from electrolysis using the electricity grid mix, no effective emission reduction might be obtained. This demonstrates that coupling electrolysis with clean electricity is a key element to achieve greater results in terms of the environmental impact reduction in hydrogen-powered vehicles. The results obtained from the environmental analysis were coherent with other studies that estimated the life cycle CO_2 emissions of fuel cell vehicles to be less than 50% of that of the traditional counterpart, with very low-use phase impacts when hydrogen is produced from renewables [62–64].

Table 9. WtW equivalent emission factors for Diesel and H_2 according to [8,34].

Emission Source	WtW Emission Factor	Unit
Diesel	3.18	kg CO_2-eq./L
Actual Hydrogen production mix	9.13	kg CO_2-eq./kg
Blue Hydrogen	3.70	kg CO_2-eq./kg
Hydrogen from Electrolysis (based on nuclear)	0.71	kg CO_2-eq./kg
Hydrogen from Electrolysis (based on grid)	29.21	kg CO_2-eq./kg
Hydrogen from Electrolysis (based on renewables)	1.87	kg CO_2-eq./kg

Table 10. Environmental impact comparison for the real work scenario.

	Fuel Consumptions	
Diesel consumption	0.55 L	
Hydrogen consumption	60.4 g	
Fuel	**Emissions (kg CO_2-eq.)**	**Difference (%)**
Diesel	1.75	-
Hydrogen (actual production mix)	0.55	−68.6
Blue Hydrogen	0.22	−87.4
Hydrogen from Electrolysis (based on nuclear)	0.043	−97.5
Hydrogen from Electrolysis (based on grid)	1.76	+0.1
Hydrogen from Electrolysis (based on renewables)	0.13	−92.6

5.4. Discussion

The degradation and environmental analysis showed that fuel cell powertrains can be a feasible solution to mitigate the impact of Non-Road Mobile Machinery on the environment in terms of equivalent CO_2 emissions. However, the applicability of these systems on a real vehicle is not straightforward. From a technical point of view, one of the major challenges is the integration of the fuel cell system on-board the vehicle. Another technical issue is regarding the on-board hydrogen storage system. Considering the simulation results for the real work scenario, to operate for 8 h straight without the need for refuelling, the hydrogen storage system should be able to stock around 4.5 kg of H_2. This storage capacity requirement can be satisfied implementing a modern type IV tank, which can store 5.8 kg of hydrogen in its gaseous form at 700 bar. The system weighs approximately 133.6 kg and has a volume of 229.6 litres [65]. Alternatively, to reduce the volume of the storage system, another possible solution is represented by metal hydride tanks. These systems can reach higher volumetric densities, up to five times higher, but have a lower gravimetric density compared to gaseous tanks [66]. Furthermore, metal hydride tanks require a heat management system to operate properly, thus a higher complexity of the whole powertrain might be introduced [67]. On the contrary, the adoption of a type IV storage tank involves safety issues that must be considered during the design stage of the vehicle [68]. However, the most insidious challenges are related to other factors that go beyond the technical issues related to the realization of a first propotype. As stated in [38], one of the most critical factors that is slowing the diffusion of fuel cell systems on vehicular applications is the early stage of development of the hydrogen refuelling network. Furthemore, the hydrogen refueling infrastructure requires an investment cost that is extremely high compared to the cost of other fuel refueling stations [69]. Another important challenge is represented by the higher purchasing cost of fuel cell vehicles with respect to their traditional and battery electric counterpart [70]. Indeed, the hydrogen tanks and the stack are the most expensive components of the fuel cell system due to the presence of expensive materials, such as platinum, used as catalyst in the stack; and carbon fibre, used to manufacture the type IV tank [39,65,71]. To reduce the cost of the fuel cell systems, DOE targets have been defined for the future [72,73]. Along with the problem related to the high cost, another aspect that must be faced is the lack of consumer awareness about fuel cell vehicles, which results in a distorted perception about their safety and performances [74]. According to [75], in vehicle users there is a combination of aversion to high purchase costs and negative perceptions of environmental benefits from adopting hydrogen that is creating a barrier between the market and fuel cell vehicles. As a consequence, overcoming the problem of a social attitude towards the acceptance of fuel cell vehicles is mandatory to promote their diffusion [76]. However, environmental policies and regulations might be important drivers for the adoption of hydrogen technologies [77,78]. In this context, the European Community is pursuing the goal of reaching a reduction of 55% in net greenhouse gas emissions for 2030 with respect to the emission levels of 1990, and of achieving carbon neutrality for 2050 [79]. However, the sector of Non-Road Mobile Machinery is at an earlier stage of decarbonisation with respect to passenger vehicles. Indeed, the actual regulation for Non-Road Mobile Machinery focuses on the emission levels of CO, HC, NO_x and particulate matter, while no limits for CO_2 are, at present, introduced [80]. Nevertheless, the limits for the aforementioned pollutants are becoming more and more stringent and are requiring more complex and bulky exhaust gas aftertreament systems. Thus, this poses technical challenges that may be drivers for the adoption of fuel cell systems, which instead do not require an aftertreatment system since they produce water at the exhaust.

6. Conclusions

Fuel cell hybrid electric powertrains represent a promising strategy to replace traditional internal combustion engines in the sector of Non-Road Mobile Machinery. Indeed, fuel cell vehicles feature zero local emission levels, high-energy density and low refuelling time, which make them more interesting than battery electric vehicles for applications where endurance and productivity are crucial with a view to be competitive on the market. In this article, the preliminary design of a fuel cell hybrid electric powertrain for a telehandler was presented. The proposed system is characterized by the presence of a PEMFC stack, with a rated power approximately equal to 50% of the electric motor nominal power, whose role is to satisfy the low-frequency component of the external load, and a battery pack, which has to handle the high frequency part of the load to avoid the fast degradation of the fuel cell system. To make a comparison between the proposed system and the traditional one, numerical models of both the powertrain were built in a MATLAB/Simulink environment. The modelling included a Multibody model to simulate the use of the telescopic arm and the extension boom. For the simulations, the authors developed a simple charge-sustaining power follower strategy to determine the power split between the fuel cell and the battery pack. The simulations were regarding two type of tests: peak power performance tests and the real work scenario simulation test. The first set of tests aimed at evaluating both the performances in terms of the acceleration and overcoming of slopes, and the behaviour of the fuel cell powertrain in terms of stress on its main components. On the contrary, the real work scenario test aimed at evaluating the fuel economy and the environmental impact of the powertrains during a typical operative scenario, characterized by the handling of loads by the mean of the telescopic arm. The results of the simulations are summarized as follows:

- The fuel cell powertrain was able to show almost the same performances of the traditional one without excessive stress on its components; indeed, a degradation analysis was conducted to address the fuel cell durability issue.
- The most stressful tests for the fuel cell powertrain were the acceleration tests, since the max change rate in the fuel cell output was equal to 15% of the rated stack power per second, and the max discharge C-rate of the battery pack was equal to 6.3.
- During the real work scenario test, the fuel cell powertrain showed a reduction in terms of the equivalent CO_2 emissions of 69% with respect to the traditional powertrain; this result was obtained considering the use of grey hydrogen for the fuel cell system and can be improved using hydrogen from a cleaner production mix.

Therefore, the authors concluded that the adoption of fuel cell systems are a feasible solution for replacing traditional internal combustion engines in telehandlers, since they can have the same performances without excessive degradation and with a reduction in terms of equivalent CO_2 emissions. However, a higher purchase cost is expected for the fuel cell vehicle with respect to the diesel-powered counterpart. Therefore, to be competitive on the market, consumers' awareness of environmental performances and safety of the proposed powertrain is mandatory. A further analysis might be regarding the development of a more detailed and optimized energy management strategy, with the aim of minimizing fuel consumption, component degradation or both of them according to a predefined objective function. As demonstrated in the literature, the adoption of optimization algorithms can enhance the powertrain performance. Other future works could investigate the possibility of introducing supercapacitors to reduce batteries' degradation. Furthermore, due to the lack of experimental data, the adoption of a monitoring device that could be installed on-board a real telehandler, to outline its realistic mission profile, could be the subject of research and attention. To immprove the simulation reliability, tests on a scaled test bench with a real fuel cell system might be conducted in the future. To promote circular economy, another subject of research could be the integration of green hydrogen production systems, powered using electricity from renewable sources, to locally produce clean hydrogen to be used in the vehicle. A typical case study might be a farm with an electrolyser powered using renewable sources. In that scenario, the farm is able to produce green hydrogen with very low equivalent CO_2 emissions using water and, for example, electricity coming from solar panels. That hydrogen can be used to power the vehicle. Furthermore, the water produced at the exhaust can be re-used in the electrolyser for the hydrogen production.

Author Contributions: Conceptualization, V.M., F.M. and A.S.; methodology, V.M., F.M. and A.S.; software, V.M., F.M. and A.S.; validation, V.M., F.M. and A.S.; formal analysis, V.M., F.M. and A.S.; investigation, V.M., F.M. and A.S.; data curation, V.M., F.M. and A.S.; writing—original draft preparation, V.M., F.M. and A.S.; writing—review and editing, V.M., F.M. and A.S.; visualization, V.M., F.M. and A.S.; supervision, V.M., F.M. and A.S. All authors have read and agreed to the published version of the manuscript.

Funding: This research received no external funding.

Institutional Review Board Statement: Not applicable.

Informed Consent Statement: Not applicable.

Data Availability Statement: The original contributions presented in the study are included in the article, further inquiries can be directed to the corresponding author.

Conflicts of Interest: The authors declare no conflicts of interest.

References

1. Anenberg, S.C.; Achakulwisut, P.; Brauer, M.; Moran, D.; Apte, J.S.; Henze, D.K. Particulate matter-attributable mortality and relationships with carbon dioxide in 250 urban areas worldwide. *Sci. Rep.* **2019**, *9*, 11552. [CrossRef]
2. McDuffle, E.E.; Martin, R.V.; Spadaro, J.V.; Burnett, R.; Smith, S.J.; O'Rourke, P.; Hammer, M.S.; van Donkelaar, A.; Bindle, L.; Shah, V.; et al. Source sector and fuel contributions to ambient PM2.5 and attributable mortality across multiple spatial scales. *Nat. Commun.* **2021**, *12*, 3594. [CrossRef]
3. Zhu, L.; Ge, X.; Chen, Y.; Zeng, X.; Pan, W.; Zhang, X.; Ben, S.; Yua, Q.; Xin, J.; Shao, W.; et al. Short-term effects of ambient air pollution and childhood lower respiratory diseases. *Sci. Rep.* **2017**, *7*, 4414. [CrossRef] [PubMed]
4. Eom, J.; Hyun, M.; Lee, J.; Lee, H. Increase in household energy consumption due to ambient air pollution. *Nat. Energy* **2020**, *5*, 976–984. [CrossRef]
5. Dai, C.; Qin, X.S.; Zhang, X.L.; Liu, B.J. Study of climate change impact on hydro-climatic extremes in the Hanjiang River basin, China, using CORDEX-EAS data. *Weather Clim. Extremes* **2022**, *38*, 100509. [CrossRef]
6. Taghizadeh-Hesary, F.; Taghizadeh-Hesary, F. The Impacts of Air Pollution on Health and Economy in Southeast Asia. *Energies* **2020**, *13*, 1812. [CrossRef]
7. McCarthy, P.; Rasul, M.G.; Moazzem, S. Analysis and comparison of performance and emissions of an internal combustion engine fuelled with petroleum diesel and different bio-diesels. *Fuel* **2011**, *90*, 2147–2157. [CrossRef]
8. Buberger, J.; Kersten, A.; Kuder, M.; Eckerle, R.; Weyh, T.; Thiringer, T. Total CO_2-equivalent life-cycle emissions from commercially available passenger cars. *Renew. Sustain. Energy Rev.* **2022**, *159*, 112158. [CrossRef]
9. Onat, N.C.; Kucukvar, M.; Tatari, O. Towards Life Cycle Sustainability Assessment of Alternative Passenger Vehicles. *Sustainability* **2014**, *6*, 9305–9342. [CrossRef]
10. European Parliament. Available online: https://www.europarl.europa.eu/news/en/headlines/economy/20221019STO44572/eu-ban-on-sale-of-new-petrol-and-diesel-cars-from-2035-explained (accessed on 3 January 2024).
11. Rietmann, N.; Hugler, B.; Lieven, T. Forecasting the trajectory of electric vehicle sales and the consequences for worldwide CO_2 emissions. *J. Clean. Prod.* **2020**, *261*, 121038. [CrossRef]
12. Martelli, S.; Mocera, F.; Somà, A. Carbon Footprint of an Orchard Tractor through a Life-Cycle Assessment Approach. *Agriculture* **2023**, *13*, 1210. [CrossRef]
13. Martelli, S.; Mocera, F.; Somà, A. New Challenges Towards Electrification Sustainability: Environmental Impact Assessment Comparison between ICE and Hybrid-Electric Orchard Tractor. In Proceedings of the 2023 JSAE/SAE Powertrains, Energy and Lubricants International Meeting, Kyoto, Japan, 29 August–1 September 2023. [CrossRef]
14. Cunanan, C.; Tran, M.-K.; Lee, Y.; Kwok, S.; Leung, V.; Fowler, M. A Review of Heavy-Duty Vehicle Powertrain Technologies: Diesel Engine Vehicles, Battery Electric Vehicles, and Hydrogen Fuel Cell Electric Vehicles. *Clean Technol.* **2021**, *3*, 474–489. [CrossRef]
15. Mocera, F.; Martini, V. Numerical Performance Investigation of a Hybrid eCVT Specialized Agricultural Tractor. *Appl. Sci.* **2022**, *12*, 2438. [CrossRef]
16. Mocera, F.; Martini, V.; Somà, A. Comparative Analysis of Hybrid Electric Architectures for Specialized Agricultural Tractors. *Energies* **2022**, *15*, 1944. [CrossRef]
17. Mocera, F.; Martelli, S.; Costamagna, M. Dynamic behaviour of a battery pack for agricultural applications. *Conf. Ser. Mater. Sci. Eng.* **2022**, *1214*, 012032. [CrossRef]
18. Mocera, F.; Somà, A. Analysis of a Parallel Hybrid Electric Tractor for Agricultural Applications. *Energies* **2020**, *13*, 3055. [CrossRef]
19. Dalboni, M.; Santarelli, P.; Patroncini, P.; Soldati, A.; Concari, C.; Lusignani, D. Electrification of a Compact Agricultural Tractor: A Successful Case Study. In Proceedings of the 2019 IEEE Transportation Electrification Conference and Expo (ITEC), Novi, MI, USA, 19–21 June 2019; IEEE: Piscataway, NJ, USA, 2019; pp. 1–6. [CrossRef]
20. Available online: https://www.farm-equipment.com/articles/19849-antonio-carraro-srx-hybrid-tractor (accessed on 15 February 2024).
21. Available online: https://www.landini.it/as/landini-rex4-full-hybrid-technical-innovation-at-eima-2022/ (accessed on 15 February 2024).
22. Martini, V.; Mocera, F.; Somà, A. Numerical Investigation of a Fuel Cell-Powered Agricultural Tractor. *Energies* **2022**, *15*, 8818. [CrossRef]
23. Di Ilio, G.; Di Giorgio, P.; Tribioli, L.; Bella, G.; Jannelli, E. Preliminary design of a fuel cell/battery hybrid powertrain for a heavy-duty yard truck for port logistics. *Energy Convers. Manag.* **2021**, *243*, 114423. [CrossRef]
24. Liukkonen, M.; Lajunen, A.; Suomela, J. Feasibility study of fuel cell-hybrid powertrains in non-road mobile machineries. *Autom. Constr.* **2013**, *35*, 296–305. [CrossRef]
25. Mocera, F.; Somà, A.; Martelli, S.; Martini, V. Trends and Future Perspective of Electrification in Agricultural Tractor-Implement Applications. *Energies* **2023**, *16*, 6601. [CrossRef]
26. Martini, V.; Mocera, F.; Somà, A. Design and Experimental Validation of a Scaled Test Bench for the Emulation of a Hybrid Fuel Cell Powertrain for Agricultural Tractors. *Appl. Sci.* **2023**, *13*, 8582. [CrossRef]
27. Ahluwalia, R.K.; Wang, X.; Star, A.G.; Papadias, D.D. Performance and cost of fuel cells for off-road heavy-duty vehicles. *Int. J. Hydrogen Energy* **2022**, *47*, 10990–11006. [CrossRef]
28. De Lorenzo, G.; Ruffo, R.M.; Fragiacomo, P. Preliminary Design of the Fuel Cells Based Energy Systems for a Cruise Ship. *World Electr. Veh. J.* **2023**, *14*, 263. [CrossRef]

29. Waseem, M.; Amir, M.; Lakshmi, G.S.; Harivardhagini, S.; Ahmad, M. Fuel cell-based hybrid electric vehicles: An integrated review of current status, key challenges, recommended policies, and future prospects. *Green Energy Intell. Transp.* **2023**, *2*, 100121. [CrossRef]
30. Louvros, P.; Trivyza, N.L.; Komianos, A.; Boulougouris, E. Fuel cell, ammomnia powered container ship: A case study. *Transp. Res. Procedia* **2023**, *72*, 2245–2252. [CrossRef]
31. Laimon, M.; Yusaf, T. Towards energy freedom: Exploring sustainable solutions for energy independence and self-sufficiency using integrated renewable energy-driven hydrogen system. *Renew. Energ.* **2024**, *222*, 119948. [CrossRef]
32. Tellez-Cruz, M.M.; Escorihuela, J.; Solorza-Feria, O.; Compañ, V. Proton Exchange Membrane Fuel Cells (PEMFCs): Advances and Challenges. *Polymers* **2021**, *13*, 3064. [CrossRef]
33. Lohse-Busch, H.; Stutenberg, K.; Duoba, M.; Liu, X.; Elgowainy, A.; Wang, M.; Wallner, T.; Richard, B.; Christenson, M. Automotive fuel cell stack and system efficiency and fuel consumption based on vehicle testing on a chassis dynamometer at min 18 °C to positive 35 °C temperatures. *Int. J. Hydrogen Energy* **2020**, *45*, 861–872. [CrossRef]
34. Ji, M.; Wang, J. Review and comparison of various hydrogen production methods based on costs and life cycle impact assessment indicators. *Int. J. Hydrogen Energy* **2021**, *46*, 38612–38635. [CrossRef]
35. Das, A.; Peu, S.D. A Comprehensive Review on Recent Advancements in Thermochemical Processes for Clean Hydrogen Production to Decarbonize the Energy Sector. *Sustainability* **2022**, *14*, 11206. [CrossRef]
36. Xu, X.; Zhou, Q.; Yu, D. The future of hydrogen energy: Bio-hydrogen production technology. *Int. J. Hydrogen Energy* **2022**, *47*, 33677–33698. [CrossRef]
37. IEA. Global Hydrogen Review. 2022. Paris. Available online: https://www.iea.org/reports/global-hydrogen-review-2022 (accessed on 4 January 2024).
38. Olabi, A.G.; Abdelkareem, M.A.; Wilberforce, T.; Alami, A.H.; Alkhalidi, A.; Hassan, M.M.; Sayed, E.T. Strength, weakness, opportunities, and threats (SWOT) analysis of fuel cells in electric vehicles. *Int. J. Hydrogen Energy* **2023**, *48*, 23185–23211. [CrossRef]
39. Cano, Z.P.; Banham, D.; Ye, S.; Hintennach, A.; Lu, J.; Fowler, M.; Chen, Z. Batteries and fuel cells for emerging electric vehicle markets. *Nat. Energy* **2018**, *3*, 279–289. [CrossRef]
40. Das, H.S.; Tan, C.W.; Yatim, A.H.M. Fuel cell hybrid electric vehicles: A review on power conditioning units and topologies. *Renew. Sustain. Energy Rev.* **2017**, *76*, 268–291. [CrossRef]
41. Pei, P.; Chang, Q.; Tang, T. A quick evaluating method for automotive fuel cell lifetime. *Int. J. Hydrogen Energy* **2008**, *33*, 3829–3836. [CrossRef]
42. Zhao, X.; Wang, L.; Zhou, Y.; Pan, B.; Wang, R.; Wang, L.; Yan, X. Energy management strategies for fuel cell hybrid electric vehicles: Classification, comparison, and outlook. *Energy Convers. Manag.* **2022**, *270*, 116179. [CrossRef]
43. Available online: https://www.merlo.com/gbr/en/p/telehandlers/medium-capacity-telehandlers/turbofarmer-42-7/ (accessed on 4 January 2024).
44. Somà, A.; Bruzzese, F.; Mocera, F.; Viglietti, E. Hybridization Factor and Performance of Hybrid Electric Telehandler Vehicle. *IEEE Trans. Ind. Appl.* **2016**, *52*, 5130–5138. [CrossRef]
45. Mathworks. *Getting Started with Simscape*; MathWorks: Natick, MA, USA, 2018.
46. Činkelj, J.; Kamnik, R.; Čepon, P.; Mihelj, M.; Munih, M. Closed-loop control of hydrualic telescopic handler. *Autum. Constr.* **2010**, *19*, 954–963. [CrossRef]
47. Golverk, A.A. The Method for Development of a Diesel Engine Universal Performance Map. *SAE Int. J. Fuels Lubr.* **1994**, *103*, 1041–1048.
48. Wang, Y.; Sun, Z.; Chen, Z. Energy management strategy for battery/supercapacitor/fuel cell hybrid source vehicles based on finite state machine. *Appl. Energy* **2019**, *254*, 113707. [CrossRef]
49. Zhang, B.; Wang, X.; Gong, D.; Xu, S. Experimental analysis of the performance of the air supply system in a 120 kW polymer electrolyte membrane fuel cell system. *Int. J. Hydrogen Energy* **2022**, *47*, 21417–21434. [CrossRef]
50. Vergori, E.; Mocera, F.; Somà, A. Battery Modelling and Simulation Using a Programmable Testing Equipment. *Computers* **2018**, *7*, 20. [CrossRef]
51. Lai, X.; Zheng, Y.; Sun, T. A comparative study of different equivalent circuit models for estimating state-of-charge of lithium-ion batteries. *Electrochim. Acta* **2018**, *259*, 566–577. [CrossRef]
52. Yu, P.; Li, M.; Wang, Y.; Chen, Z. Fuel Cell Hybrid Electric Vehicles: A Review of Topologies and Energy Management Strategies. *World Electr. Veh. J.* **2022**, *13*, 172. [CrossRef]
53. Iqbal, M.; Laurent, J.; Benmouna, A.; Becherif, M.; Ramadan, H.S.; Claude, F. Ageing-aware load following control for composite-cost optimal energy management of fuel cell hybrid electric vehicle. *Energy* **2022**, *254*, 124233. [CrossRef]
54. Xu, L.; Mueller, C.D.; Li, J.; Ouyang, M.; Hu, Z. Multi-objective component sizing based on optimal energy management strategy of fuel cell electric vehicles. *Appl. Energy* **2015**, *157*, 664–674. [CrossRef]
55. Chen, H.; Zhao, X.; Zhang, T.; Pei, P. The reactant starvation of the proton exchange membrane fuel cells for vehicular applications: A review. *Energy Convers. Manag.* **2019**, *182*, 282–298. [CrossRef]
56. Mocera, F.; Vergori, E.; Somà, A. Battery Performance Analysis for Working Vehicle Applications. *IEEE Trans. Ind. Appl.* **2020**, *56*, 644–653. [CrossRef]

57. Yang, H.; Sun, Y.; Xia, C.; Zhang, H. Research on Energy Management Strategy of Fuel Cell Electric Tractor Based on Multi-Algorithm Fusion and Optimization. *Energies* **2022**, *15*, 6389. [CrossRef]
58. Jia, C.; Zhou, J.; He, H.; Li, J.; Wei, Z.; Li, K.; Shi, M. A novel energy management strategy for hybrid electric bus with fuel cell health and battery thermal- and health-constrained awareness. *Energy* **2023**, *271*, 127105. [CrossRef]
59. Garland, N.; Benjamin, T.; Kopasz, J. DOE Fuel Cell Program: Durability Technical Targets and Testing Protocols. *ECS Trans.* **2007**, *11*, 923. [CrossRef]
60. Available online: https://www.energy.gov/eere/fuelcells/doe-technical-targets-fuel-cell-systems-and-stacks-transportation-applications (accessed on 12 February 2024).
61. Nordelöf, A.; Messagie, M.; Tillman, A.M.; Ljunggren Söderman, M.; van Mierlo, J. Environmental impacts of hybrid, plug-in hybrid, and battery electric vehicles—what can we learn from life cycle assessment? *Int. J. Life Cycle Asses.* **2014**, *19*, 1866–1890. [CrossRef]
62. Sinha, P.; Brophy, B. Life cycle assessment of renewable hydrogen for fuel cell passenger vehicles in California. *Sustain. Energy Technol. Assess.* **2021**, *45*, 101188. [CrossRef]
63. Teimouri, A.; Kabeh, K.Z.; Changizian, S.; Ahmadi, P.; Mortazavi, M. Comparative lifecycle assessment of hydrogen fuel cell, electric, CNG, and gasoline-powered vehicles under real driving conditions assessment. *Int. J. Hydrogen Energy* **2022**, *47*, 37990–38002. [CrossRef]
64. Teng, Z.; Tan, C.; Liu, P.; Han, M. Analysis on carbon emission reduction intensity of fuel cell vehicles from a life-cycle perspective. *Front. Energy* **2023** . [CrossRef]
65. Hua, T.Q.; Roh,H.-S.; Ahluwalia, R.K. Performance assessment of 700-bar compressed hydrogen storage for light duty fuel cell vehicles. *Int. J. Hydrogen Energy* **2017**, *42*, 25121–25129. [CrossRef]
66. Li, Y.; Teliz, E.; Zinola, F.; Díaz, V. Design of a AB5-metal hydride cylindrical tank for hydrogen storage. *Int. J. Hydrogen Energy* **2021**, *46*, 33889–33898. [CrossRef]
67. Nguyen, H.Q.; Shabani, B. Review of metal hydride hydrogen storage thermal management for use in the fuel cell systems. *Int. J. Hydrogen Energy* **2021**, *46*, 31699–31726. [CrossRef]
68. Guo, L.; Su, J.; Wang, Z.; Shi, J.; Guan, X.; Cao, W.; Ou, Z. Hydrogen safety: An obstacle that must be overcome on the road towards future hydrogen economy. *Int. J. Hydrogen Energy* **2024**, *51*, 105–1078. [CrossRef]
69. Isaac, N.; Saha, A.K. A Review of the Optimization Strategies and Methods Used to Locate Hydrogen Fuel Refueling Stations. *Energies* **2023**, *16*, 2171. [CrossRef]
70. Moon, S.; Lee, Y.-J.; Lee, D.-J. A cost-effectiveness analysis of fuel cell electric vehicles considering infrastructure costs and greenhouse gas emissions: An empirical case study in Korea. *Sustain. Energy Technol. Assess.* **2022**, *54*, 102777. [CrossRef]
71. Usai, L.; Hung, C.R.; Vasquez, F.; Windsheimer, M.; Burheim, O.S.; Strømman, A.H. Life cycle assessment of fuel cell systems for light duty vehicles, current state-of-the-art and future impacts. *J. Clean. Prod.* **2021**, *280*, 125086. [CrossRef]
72. James, B.D. Fuel Cell Cost and Performance Analysis. Available online: https://www.hydrogen.energy.gov/docs/hydrogenprogramlibraries/pdfs/review22/fc353_james_2022_o-pdf.pdf?Status=Master (accessed on 13 February 2024).
73. Houchins, C.; James, B.D. Hydrogen Storage Cost Analysis. Available online: https://www.hydrogen.energy.gov/docs/hydrogenprogramlibraries/pdfs/review22/st235_houchins_2022_p-pdf.pdf?Status=Master (accessed on 13 February 2024).
74. Rawat, A.; Garg, C.P.; Sinha, P. Analysis of the key hydrogen fuel vehicles adoption barriers to reduce carbon emissions under net zero target in emerging market. *Energy Policy* **2024**, *184*, 113847. [CrossRef]
75. Trencher, G.; Wesseling, J. Roadblocks to fuel-cell electric vehicle diffusion: Evidence from Germany, Japan and California. *Transp. Res. D Transp. Environ.* **2022**, *112*, 103458. [CrossRef]
76. Al-Amin, A.Q.; Ambrose, A.F.; Masud, M.M.; Azam, M.N. People purchase intention towards hydrogen fuel cell vehicles: An experiential enquiry in Malaysia. *Int. J. Hydrogen Energy* **2016**, *41*, 2117–2127. [CrossRef]
77. Vallejos-Romero, A.; Cordoves-Sánchez, M.; Cisternas, C.; Sáez-Ardura, F.; Rodríguez, I.; Aledo, A.; Boso, Á.; Prades, J.; Álvarez, B. Green Hydrogen and Social Sciences: Issues, Problems, and Future Challenges. *Sustainability* **2023**, *15*, 303. [CrossRef]
78. Quarton, C.J.; Samsatli, S. How to incentivise hydrogen energy technologies for net zero: Whole-system value chain optimisation of policy scenarios. *Sustain. Prod. Consum.* **2021**, *27*, 1215–1238. [CrossRef]
79. Available online: https://eur-lex.europa.eu/legal-content/EN/TXT/?uri=CELEX:32021R1119 (accessed on 14 February 2024).
80. Available online: https://eur-lex.europa.eu/legal-content/EN/TXT/?uri=CELEX:32016R1628 (accessed on 14 February 2024).

World Electric Vehicle Journal

MDPI

Article

A Versatile Control Method for Multi-Agricultural Machine Cooperative Steering Applicable to Two Steering Modes

Weizhen Zhu, Yuhao Zhang, Weiwei Kong *, Fachao Jiang and Pengxiao Ji

College of Engineering, China Agricultural University, Beijing 100083, China; sy20223071562@cau.edu.cn (W.Z.); davidzhanghao@me.com (Y.Z.); jfachao@cau.edu.cn (F.J.); pengxiaoji@cau.edu.cn (P.J.)
* Correspondence: kongweiwei@cau.edu.cn

Abstract: This article aims to address the unnecessary stopping and low efficiency issues present in existing multi-machine cooperative steering control methods. To tackle this challenge, a novel cooperative control approach for multiple agricultural machines is proposed, considering two typical steering modes of farm machinery. This approach encompasses a multi-machine cooperative control framework suitable for both steering modes. Based on the established lateral and longitudinal kinematics models of the farm machines, the method includes a path-tracking controller designed using the pure pursuit and Stanley algorithms, a formation-keeping controller based on PID control, and a T-turn cooperative-steering controller based on a problem-solving approach. To assess the method's viability, a collaborative simulation platform utilizing CarSim and Simulink was constructed, which conducted simulations for both U-turn and T-turn cooperative steering controls. The simulation results indicate that the proposed control framework and methodology can effectively ensure no collision risk during the U-turn and T-turn cooperative steering processes for three farm machines, eliminating stopping in T-turn, enhancing safety, and improving fuel economy. Compared with traditional sequential control methods, the proposed approach reduced operation time by 17.47 s and increased efficiency by 15.29% in the same scenarios.

Keywords: agricultural vehicle steering mode; multiple agricultural machine cooperative control; path tracking; formation maintenance

Citation: Zhu, W.; Zhang, Y.; Kong, W.; Jiang, F.; Ji, P. A Versatile Control Method for Multi-Agricultural Machine Cooperative Steering Applicable to Two Steering Modes. *World Electr. Veh. J.* **2024**, *15*, 126. https://doi.org/10.3390/wevj15040126

Academic Editor: Joeri Van Mierlo

Received: 18 February 2024
Revised: 19 March 2024
Accepted: 20 March 2024
Published: 22 March 2024

1. Introduction

With the intensification, scaling, and industrialization of agriculture in China, as well as the rising demands for efficiency and the increasing complexity of tasks in agricultural operations, traditional automatic navigation techniques for agricultural machinery are no longer sufficient to meet the challenges of safety and break through efficiency bottlenecks [1]. Therefore, the development of intelligent agricultural machinery technology, including multi-machine cooperative systems, has become an urgent need for the evolution of China's agriculture [2–4].

The multi-machine cooperative technology has garnered extensive recognition from both domestic and international researchers. Its primary components encompass positioning and communication technology, path planning technology, and control technology [5]. The positioning and communication sector is responsible for acquiring and exchanging critical data such as the position and heading of both the subject vehicle and others in the vicinity. To this end, Zhu Z et al. equipped farm machinery with potentiometers, magnetic speed sensors, and RTK-GPS to ascertain front wheel steering angles, velocities, heading angles, and positional coordinates [6]. Gerasimos G. Rigatos proposed a multi-machine collaborative control approach based on state estimation. This method integrates information from multiple sensors using a derivative-free nonlinear filtering technique. The accurate estimation of the agricultural machinery's position and its motion characteristics is achieved by replacing the extended information filter with a recursive standard information filter for distributed state estimation [7]. Li S et al. designed communication protocol frames for inter-vehicle communication within a master–slave coordinated system among agricultural vehicles, enabling the exchange of status information [8]. Wenju Mao et al. developed a robotic navigation system for orchard-harvesting machines featuring a dual master–slave mode. The communication protocol data frame format between orchard-harvesting robots was elucidated for the transmission of GNSS coordinates, velocity, navigation modes, and other information pertaining to the harvesting and transport machines [9]. Path planning technology considers the optimal path for a multi-machine system to avoid collisions, planning and calculating the route for each agricultural machine accordingly. Martin Andreas Falk Jensen et al. incorporated optimization criteria such as time and operational distance, utilizing the Dijkstra algorithm to address path planning for a multi-machine system in field and intra-field transportation scenarios [10]. Addressing the challenges of heterogeneous agricultural machinery fleets, which vary in speed, fuel consumption, turning radius, and fuel tank size, Jesus Conesa-Muñoz et al. proposed a practical multi-machine cooperative path planning method. This approach deliberates over various optimization standards such as distance, time, and input costs, ensuring a comprehensive strategy for efficient path planning [11]. Timo Blender et al. employed a centralized entity, Opti-Visor, to manage the path planning and optimization for a fleet of agricultural machines. The architecture and functionality of Opti-Visor within the agricultural machine fleet system were delineated [12]. Control technology ensures the synchronized operation of a multi-machine cooperative system in prescribed formations, at designated speeds, and with appropriate longitudinal and lateral spacing as per mission specifications. A cooperative navigation control method for a harvesting fleet, premised upon a leader–follower structure, was proposed by Bai X et al. [13]. To manage headland turning maneuvers, Zhang C et al. devised control strategies and conducted field tests with real machinery, thereby laying the groundwork for operational efficacy in agricultural tasks requiring such complex turns [14].

Current research on multi-machine cooperative localization, communication technology, and path planning has reached a relatively mature stage. However, in the domain of control technology research, most studies are focused on employing various methodologies to address the challenges associated with the collaborative control of two agricultural machines. Noboru Noguchi proposed two fundamental motion control algorithms for master–slave agricultural machine systems: the GOTO algorithm and the FOLLOW algorithm [15]. In response to uncertainties such as variations in workload and soil conditions during agricultural operations, leading to issues of instability, delayed response, and control challenges in the cooperative following of agricultural machinery, Xu Guangfei et al. proposed a hierarchical control architecture for master–slave multi-agricultural machine following [16]. Wang Zhiqing designed the speed control system, heading–following control system, and following distance control system for two agricultural machines, as well as a collaborative control system for the lead and follow vehicles [17]. Chi Zhang et al. proposed a multi-tractor system for field operations aimed at reducing total operation time, improving work efficiency, and presenting strategies for lateral and longitudinal spacing control [18]. While a minority of studies address three or more agricultural machines, Stavros G. Vougioukas proposed a distributed control framework for multiple agricultural machines, wherein each machine was equipped with a nonlinear model predictive tracking controller. Additionally, it received motion trajectories from other machines, incorporating them into its own control considerations [19]. Zhang Wenyu et al. designed a longitudinal relative position collaborative control method suitable for master–slave navigation in coordinated harvesting and unloading operations [20]. Zheng Xinyao addressed the steering control issue for the follower machine by proposing a fuzzy adaptive PID variable damping steering control method. This approach aims to ensure precise and rapid tracking of the desired steering angle command for the follower machine's steering wheel [21]. Gou ABE et al. developed a control algorithm based on a laser scanner to recognize the relative distance and direction to the lead vehicle and follow it [22].

In summary, existing research on multi-machine cooperative control is limited to linear path coordination, and the rare studies on headland turn control typically focus on only one type of turning method and often employ time-sequential control [23,24]. This approach can lead to unnecessary stopping and waiting, reducing operational efficiency. The main contributions of the paper are as follows:

(1) This article proposes a versatile multi-machine control architecture, applicable to two typical agricultural machine steering modes, including U-turn and T-turn.
(2) Building upon the control architecture, a cooperative control approach for multiple agricultural machines in both straight-line and headland turn steering scenarios is proposed. This method, while ensuring the safety of agricultural machines, aims to enhance operational efficiency and fuel economy.

The rest of the paper is organized as follows: Section 2 provides a detailed description of the problem addressed in this paper, while Section 3 presents the overall control architecture, establishes the longitudinal and lateral kinematic models for agricultural machine, and designs a multi-robot cooperative controller based on this foundation. Section 4 discusses the simulation results; Section 5 concludes the article and discusses future work.

2. Problem Description

This article focuses on a cohort of three homogenized machines equipped with implements, aimed at maintaining formation during straight-line operations, as well as ensuring safety and efficiency during headland turns. The lead machine in this group is driven by a human, operating at a constant speed, while the second and third tractors are autonomous agricultural machines, programmed to follow their respective preceding vehicle.

During straight-line operations, the following agricultural machines maintain the formation, as depicted in Figure 1, while working in tandem with the lead agricultural machine. The green lines in the Figure 1 refer to the operation lines of agricultural machine.

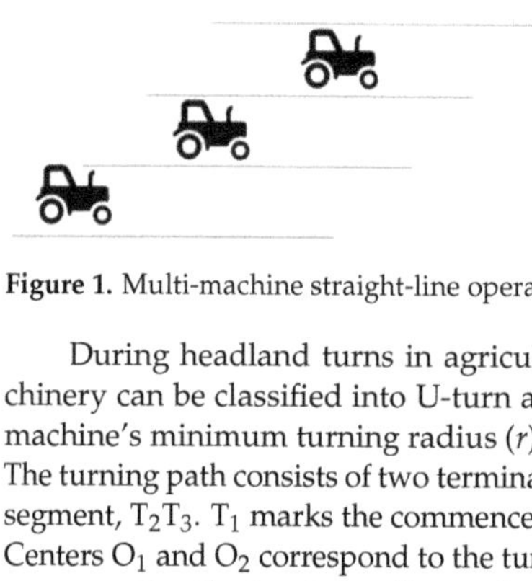

Figure 1. Multi-machine straight-line operation formation.

During headland turns in agricultural operations, the steering method of the machinery can be classified into U-turn and T-turn, based on the relationship between the machine's minimum turning radius (r) and the working width (w), as shown in Figure 2. The turning path consists of two terminal circular arcs, T_1T_2 and T_3T_4, and a connecting line segment, T_2T_3. T_1 marks the commencement of the turn, while T_4 indicates the completion. Centers O_1 and O_2 correspond to the turning circular arcs T_1T_2 and T_3T_4, respectively, with r representing the turning radius, and w indicating the working width. A U-turn approach is used when the working width w is greater than or equal to twice the turning radius, signified as $w \geq 2r$; conversely, a T-turn method is preferred when the working width w is less than double the turning radius, denoted as $w < 2r$ [25].

Addressing straight-line conditions, control measures are implemented on the agricultural machines to maintain a designated formation among the three machines while ensuring operational safety. Concerning headland turn coordination, the aim is to minimize the overall turning time without any risk of collision, and to eliminate stopping and waiting during the turn, thereby enhancing turning efficiency and fuel economy.

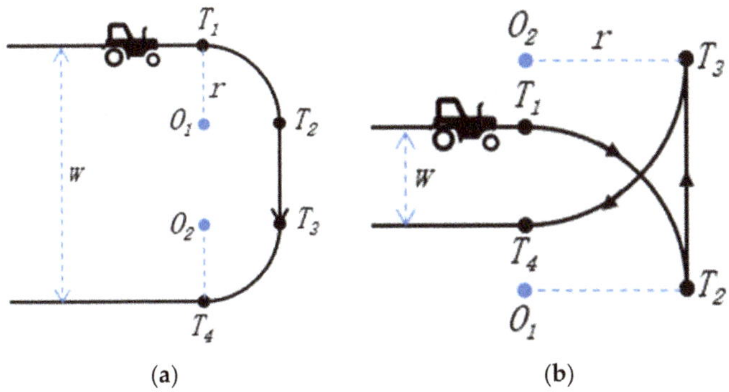

Figure 2. Agricultural machine steering model: (**a**) U-turn and (**b**) T-turn.

3. Methodology

3.1. Integral Control Framework

The multi-machine cooperative control architecture proposed in this study primarily comprises a comprehensive kinematic model for universal agricultural machinery, a path-tracking controller, a formation-maintenance controller, a collision risk detection module, and a controller to optimize cooperative steering during T-turns. This framework ensures that the steering actions performed during U-turns do not elevate the risk of collision; thus, an additional controller for cooperative steering optimization during U-turns is unnecessary. It is sufficient to confirm the safety of the agricultural machinery throughout these maneuvers.

The kinematic model of agricultural machinery, encompassing both lateral and longitudinal dynamics, forms the essential groundwork for the associated lateral and longitudinal control systems. A path-tracking controller is derived based on the lateral kinematic model of the agricultural machinery, enabling machinery operation along pre-planned straight and headland turn paths. Furthermore, based on the longitudinal dynamic model and data such as speed and spacing from the onboard millimeter-wave radar, a multi-machine formation-keeping controller for straight trajectories is designed. This controller generates the necessary speeds to maintain formation, which serves as the speed input for the path-tracking controller on straight paths.

Additionally, a collision risk detection module based on the Separating Axis Theorem was designed to accommodate multi-machine coordinated steering under two steering modes. This module verifies the safety of three agricultural machines during the cooperative steering, providing collision-free minimum spacing for the formation-keeping controllers of both steering types. For the multi-machine T-turn cooperative steering, a steering optimization controller was engineered. Its output determines the speed under the cooperative steering path of the path-tracking controller, enabling three agricultural machines to execute the cooperative steering without topping and waiting, thus minimizing the overall turning time. The integrated control architecture is illustrated in Figure 3, and the control flow chart is shown in Figure 4. v and δ in Figure 4 represent the desired speed and the desired front wheel angle that the formation-keeping controller and pure pursuit controller outputs to the kinematic model.

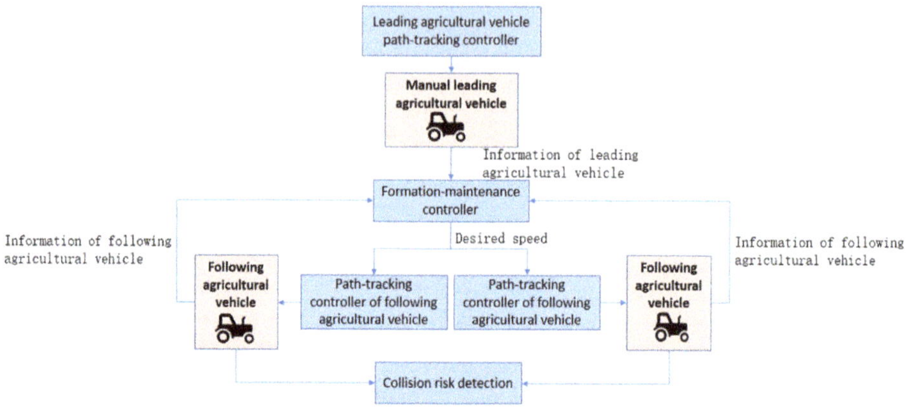

Figure 3. Integrated control architecture.

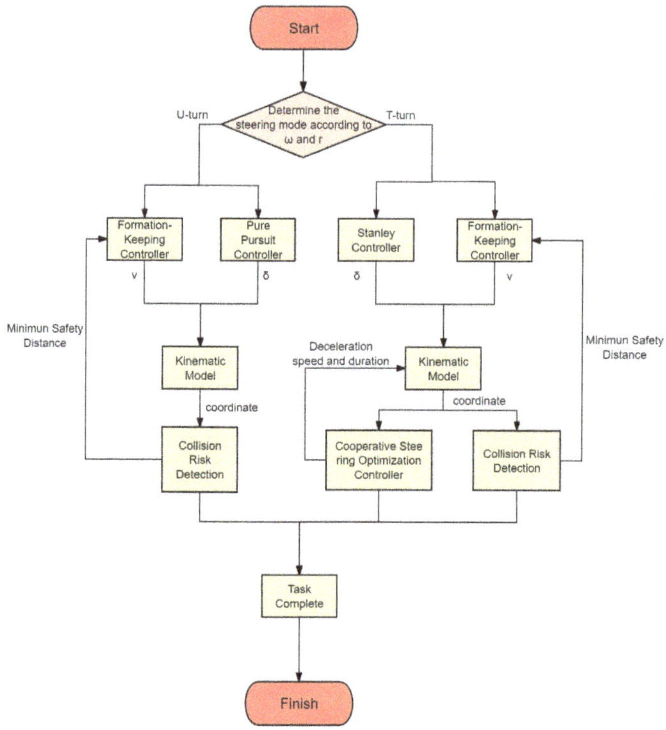

Figure 4. Control flow chart.

3.2. *Kinematic Model*

3.2.1. Lateral Kinematics

The lateral kinematic model provides a basis for the design of the path-tracking controller. Assuming there is no interaction between the machine and the ground, and disregarding motions such as roll, pitch, and sideslip, the agricultural machinery can be simplified to a two-wheeled machine model for kinematic analysis, as illustrated in Figure 5. Within this simplified model, point M represents the point on the path that is closest to the control point O.

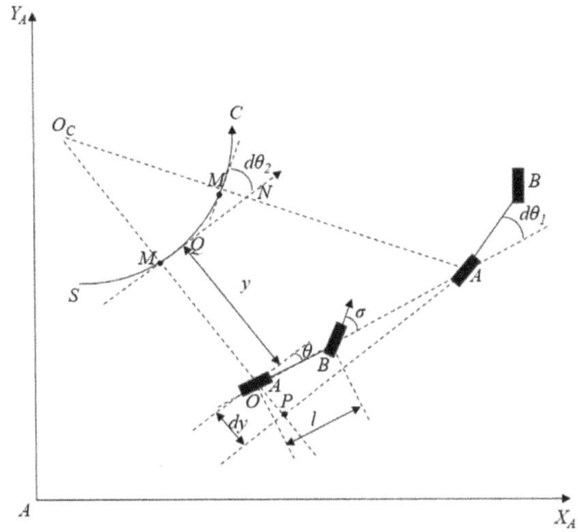

Figure 5. Lateral kinematics model diagram.

Based on the schematic of the kinematic model demonstrated above, along with geometric knowledge and physical principles, it is possible to derive a set of differential equations representing the kinematics of agricultural machinery, which are presented as follows:

$$\begin{cases} \frac{ds}{dt} = \frac{v\cos\theta}{1-c(s)y} \\ \frac{dy}{dt} = v\sin\theta \\ \frac{d\theta}{dt} = v\left(\frac{\tan\delta}{l} - \frac{c(s)\cos\theta}{c(s)y}\right) \end{cases} \quad (1)$$

where $c(s)$ is the curvature of the following curve path at point M; v is the speed of the agricultural machine; l is the wheelbase of the agricultural machine; s is the arc length of point M along curve C; y is the lateral deviation of the agricultural machine from the set path; θ is the heading deviation angle of the agricultural machine; and δ is the steering wheel deviation angle of the agricultural machine.

The kinematic model for agricultural machinery along a straight-path trajectory is described as follows:

$$\begin{cases} \frac{ds}{dt} = v\cos\theta \\ \frac{dy}{dt} = v\sin\theta \\ \frac{d\theta}{dt} = v\frac{\tan\delta}{l} \end{cases} \quad (2)$$

Performing elementary calculations yields the system of differential equations for the lateral kinematic model of the agricultural machine, as follows:

$$\begin{cases} \frac{dy}{ds} = \tan\theta \\ \frac{d\theta}{ds} = \frac{\tan\delta}{l\cos\theta} \end{cases} \quad (3)$$

3.2.2. Longitudinal Kinematics

The longitudinal kinematic model provides a foundation for the design of formation-maintenance controllers. The longitudinal motion state of the following agricultural machinery during cooperative operation can be represented as:

$$\frac{N_g N_0 \eta_t}{r_w} M_e - K_P P_b = m\dot{v}_f + C_A {v_f}^2 + mgf \quad (4)$$

where N_g is the engine power transmission ratio; N_0 is the engine-to-wheel transmission ratio; η_t is the mechanical efficiency of the transmission system; M_e is the engine output torque; K_p is the scaling factor; P_b is the brake pressure; r_w is the wheel radius; v_f is the speed of the following agricultural machine; m is the weight of the following agricultural machine; C_A is the aerodynamic drag coefficient; g is the acceleration due to gravity; and f is the rolling resistance coefficient.

During the following process, a fixed spacing model is employed, from which the desired following distance, denoted as d_d, can be derived using the collision risk detection module. For detailed derivation, refer to Section 3.3.2. The derived longitudinal kinematic model for agricultural machinery following is as follows:

$$\Delta\dot{d} = \Delta v \tag{5}$$

$$\Delta\dot{v} = \frac{1}{m}\left(C_A {v_f}^2 + mgf\right) - \frac{N_g N_0 \eta_t}{m r_w} M_e \tag{6}$$

where Δd is the following distance error; d_r is the actual following distance; Δv is the following speed error; and v_m is the speed of the leading machine.

3.3. Controller Design

In accordance with the integral control framework and the kinematic models established in Section 2, the multi-machine cooperative control system is delineated into three distinct components: the linear formation-keeping controller, the headland cooperated steering optimization controller, and the path-tracking controller design. The formation-keeping controller and the headland cooperated steering optimization controller function to furnish the longitudinal velocity inputs requisite for the path-tracking controller, which undertakes the lateral control tasks of the agricultural machinery.

3.3.1. Straight-Line Formation-Keeping Controller

The functionality of the formation-keeping controller lies in adjustments made to the following farm machine's speed, ensuring that the follower maintains a set longitudinal distance from the leader during field operations.

The controller employs a fixed-spacing strategy PID controller, in which the spacing error 'e_i' is fed into the PID as an error term. This error is then subjected to a linear combination of proportional, integral, and differential calculations to produce the control variable. The model is illustrated in Figure 6.

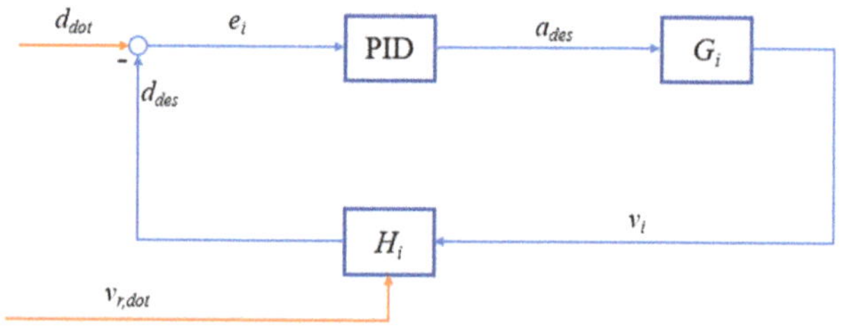

Figure 6. PID controller model.

After being processed by the PID controller, the error 'e_i' yields the desired acceleration for the subject machine. The module 'H_i' represents the spacing strategy, which determines the desired spacing between the subject machine and the lead machine. The 'G_i' module is the longitudinal machine controller, which provides the subject machine's velocity as the input to the spacing strategy module.

3.3.2. Headland-Turn Cooperative Steering Optimization Controller

(1) Collision Risk Detection

A safety model for agricultural machinery is structured through the implementation of an Oriented Bounding Box (OBB) [26] to identify zones susceptible to collision. Given that the outline of the agricultural machine typically projects a nearly rectangular shape on the ground, a rectangle defined by the points P_1 P_2 P_3 P_4 serves as the enclosing surface of the safety model, as illustrated in Figure 7. The dimensions of this rectangle are the combined aggregate of the length and width of the machine's ground projection including its implements, with an additional minimum safety margin incorporated. The heading angle of the machine is represented by θ, and a minimum safety distance of 0.5 m is established to uphold the standard of operational safety.

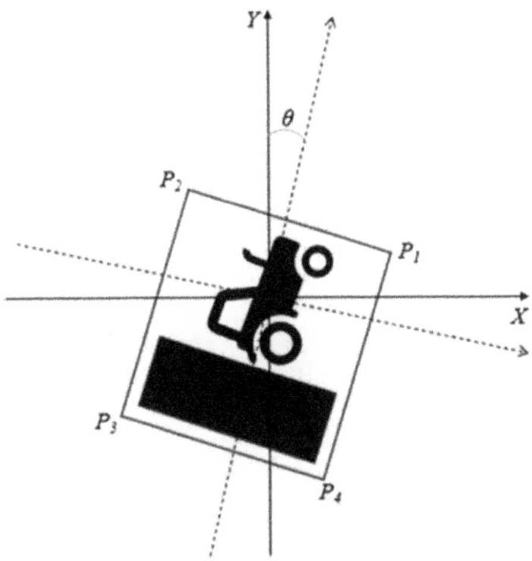

Figure 7. Agricultural machine safety model.

Utilizing the Separating Axis Theorem [27], a model for detecting the safety state of agricultural machinery is established, as illustrated in Figure 8. In this model, O_A and O_B are identified as the geometric centers of the OBB for machine 1 and 2, respectively.

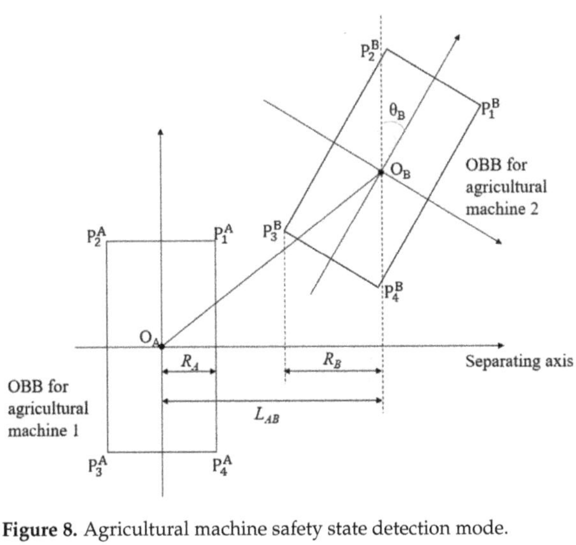

Figure 8. Agricultural machine safety state detection mode.

A separated axis coordinate system is established using O_A as the origin, where the separating axis is parallel to the lateral symmetry axis of the OBB of machine 1. L_{AB} denotes the projection length of the geometric center distance between the OBB of the two machines on the separating axis. R_A and R_B are the half-lengths of the maximum projection lengths on the separating axis of machine 1 and 2's OBB section, which can be referred to as the projection radius. The formulas for calculating R_A, R_B, and L_{AB} are as follows:

$$\begin{cases} R_A = W_A/2 \\ R_B = \frac{W_B cos\theta_B}{2} + \frac{L_B sin\theta_B}{2} \\ L_{AB} = x_B - x_A \end{cases} \quad (7)$$

where W_A is the width of the OBB for agricultural machine 1; θ_B is the heading angle of agricultural machine 2 relative to agricultural machine 1; L_B is the length of the OBB for agricultural machine 2; and W_B is the width of agricultural machine 2.

According to the Separating Axis Theorem, if the combined projection radius of the agricultural machine OBB sections on the separating axis (denoted as L_{AB}) exceed the sum of the projection lengths of the geometric center distances on the separating axis (denoted as R_A + R_B), it indicates that there is no collision between the machines, hence, no risk of collision.

(2) U-turn Cooperative Steering

As the agricultural machinery reaches the headland and begins the coordinated steering of U-turns, the distance between the machines widens. Therefore, to achieve effective control, it is necessary to minimize the spacing as much as possible while ensuring there is no risk of collision during straight-line operations.

(3) T-turn Cooperative Steering

In T-turn cooperative steering, the large turning radius of the agricultural machine makes it unable to directly turn into the next work path without reversing adjustments. This reduces the distance between machines and increases the risk of collision. Hence, the two following machines must decelerate in advance during straight-line operations to avoid stopping and waiting in the entire cooperative steering process. After the lead agricultural machine and the first following machine have completed the turn, to ensure the distance between the machines is within a safe range, both the lead and the first following agricultural machine need to decelerate. They should continue at a reduced speed until the minimum safe distance is restored, after which they can resume work at the original speed. The control logic is as follows: First, the formation-keeping controller shortens the distance to the minimum safety distance. Once the three machines reach a stable speed, the T-turn headland-steering controller calculates the deceleration, target speeds, and the duration of target speeds for the two following machines, so they can complete the turn without stopping and waiting and without the risk of collision. After completing the turn, the controller instructs the lead agricultural machine and the first following machine to decelerate until they reach the minimum safe distance, after which they resume working speed to continue operations.

Based on the above description, the design of the headland steering optimization controller is as follows:

$$S_{ahead} - S_{ego} \geq L \quad (8)$$

where:

$$S_{ahead} = \int_0^{t_2} v_{ahead} dt \quad (9)$$

$$S_{ego} = \int_0^{t_1} (-at + v_{ego}) dt + \int_{t_2}^{t_2} v_{desire} dt \quad (10)$$

$$L = L_{ABmin} tan\theta_B \quad (11)$$

And to ensure the comfort during the deceleration process of the agricultural machines, the acceleration 'a', which is measured in meters per second squared (m/s^2), needs to satisfy:

$$a \leq 1.5 \tag{12}$$

In Equations (9)–(11), S_{ahead} denotes the travel distance of the preceding machine during the deceleration process, measured in meters (m); S_{ego} signifies the travel distance of the ego vehicle, which refers to the machine under consideration or the machine executing the maneuver, also measured in meters (m).; L_{ABmin} is the minimum safe distance for the agricultural machine; V_{ahead} is the lead machine speed; V_{ego} is the machine speed before it decelerates; V_{desire} is the machine speed after it decelerates; t_1 is the duration of the deceleration phase; and t_2 is the duration of the V_{desire}.

Based on the above controllers, the deceleration, target speed, and the duration of the target speed during the T-type coordinated steering process can be determined. These values serve as inputs for the following agricultural machine.

3.3.3. Path-Tracking Controller

The primary function of the path-tracking controller is to ascertain that the lead agricultural machine and the two following machines accurately execute their operations along designated pathways, thereby maintaining a preset lateral distance between each machine, equivalent to the working row width of the machines. The controller employs the actual positional data of the agricultural machinery along with the lateral and heading error from the set path as input parameters. Through a carefully designed control algorithm, it generates the desired steering angle for the wheels. This process ensures that the lateral distance between each following machine and the leading machine is consistently maintained within the defined numerical range. The integrity of this system is critical to achieve precise alignment and coordination in agricultural operations, enabling the efficient utilization of the machinery within the prescribed operational parameters.

(1) U-turn Cooperative Steering

For the comparatively straightforward U-turn cooperative steering, the path-tracking controller utilizes a pure pursuit algorithm, as illustrated in Figure 9. In this schematic, point A represents the center of the rear axle, B is the center of the front axle, and C designates the target point on the path closest to point A.

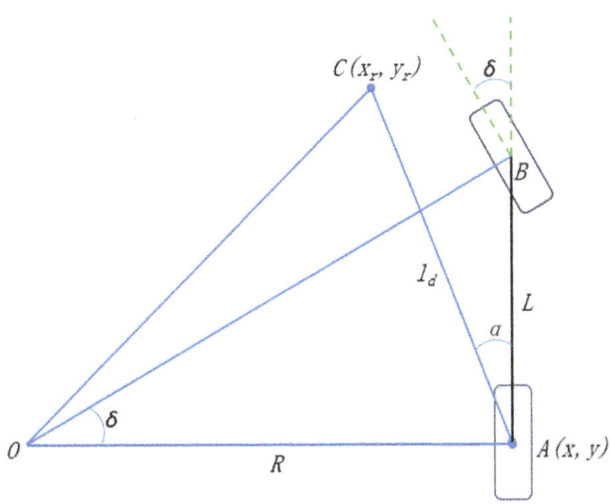

Figure 9. Schematic diagram of pure pursuit algorithm.

The geometric relationships in conjunction with the formula for calculating the turning radius yield the following result:

$$\begin{cases} \frac{1}{R} = \frac{2sin\alpha}{l_d} \\ tan\delta = \frac{L}{R} \end{cases} \quad (13)$$

The target steering angle for the front wheels can be derived as follows:

$$\delta = arctan\left(\frac{2Lsin\alpha}{l_d}\right) \quad (14)$$

where R is the steering radius; L is the wheelbase; δ is the front wheel angle; α is the angle between the machine body and the look-ahead point; and l_d is the look-ahead distance.

'e' is defined as the lateral error of the look-ahead point, then

$$e = l_d sin\alpha \quad (15)$$

The control law for the pure pursuit algorithm is derived as follows:

$$\delta = arctan\left(\frac{2Le}{l_d^2}\right) \quad (16)$$

(2) T-turn Cooperative Steering

For the T-turn cooperative steering, which involves both forward and reverse movements, the path-tracking controller adopts the Stanley algorithm, which can better handle curved paths and lateral errors. The principles of this algorithm are depicted in Figure 10. Analogous to the pure pursuit algorithm, point A marks the center of the rear axle, and point B marks the center of the front axle. Point P represents the nearest path point to the center of the front axle.

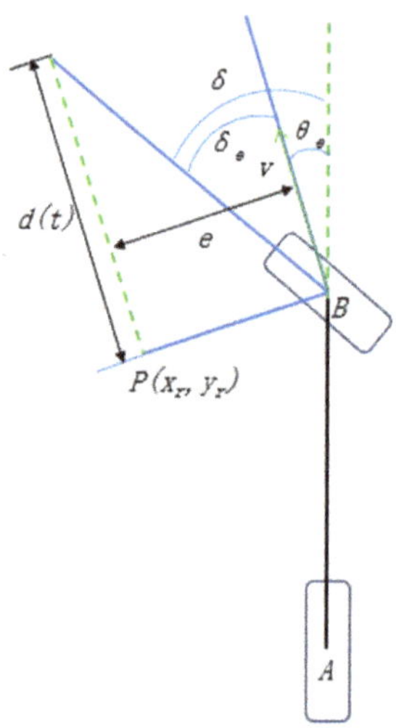

Figure 10. Schematic diagram of Stanley algorithm.

Neglecting the lateral tracking error 'e', the angle of the front wheel aligns with the tangent to the given path, which results in the following equation:

$$\delta_{\theta_e} = \theta_e(t) \quad (17)$$

Disregarding the heading error (θ_e), a more substantial lateral tracking error results in a greater steering angle of the front wheel. Assuming that the vehicle's anticipated trajectory intersects with the tangent of the nearest point on the given path at a distance $d(t)$ ahead of the front wheel, the following nonlinear proportional function is deduced based on the geometric relationships:

$$\delta_e(t) = arctan\frac{ke(t)}{v(t)} \tag{18}$$

Considering both control factors, the control law for the front wheel steering angle is as follows:

$$\delta(t) = \theta_e(t) + \frac{ke(t)}{v(t)} \tag{19}$$

Based on the linear bicycle kinematics model and geometric relationships, the rate of change of lateral error can be obtained as follows:

$$\dot{e}(t) = \frac{-ke(t)}{\sqrt{1+\left(\frac{ke(t)}{v(t)}\right)^2}} \tag{20}$$

When the lateral tracking error $e(t)$ is small, it can be deduced that:

$$e(t) = e(0) \times e^{-kt} \tag{21}$$

Therefore, the lateral error exponent converges to $e(t) = 0$, where the parameter k determines the convergence speed of the lateral error. For any lateral error, the differential equation monotonically approaches 0.

4. Simulation Validation

To validate the model and method proposed in this article, a simulation scenario was established in CarSim, as shown in Figure 11, and the proposed control algorithm was constructed in Simulink to set up a CarSim/Simulink joint simulation platform. The platform was then used to conduct a simulated validation of U-turn and T-turn cooperative steering control with one lead agricultural machine and two following agricultural machines.

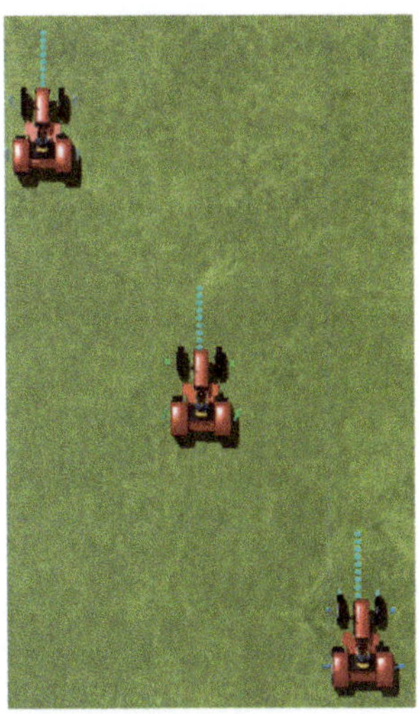

Figure 11. CarSim simulation scenario diagram.

4.1. Scenario Design

The YTO-LX800 tractor is chosen for this study, and its parameters are presented in Table 1. It is equipped with implements of dimensions 8000 × 2000 mm and 4000 × 2000 mm for U-turn and T-turn scenarios, respectively, determined based on the relationship between turning radius and working width. The safety model can be approximated as a 7 × 7 m rectangle.

Table 1. Parameters of YTO-LX800.

Parameter		Value
Model		YTO-LX800
Dimensions (mm)		4250 × 2090 × 2850
Minimum turning radius (m)		4
Wheelbase (mm)		2342
Total vehicle weight (kg)		2725
Speed range (km/h)	Forward	1.92 to 31.72
	Reverse	5 to 15.01
Maximum traction force (kN)		≥22.4

A field block, MNPQ, is selected, approximately 100 m in length, and the routes for three tractors to perform cooperative U-turn and T-turn maneuvers are planned, as shown in Figure 12. Different colors in Figure 12 represent the paths of different farm machines. The path AB represents the path of the lead machine, while A_1B_1 and A_2B_2 correspond to the paths of the following machines. The coordinates of the path points are entered into the path-tracking controller. It is stipulated that the machines maintain a stable working speed of 10 km/h, while the speed during reverse motion is set to 5 km/h. Furthermore, the acceleration during the entire process is to be confined within the limits of $\pm 1.5\ \text{m/s}^2$.

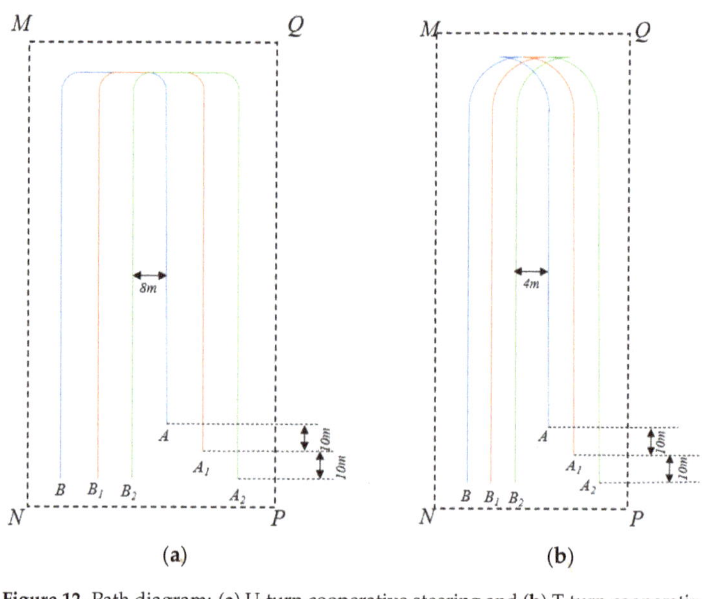

Figure 12. Path diagram: (**a**) U-turn cooperative steering and (**b**) T-turn cooperative steering.

In addition, due to the significant curvature changes in the T-turn cooperative steering and the involvement of forward and reverse path transitions, it is necessary to divide the T-turn into three segments and track each segment separately, as shown in Figure 13. Serial number 1–3 in Figure 13 indicates the sequence of the segmented path of agricultural machine. The switching conditions between different paths are determined based on factors such as the position and heading angle of the tractors.

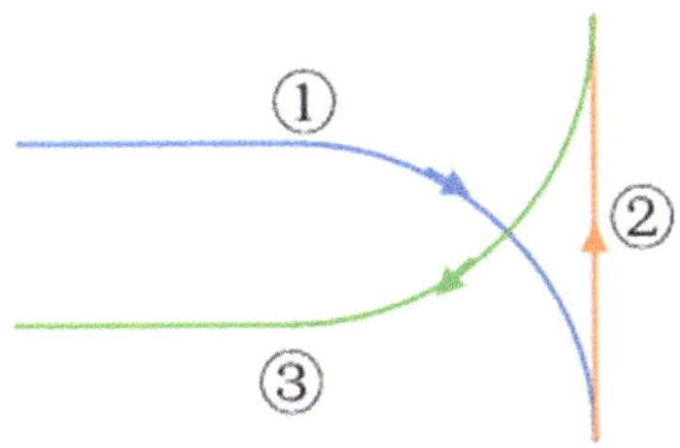

Figure 13. Path segmentation diagram.

A joint simulation platform with Carsim and Simulink is built, where U-turn and T-turn cooperative scenarios for straight-line and steering path simulations are constructed within Carsim. Meanwhile, in Simulink, the various controllers, as described in Section 3, are established for simulation validation.

4.2. Analysis of Results

4.2.1. U-Type Cooperative Steering

Under the U-turn cooperative steering scenario, the path tracking trajectories of three agricultural machines are shown in Figure 14, where VL represents the lead machine, while VF1 and VF2 are the first and second following machines, respectively. On each machine's trajectory, squares are used to denote the position of the machine at 16 different points in time, with the number 1–16 indicating the sequence of moments. Establishing VF2's initial position as the coordinate origin, with the travel direction along the positive X-axis and the steering direction along the positive Y-axis, the initial coordinates for VL, VF1, and VF2 are defined as (20, 12), (10, 6), and (0, 0), respectively. The collaborative operation task is deemed completed and the simulation is terminated upon VF2 achieving a new X-coordinate of 0 after steering, corresponding to the coordinates (0, 18).

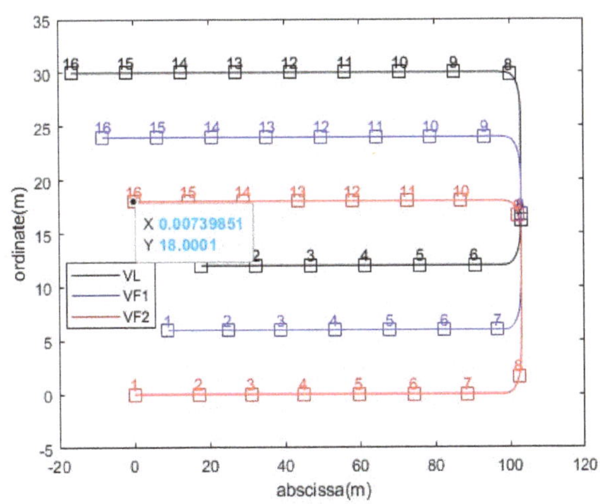

(a)

Figure 14. *Cont.*

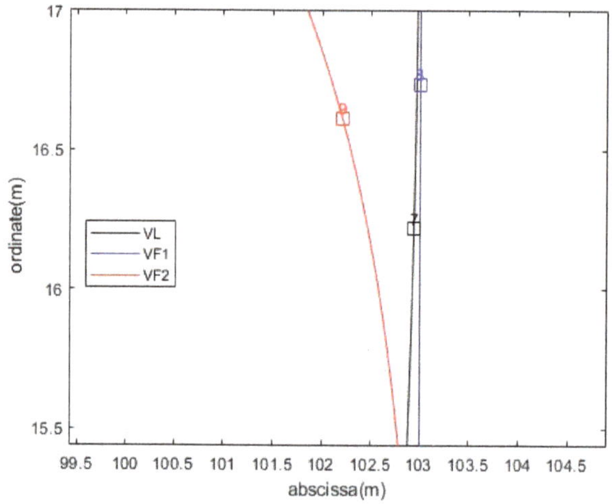

(**b**)

Figure 14. U-type cooperative steering trajectory diagram: (**a**) global trajectory and (**b**) local trajectory.

The longitudinal speeds are depicted in Figure 15, where the two following agricultural vehicles initially accelerate to reduce the distance to the leading vehicle and then reach a stable state at approximately 30 s. The simulation concludes at the 78.353 s mark upon the completion of the operational task. The inter-vehicle spacing is illustrated in Figure 16, with VL/VF1 representing the distance between the leading agricultural vehicle and the first following agricultural vehicle, and VF1/VF2 indicating the spacing between the first and second agricultural vehicles. Additionally, the red dashed line in Figure 16 represents the minimum safety distance, and the results from the collision risk detection module consistently indicate the absence of collision risks throughout the entire process.

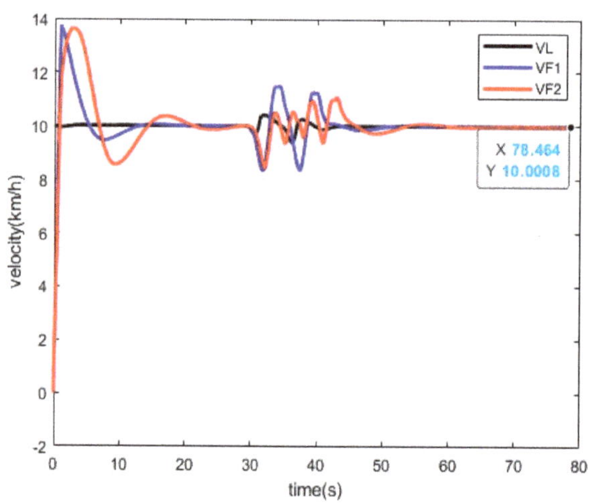

Figure 15. U-type cooperative steering longitudinal speed diagram.

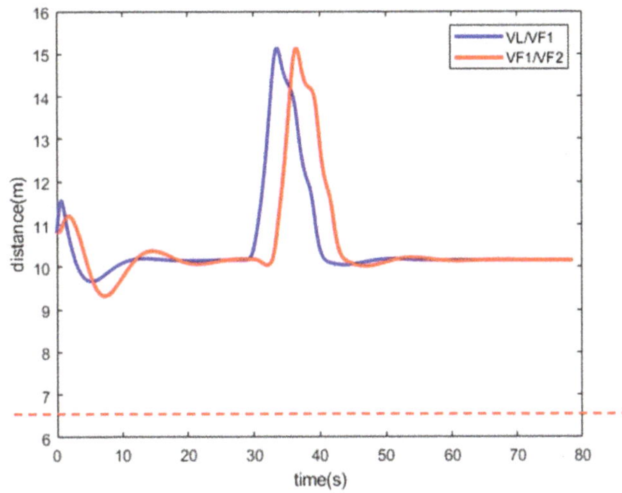

Figure 16. U-type cooperative steering agricultural vehicle distance diagram.

Based on the trajectory and spacing charts from the U-turn cooperative steering simulation, as well as the results of the collision risk detection module, it can be discerned that throughout the whole process, the distances between the agricultural machines were always above the minimum safety gap. Moreover, at any given moment, the plotted paths of the machines did not conflict, and there was no risk of collision.

The simulation results indicate that the method proposed in this study for the multiple agricultural machine cooperative operation and U-turn cooperative steering enables the leader and follower machines to work according to the planned paths while maintaining a certain formation. Furthermore, throughout the entire operation and cooperative steering process, there is no risk of collision between the machines.

4.2.2. T-Type Cooperative Steering

In a T-turn cooperative steering scenario, the path tracking trajectories of three agricultural machines are depicted in Figure 17. The squares and numerals within the figure carry the same meaning as in the U-turn cooperative steering scenario, representing the location of the machines at various time intervals and the sequential order of these moments, respectively. Similarly, stipulating VF2's initial position as the coordinate origin, with the travel direction along the positive X-axis and the steering direction along the positive Y-axis, the initial coordinates for VL, VF1, and VF2 are set at (20, 6), (10, 3), and (0, 0), respectively. The collaborative operation task is considered complete and the simulation is halted upon VF2 reaching a new X-coordinate of 0 after steering, corresponding to the coordinates (0, 9).

As illustrated in Figure 18, the longitudinal speed of the agricultural machinery begins with an acceleration by the two following machines to reduce the distance to the lead vehicle, subsequently stabilizing at 10 km/h. Once stable, the headland cooperative steering controller calculates the required deceleration, target velocity, and the duration for maintaining the target velocity to ensure a collision-free turning process. This results in a controlled deceleration of the two following machines, corresponding to the section between the blue lines in the diagram. In particular, the second following machine must sustain the target speed for a longer period to increase the gap from the first following machine. After the lead agricultural machine and the first trailing machine have completed the turn, they decelerate to wait for the following machine, corresponding to the section between the green lines in the diagram. The three machines then shorten the interval between them to the minimum safety distance before resuming a stable speed and state. At 96.73 s, the operational task is completed, and the simulation is terminated.

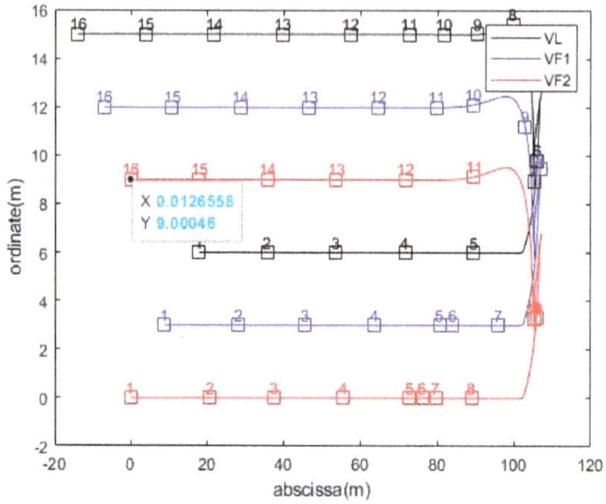

(**a**)

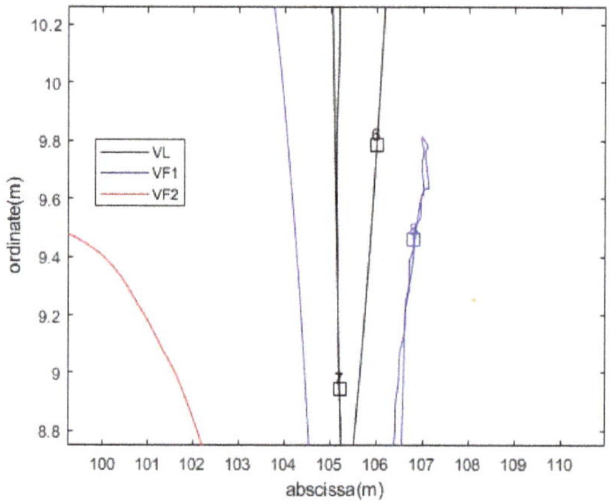

(**b**)

Figure 17. T-turn cooperative steering trajectory diagram: (**a**) global trajectory and (**b**) local trajectory.

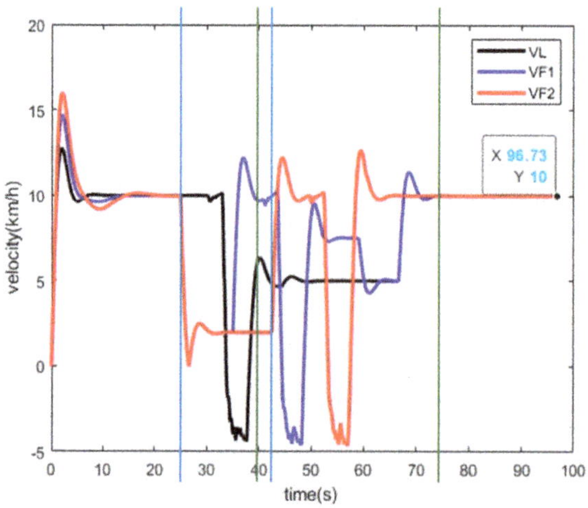

Figure 18. T-type cooperative steering longitudinal speed diagram.

Figure 19 illustrates the relationship between the spacing of the three agricultural machines during the T-turn collaborative turning process, and their correlation with the minimum safety distance, corresponding to the red dashed line.

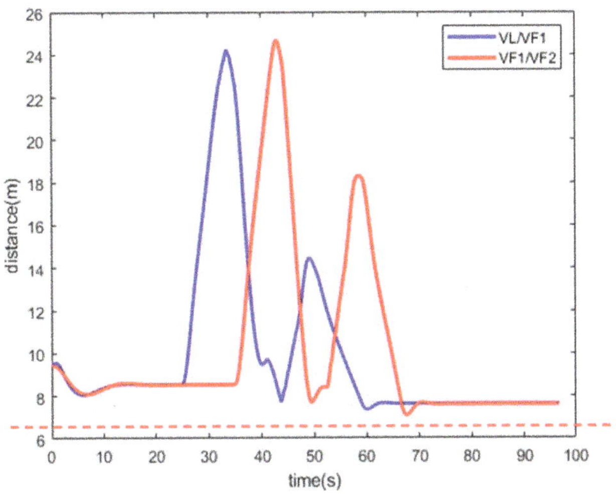

Figure 19. T-type cooperative steering agricultural vehicle distance diagram.

In addition, under the same scenario as the T-turn cooperative turning, a simulation of sequential cooperative steering control is conducted, involving a straight-line segment of 200 m at a forward speed of 10 km/h and a reverse speed of 5 km/h. The steering controller halts the rear machine once the front machine initiates steering and waits until the front vehicle has fully completed the turn. The rear machine begins its turn after the front machine reaches point T4 in Figure 2b. The simulation results are illustrated in Figures 20 and 21. The squares and numerals within the figure carry the same meaning as in the T-turn cooperative steering scenario, representing the location of the machines at various time intervals and the sequential order of these moments, respectively.

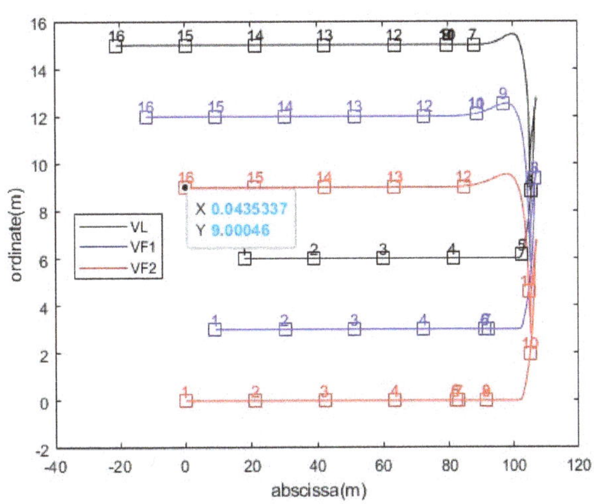

Figure 20. T-turn cooperative steering comparison method trajectory diagram.

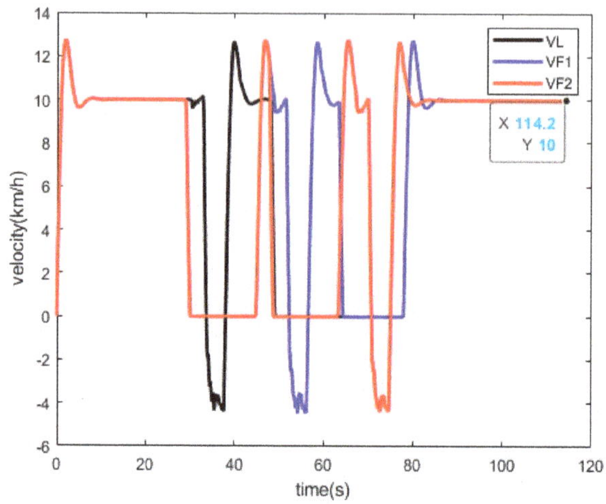

Figure 21. T-turn cooperative steering comparison method longitudinal speed diagram.

The initial coordinates for the comparative methods are the same as those mentioned earlier for the T-turn steering. Similarly, when VF2 completes the operational task, reaching coordinates (0, 9), it is considered that the three agricultural machines have completed the task, and the simulation is terminated, corresponding to a time of 114.2 s. Therefore, it can be considered that the control method proposed in this paper, compared to the sequential control method, reduces the operation time per 200 m by 17.47 s, thereby improving efficiency by 15.29%.

5. Conclusions

Addressing the issue of unnecessary stopping and waiting, as well as reduced operational efficiency due to the sequential control in existing multi-machine cooperative headland steering, this paper proposes a novel method for multi-machine cooperative straight-line and headland steering that accounts for both U-turn and T-turn cooperative maneuvers. The method comprises the design of a path-following controller, a formation-keeping controller, and a headland cooperative steering controller. The simulation results indicate that during U-turn cooperative steering, three agricultural machines can successfully navigate a pre-planned path and maintain formation while collaborating and steering without any risk of collision. In the case of T-turn cooperative steering, the three machines can complete the process without collision risks, unnecessary stopping, or waiting. Compared to sequential control, the operational time is reduced by 17.47 s per 200 m, enhancing efficiency by 15.29%.

However, this paper has some apparent limitations: (1) The kinematic model of agricultural machinery is simplified to a two-wheel model with front-wheel steering, without considering actual position feedback. (2) The cooperative steering control is only addressed for the same model of agricultural machines, neglecting coordinated steering control for machines with different turning radii or operating widths. Further discussions and investigations addressing these limitations will be pursued in subsequent research endeavors.

Author Contributions: Conceptualization, W.Z. and Y.Z.; methodology, W.Z.; software, W.Z.; validation, W.Z., Y.Z. and W.K.; formal analysis, P.J.; investigation, Y.Z.; resources, F.J.; data curation, W.Z.; writing—original draft preparation, W.Z.; writing—review and editing, Y.Z.; visualization, W.Z.; supervision, Y.Z.; project administration, W.K. and F.J.; funding acquisition, W.K. All authors have read and agreed to the published version of the manuscript.

Funding: This research was sponsored by the National Key Research and Development Program of China (2022YFB2503203) and the Beijing Nova Program (20220484040).

Data Availability Statement: The original contributions presented in the study are included in the article, further inquiries can be directed to the corresponding author.

Acknowledgments: The authors would like to thank the National Key Research and Development Program of China (2022YFB2503203) and the Beijing Nova Program (20220484040).

Conflicts of Interest: The authors declare no conflicts of interest.

References

1. Zhang, M.; Ji, Y.; Li, S.; Cao, R.; Xu, H.; Zhang, Z. Research Progress of Agricultural Machinery Navigation Technology. *Trans. Chin. Soc. Agric. Mach.* **2020**, *51*, 1–18.
2. Jin, S.; Liu, G.; Yan, X.; Shi, H.; Cheng, L. Necessity and Application Prospects of Precision Agriculture Development in China. *Zhejiang Agric. Sci.* **2010**, *2*, 414–416. [CrossRef]
3. Hu, J.; Gao, L.; Bai, X.; Li, T.; Liu, X. Review of research on automatic guidance of agricultural vehicles. *Trans. Chin. Soc. Agric. Eng.* **2015**, *31*, 1–10. [CrossRef]
4. Chen, X.; Wen, H.; Zhang, W.; Pan, F.; Zhao, Y. Advances and progress of agricultural machinery and sensing technology fusion. *Smart Agric.* **2020**, *2*, 1–16. [CrossRef]
5. Ju, C.; Kim, J.; Seol, J.; Son, H.I. A review on multirobot systems in agriculture. *Comput. Electron. Agric.* **2022**, *202*, 107336. [CrossRef]
6. Zhu, Z.; Song, Z.; Xie, B.; Chen, J.; Wutian, C.; Mao, E. Automatic Control System of Tractors Platooning. *Trans. Chin. Soc. Agric. Mach.* **2009**, *40*, 149–154.
7. Rigatos, G.G. Derivative-free distributed nonlinear Kalman filtering for cooperating agricultural robots. In Proceedings of the 2013 IEEE International Symposium on Industrial Electronics, Taipei, Taiwan, 28–31 May 2013; pp. 1–6. [CrossRef]
8. Li, S.; Xu, H.; Ji, Y.; Cao, R.; Zhang, M.; Li, H. Development of a following agricultural machinery automatic navigation system. *Comput. Electron. Agric.* **2019**, *158*, 335–344. [CrossRef]
9. Mao, W.; Liu, H.; Hao, W.; Yang, F.; Liu, Z. Development of a Combined Orchard Harvesting Robot Navigation System. *Remote Sens.* **2022**, *14*, 675. [CrossRef]
10. Martin, A.; Dionysis, B.; Claus, G.; Morten, R.; Kasper, L. In-field and inter-field path planning for agricultural transport units. *Comput. Ind. Eng.* **2012**, *63*, 1054–1061. [CrossRef]
11. Conesa-Muñoz, J.; Bengochea-Guevara, J.M.; Andujar, D.; Ribeiro, A. Efficient Distribution of a Fleet of Heterogeneous Vehicles in Agriculture: A Practical Approach to Multi-path Planning. In Proceedings of the 2015 IEEE International Conference on Autonomous Robot Systems and Competitions, Vila Real, Portugal, 8–10 April 2015; pp. 56–61. [CrossRef]
12. Blender, T.; Buchner, T.; Fernandez, B.; Pichlmaier, B.; Schlegel, C. Managing a Mobile Agricultural Robot Swarm for a seeding task. In Proceedings of the IECON 2016-42nd Annual Conference of the IEEE Industrial Electronics Society, Florence, Italy, 23–26 October 2016; pp. 6879–6886. [CrossRef]
13. Bai, X.; Wang, Z.; Hu, J.; Gao, L.; Xiong, F. Harvester Group Corporative Navigation Method Based on Leader-Follower Structure. *Trans. Chin. Soc. Agric. Mach.* **2017**, *48*, 14–21. [CrossRef]
14. Zhang, C.; Noguchi, N. Development of a multi-robot tractor system for agriculture field work. *Comput. Electron. Agric.* **2017**, *142*, 79–90. [CrossRef]
15. Noguchi, N.; Will, J.; Reid, J.; Zhang, Q. Development of a master–slave robot system for farm operations. *Comput. Electron. Agric.* **2004**, *44*, 1–19. [CrossRef]
16. Xu, G.; Chen, M.; Miao, H.; Yao, W.; Diao, P.; Wang, W. Control Method of Agricultural Machinery Master-Slave Following Operation Based on Model Predictive Control. *Trans. Chin. Soc. Agric. Mach.* **2020**, *51*, 11–20. [CrossRef]
17. Wang, Z. Design and Realization on Autonomous Following Control System for Agricultural Vehicles. Master's Thesis, Nanjing Agricultural University, Nanjing, China, 2014.
18. Zhang, C.; Noguchi, N.; Yang, L. Leader–follower system using two robot tractors to improve work efficiency. *Comput. Electron. Agric.* **2016**, *121*, 269–281. [CrossRef]
19. Stavros, G.; Vougioukas, A. distributed control framework for motion coordination of teams of autonomous agricultural vehicles. *Biosyst. Eng.* **2012**, *113*, 284–297. [CrossRef]
20. Zhang, W.; Zhang, Z.; Luo, X.; He, J.; Hu, L.; Yue, B. Position-velocity coupling control method and experiments for longitudinal relative position of harvester and grain truck. *Trans. Chin. Soc. Agric. Eng.* **2021**, *37*, 1–11. [CrossRef]
21. Zheng, X. Master-Slave Cooperative Control Method Harvester-Grain Carrier. Master's Thesis, Shenyang University of Technology, Shenyang, China, 2021. [CrossRef]
22. Abe, G.; Mizushima, A.; Noguchi, N. Mizushima and Noboru Noguchi. Study on a Straight Follower Control Algorithm based on a Laser Scanner. *J. Jpn. Soc. Agric. Mach.* **2005**, *67*, 65–71. [CrossRef]
23. Iida, M.; Kudou, M.; Ono, K.; Umeda, M. Automatic following control for agricultural vehicle. In Proceedings of the 6th International Workshop on Advanced Motion Control. Proceedings (Cat. No.00TH8494), Nagoya, Japan, 30 March–1 April 2000; pp. 158–162. [CrossRef]

24. Zhang, X.; Geimer, M.; Grandl, L.; Kammerbauer, B. Method for an electronic controlled platooning system of agricultural vehicles. In Proceedings of the 2009 IEEE International Conference on Vehicular Electronics and Safety (ICVES), Pune, India, 11–12 November 2009; pp. 156–161. [CrossRef]
25. Zhai, Z.; Wang, X.; Wang, L.; Zhu, Z.; Du, Y.; Mao, E. Collaborative Path Planning for Autonomous Agricultural Machinery of Master-Slave Cooperation. *Trans. Chin. Soc. Agric. Mach.* **2021**, *52*, 542–547. [CrossRef]
26. Hu, Z.; Qin, Q. Algorithm for Finding Minimum Volume Oriented Bounding Boxes Based on Convex Hull. *J. Hunan Univ. Nat. Sci.* **2019**, *46*, 105–111. [CrossRef]
27. Liu, C.; Jiang, X.; Shi, H. Improved Collision Detection Algorithm Based on Oriented Bounding Box. *Comput. Technol. Dev.* **2018**, *28*, 43–48. [CrossRef]

World Electric Vehicle Journal

MDPI

Article

Research on Energy Management Strategy for Series Hybrid Tractor under Typical Operating Conditions Based on Dynamic Programming

Xianghai Yan [1,2], Yifan Zhao [1], Xiaohui Liu [1], Mengnan Liu [2], Yiwei Wu [1,2] and Jingyun Zhang [1,2,*]

1 College of Vehicle and Traffic Engineering, Henan University of Science and Technology, Luoyang 471003, China; 9905167@haust.edu.cn (X.Y.); 220320030288@stu.haust.edu.cn (Y.Z.); 220320030334@stu.haust.edu.cn (X.L.); wyw5156@haust.edu.cn (Y.W.)

2 State Key Laboratory of Intelligent Agricultural Power Equipment, Luoyang 471003, China; liumengnan27@163.com

* Correspondence: 9905070@haust.edu.cn

Abstract: In response to the issues of hybrid tractors' energy management strategies, such as reliance on experience, difficulty in achieving optimal control, and incomplete analysis of typical operating conditions of tractors, an energy management strategy based on dynamic programming is proposed in combination with various typical operating conditions of tractors. This is aimed at providing a reference for the modeling and energy management strategies of series hybrid tractors. Taking the series hybrid tractor as the research object, the tractor dynamics models under three typical working conditions of plowing, rotary tillage, and transportation were established. With the minimum total fuel consumption of the tractor as the optimization target, the engine power as the control variable, and the state of charge of the power battery as the state variable, an energy management strategy based on a dynamic programming algorithm was established and simulation experiments were conducted. The simulation results show that, compared with the power-following energy management strategy, the energy management strategy based on the dynamic programming algorithm can reasonably control the operating state of the engine. Under the three typical working conditions of plowing, rotary tillage, and transportation, the battery SOC consumption increased by approximately 8.37%, 7.24%, and 0.77%, respectively, while the total fuel consumption decreased by approximately 25.28%, 21.54%, and 13.24%, respectively.

Keywords: hybrid tractor; energy management strategy; dynamic programming; minimum total fuel consumption

Citation: Yan, X.; Zhao, Y.; Liu, X.; Liu, M.; Wu, Y.; Zhang, J. Research on Energy Management Strategy for Series Hybrid Tractor under Typical Operating Conditions Based on Dynamic Programming. *World Electr. Veh. J.* **2024**, *15*, 156. https://doi.org/10.3390/wevj15040156

Academic Editor: Peter Van den Bossche

Received: 12 March 2024
Revised: 25 March 2024
Accepted: 4 April 2024
Published: 9 April 2024

1. Introduction

Tractors occupy an important position in the development of modern agriculture and are one of the main types of agricultural power machinery. However, traditional diesel tractors have problems such as high pollution emissions and poor fuel economy [1–3]. Facing increasingly strict emission standards around the world, the agricultural machinery industry is also increasingly urgent in its demand for environmentally friendly and energy-saving agricultural tractors [4,5]. Pure electric tractors can achieve zero pollution, but, due to current battery technology limitations, they have a short continuous operation time and are unable to perform high-load agricultural production for extended periods of time [6,7]. The series hybrid tractor is equipped with an engine and a generator on the basis of the pure electric tractor. By reasonably controlling the operating state of the engine, it can achieve the same power performance as the traditional diesel tractor while reducing fuel consumption [8–10].

The energy management strategy has an important impact on the fuel economy, power performance, and lifespan of the power source of the hybrid tractor [11,12]. Currently, energy management strategies are basically divided into three categories: rule-based energy management strategies, learning-based energy management strategies, and optimization-based energy management strategies [13]. Rule-based energy management strategies are simple to develop and highly feasible, and were the first energy management strategies applied to hybrid vehicles [14]. Chen et al. [15] designed an adaptive fuzzy energy management strategy for extended-range electric vehicles by using BP neural network optimized by an improved genetic algorithm, which effectively improved the fuel economy of the whole vehicle. Yang et al. [16] designed an energy management strategy that combines constant temperature control, power following, and fuzzy rules. This strategy reduces equivalent hydrogen consumption while increasing the lifespan of the system. Zou et al. [17] proposed an energy management strategy for fuel cell hybrid vehicles that utilizes fuzzy logic to optimize the power following control strategy. This control strategy optimizes the output power of the hydrogen fuel cell while reducing hydrogen consumption. However, rule-based energy management strategies usually require a great deal of debugging to determine suitable parameters, relying on the developer's design experience, and it is difficult to achieve optimal control [18].

Learning-based energy management strategies are control strategies with adaptive learning capabilities and good robustness [19]. Xu et al. [20] proposed a supervised learning-based driving cycle pattern recognition method that can accurately predict road conditions and improve the fuel economy of hybrid vehicles. Wu et al. [21] proposed an energy management strategy based on deep deterministic policy gradients, which has near-global optimal dynamic programming performance and can achieve optimal energy allocation for vehicles in continuous spaces. Wang et al. [22] combined computer vision with deep reinforcement learning, enabling the algorithm to autonomously learn the optimal control strategy using visual information collected from on-board cameras, which resulted in reduced fuel consumption and achieved performance at 96.5% of the global optimum dynamic programming. However, learning-based energy management strategies require a large amount of data for training, have high computational requirements, and have relatively complex control strategies [23].

Optimization-based energy management strategies use cost functions as optimization objectives and measure the optimization effect by minimizing the cost function [24]. Zhao et al. [25] proposed an energy management strategy based on the principle of maximizing external energy efficiency, which significantly reduces equivalent hydrogen consumption and improves the overall efficiency of the hybrid tractor. Dou et al. [26] proposed an energy management strategy based on an equivalent fuel consumption minimization algorithm. This strategy can adaptively distribute the required torque based on the load condition, resulting in better fuel consumption compared to rule-based energy management strategies. Curiel-Olivares et al. [27] proposed a model-predictive-based energy management strategy for series hybrid tractors. This strategy outperforms rule-based energy management strategies in terms of fuel consumption while also optimizing the operating state of the battery.

This article takes a series diesel–electric hybrid tractor as the research object and proposes a globally optimal hybrid energy management strategy based on dynamic programming (DP) [28–30]. By reasonably controlling the operating state of the engine and optimizing its output power, it is possible to reduce the total fuel consumption of tractors under three typical working conditions of plowing, tilling, and transportation while ensuring the tractor's power performance. The remainder of this article is organized as follows. Section 2 introduces the topological structure and main performance parameters of the power system for series hybrid tractors. Section 3 explains the simulation models of various components of series hybrid tractors. Section 4 designs two energy management strategies, namely those based on dynamic programming and power following (PF). Section 5 verifies and analyzes the energy management strategies through simulation experiments. Section 6 discusses the results of the simulation experiments and outlines future research directions. Section 7 summarizes the research content and experimental results of this paper.

2. Structural Parameters of Tractor's Power System

Figure 1 shows the topological structure of the power system of a series diesel–electric hybrid tractor. This tractor uses a drive motor as the power source, and the torque output by the drive motor is transmitted to the drive wheels and power takeoff (PTO) through the transmission system. The power battery delivers power to the drive motor through a power converter. When the power battery's charge is insufficient, the engine drives the generator to generate electricity, which is then delivered to the power battery through the power converter to charge the battery. The main component parameters of the series diesel–electric hybrid tractor are shown in Table 1, including the rated power and rated speed of the diesel engine and the drive motor, as well as the rated capacity and rated voltage of the power battery and other specification parameters.

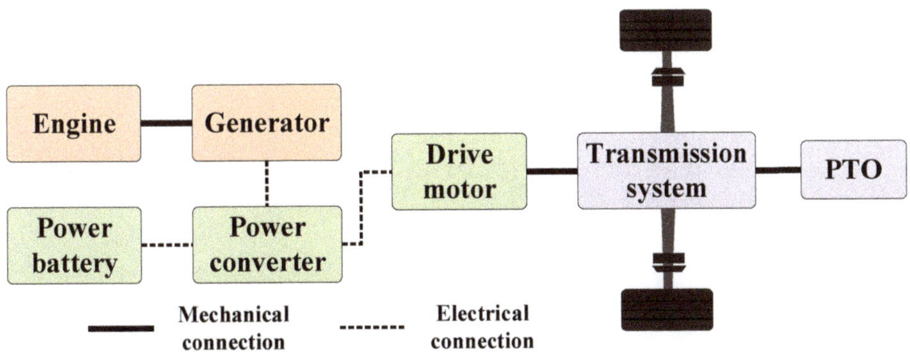

Figure 1. Topological structure diagram of the power system for a series hybrid tractor.

Table 1. Parameters of main components of hybrid tractor.

Component	Parameter	Value (Unit)
Diesel engine	Rated power Rated speed Maximum torque speed	85 (kW) 2300 (r/min) 1500~1700 (r/min)
Drive motor	Rated power Rated speed Rated torque	63 (kW) 2000 (r/min) 300 (N·m)
Power battery	Rated capacity Rated voltage SOC	70 (A·h) 330 (V) 0.25~0.90

3. Hybrid Tractor Model Construction

3.1. Driver Model

Based on the principle of forward modeling, this article uses the difference between the target vehicle speed and the current vehicle speed as input, and the acceleration pedal opening and the brake pedal opening as output to build a driver model based on PI control. The principle of the driver model is shown in the following equation [31]:

$$O_p = k_p e + k_i \int e dt \tag{1}$$

$$e = v_{ref} - v_{act} \tag{2}$$

where k_p is the proportional coefficient, k_i is the integral coefficient; e is the difference between the target velocity of the tractor and the current velocity of the tractor, km/h; O_p is the pedal opening, where $O_p \in (0,1)$ indicates the accelerator pedal opening, and

$O_p \in (-1,0)$ indicates the brake pedal opening; v_{ref} is the target velocity of the tractor, km/h, and v_{act} is the current velocity of the tractor, km/h.

3.2. Generator Set Model

In the structure of a series hybrid system, the engine drives the generator to generate electricity through mechanical connection, and the engine and generator are not connected to the transmission system, relatively independent of the power system of the entire vehicle. Therefore, the engine and generator are usually considered as a whole, namely the generator set. The research on hybrid energy management strategies focuses on analyzing the fuel economy of tractors, so numerical modeling methods are adopted, considering only the input and output relationships of the generator set, as shown in the following equation [32]:

$$P_e = \frac{n_e T_e}{9549} \tag{3}$$

$$P_G = P_e \eta_G \tag{4}$$

where P_e is the engine power, kW; n_e is the engine speed, r/min; T_e is the engine torque, N·m; P_G is the generator set power, kW; η_G is the generator efficiency.

Under the entire set of operating conditions, the total fuel consumption of the engine is represented by the following equation:

$$b_e = f(n_e, T_e) \tag{5}$$

$$E = \int \frac{P_e b_e}{3600 \rho_f} dt \tag{6}$$

where E is the total fuel consumption of the engine, L; b_e is the fuel consumption rate of the engine, g/kWh; and ρ_f is the density of diesel fuel, g/L.

In order to achieve optimal fuel economy for the engine, the engine MAP and optimal operating line (OOL) were fitted using engine bench test data, as shown in Figure 2.

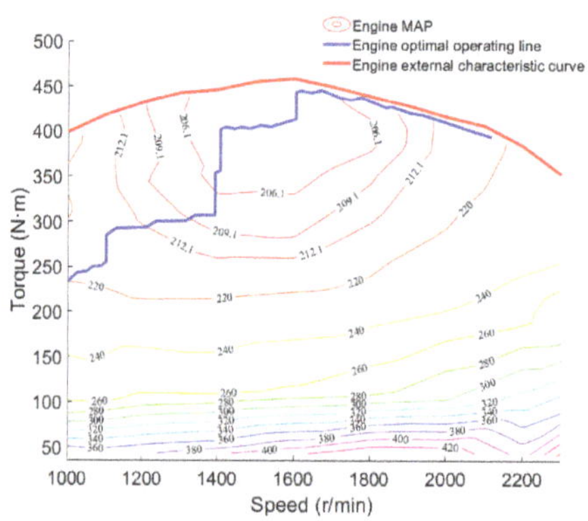

Figure 2. Diesel engine MAP.

Figure 2 includes the engine fuel consumption rate MAP, optimal operating line of the engine, and the external characteristic curve of the engine. The fitted engine characteristic curve in Figure 2 can provide data support for the engine to operate in the optimal state and accurately obtain the current fuel consumption rate of the engine.

3.3. Drive Motor Model

The drive motor model employs a numerical modeling approach. The corresponding relationship between motor speed and motor torque is determined through drive motor bench test data using a look-up table method. The torque control mode is adopted, and the motor output is controlled through the accelerator pedal opening signal. The mathematical modeling principle of the drive motor is shown in Equation (7). Based on the drive motor bench test data, the drive motor's external characteristic curve is obtained, and then the motor efficiency MAP is obtained through interpolation fitting, as shown in Figure 3 [33]:

$$\begin{cases} T_{m\_\max} = f(n_m) \\ T_m = k_{ac} T_{m\_\max} \\ \eta_m = f(n_m, T_m) \\ P_m = \frac{n_m T_m}{9549 \eta_m} \end{cases} \tag{7}$$

where n_m is the drive motor speed, r/min; T_m is the drive motor torque, N·m; $T_{m\_\max}$ is the maximum torque at the current drive motor speed; k_{ac} is the accelerator pedal opening; η_m is the drive motor efficiency; and P_m is the drive motor power, kW.

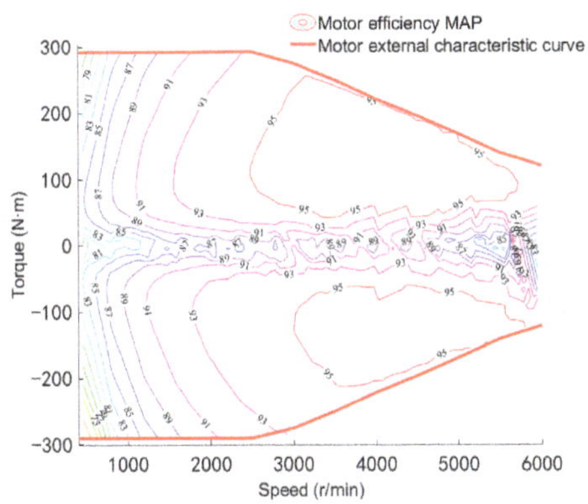

Figure 3. Motor efficiency MAP.

3.4. Transmission System Model

The power source of a series hybrid tractor comes from the drive motor, and the torque generated by the drive motor acts on the drive wheels and the PTO through the transmission system. The transmission system model is shown in the following equation:

$$F_{tr} = \frac{T_m i_g i_0 \eta_T}{R_w} - F_{br} \tag{8}$$

$$F_{br} = k_{br} F_{br\_\max} \tag{9}$$

where F_{tr} is the forward traction force acting on the tractor through the transmission system by the drive motor torque, N; i_g is the gear ratio of the transmission; i_0 is the gear ratio of the final drive; η_T is the efficiency of the transmission system; R_w is the radius of the drive wheel, m; F_{br} is the braking force of the brake, N; k_{br} is the brake pedal opening; $F_{br\_\max}$ is the maximum braking force of the brake, N.

The drive motor speed can also be calculated based on the tractor's current vehicle speed and transmission system parameters, as shown in Equation (10):

$$n_m = \frac{v_{act} i_g i_0}{0.377 R_w} \tag{10}$$

3.5. Tractor Plowing Condition Dynamics Model

Under the plowing condition of a tractor, its driving resistance is mainly determined by the plowing resistance and the rolling resistance, which are calculated as shown in the following equation:

$$F_t = F_{tr} - \left(F_L + F_f\right) \tag{11}$$

$$F_L = Zbhk \tag{12}$$

$$F_f = mgf\cos\alpha \tag{13}$$

where F_t is the driving force, N; F_L is the plowing resistance, N; F_f is the rolling resistance, N; Z is the number of plowshares; b is the width of a single plowshare, cm; h is the plowing depth, cm; k is the specific resistance of the soil, N/cm^2; m is the operating mass of the tractor, kg; g is the acceleration of gravity, m/s^2; f is the rolling resistance coefficient; α is the slope angle, (°).

The current vehicle speed of the tractor can also be calculated based on the driving force, as shown in Equation (14):

$$v_{act} = \int \frac{F_t}{3.6m} dt \tag{14}$$

3.6. Tractor Rotary Tillage Condition Dynamics Model

When performing rotary tillage operations, the series hybrid tractor can neglect the effects of air resistance and acceleration resistance. Due to the complexity of the formula for calculating rotary tillage power and the many influencing factors, this paper uses empirical formulas for calculation. The power balance is shown in the following equation [34,35]:

$$P_m = P_{drive} + P_r \tag{15}$$

$$P_{drive} = \frac{v_{act}\left(F_f + F_i\right)}{3600\eta_T} \tag{16}$$

$$F_i = mg\sin\alpha \tag{17}$$

$$P_r = \frac{3.6kBhv_{act}}{\eta_r} \tag{18}$$

where P_{drive} is the tractor's driving power, kW; F_i is the slope resistance, N; P_r is the power of the rotary tiller, kW; B is the width of the rotary tillage area; η_r is the transmission efficiency of the rotary tiller unit.

3.7. Tractor Transportation Condition Dynamics Model

When a tractor is performing transportation operations, the relationship between the driving force and the driving resistance is balanced as shown in the following equation:

$$F_t = F_{tr} - \left(F_f + F_i + F_{ac} + F_{ar}\right) \tag{19}$$

$$F_{ac} = m\delta a \tag{20}$$

$$F_{ar} = \frac{C_D A v_{act}^2}{21.15} \tag{21}$$

where F_{ac} is the acceleration resistance, N; F_{af} is the air resistance, N; δ is the tractor mass conversion coefficient; a is the tractor acceleration, m/s^2; C_D is the wind resistance coefficient of the tractor; A is the windward area of the tractor, m^2.

3.8. Power Battery Model

In the research on energy management strategies for hybrid electric vehicles, the battery model mainly reflects the interrelationship between the battery power and the power of other power systems in the vehicle. Therefore, this article treats the power battery as an ideal voltage source, ignores the temperature's impact on the battery's voltage, and adopts the Rint model to model the power battery. The dynamic equations for the battery's state of charge (SOC) and battery power are shown in the following equation:

$$\dot{SOC} = -\frac{U_{oc} - \sqrt{U_{oc}^2 - 4R_{\text{int}}P_B}}{2Q_B R_{\text{int}}} \tag{22}$$

$$P_B = \begin{cases} (P_m + P_e)\eta_B, (P_m + P_e) < 0 \\ \frac{(P_m+P_e)}{\eta_B}, (P_m + P_e) > 0 \end{cases} \tag{23}$$

where U_{oc} is the open-circuit voltage of the power battery, V; R_{int} is the internal resistance of the power battery, Ω; P_B is the battery power, kW; Q_B is the rated capacity of the battery, A·h; η_B is the charge and discharge efficiency of the power battery. When (P_m + P_e) is less than 0, the power battery is charging; when (P_m + P_e) is greater than 0, the power battery is discharging.

3.9. Tractor Simulation Model

By studying the working characteristics and structural composition of a series hybrid tractor and combining the modeling requirements of the energy management strategy, a tractor simulation model is built based on Matlab/Simulink. The model includes a driver model, a drive motor model, a transmission system model, a power battery model, tractor (plowing, rotary tillage, and transportation) dynamics models, a generator set model, and an energy management strategy model. The specific model structure is shown in Figure 4. Based on the difference e between the target vehicle speed and the current vehicle speed under the current operating conditions, the driver model outputs the acceleration pedal angle and brake pedal angle (k_{ac}, k_{br}). By controlling the transmission system model, the dynamic model of the tractor under various operating conditions, and the drive motor model through the acceleration pedal angle and brake pedal angle, the drive motor power P_m is obtained. Meanwhile, the power P_G of the generator set model is allocated according to the established control strategy through the energy management strategy model. Finally, the battery power P_B and SOC are calculated through the power battery model.

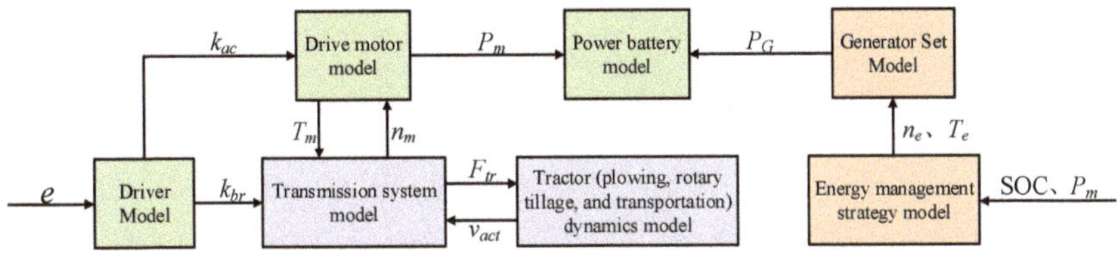

Figure 4. Simplified diagram of tractor simulation model.

4. Energy Management Strategy Design

4.1. Energy Management Strategy Based on Dynamic Programming

4.1.1. Dynamic Programming Energy Management Strategy Model

For a series hybrid tractor, the optimization goal of its energy management strategy is to reasonably control the operating state and output power of the engine so that the tractor achieves the optimal fuel consumption under the current working conditions. The DP algorithm is a multi-stage decision optimization algorithm that divides the multi-stage decisionmaking process into multiple single-stage problems based on the Bellman optimality principle. By defining appropriate control variables, state variables, and objective functions, the DP algorithm uses reverse calculation to solve the multiple single-stage problems to obtain the optimal control.

During the entire set of operating conditions of a tractor, its overall state changes over time. Therefore, when establishing a dynamic programming algorithm, the entire set of operating conditions is divided into N stages with a 1 s interval based on the tractor's operating conditions. State variables reflect the change process of the controlled object. For a series hybrid tractor, the state of charge (SOC) of the power battery can represent the state changes of the tractor under the entire set of operating conditions. Therefore, the SOC of the power battery is selected as the state variable. During the operation of the tractor, the main factor affecting the change in the SOC of the power battery is the output power of the engine. Therefore, the engine power is selected as the control variable.

The state variables and control variables are discretized as shown in Equation (24):

$$\begin{cases} x(k) = [SOC(k)]^T \\ SOC(k) \in \{SOC_1, SOC_2, \cdots, SOC_N\} \\ u(k) = [P_e(k)]^T \\ u_m(k) \in \{u_{m1}, u_{m2}, \cdots, u_{mj}\} \end{cases} \tag{24}$$

where N is the dimensions of the discrete space; j is the number of discrete points.

From Equation (22), the state transition equation can be obtained as

$$SOC(k+1) = SOC(k) + \frac{U_{oc}(t) - \sqrt{U_{oc}^2(t) - 4R_{\text{int}}(t)P_B(t)}}{2Q_B R_{\text{int}}(t)} \tag{25}$$

Taking the total fuel consumption of the engine under the entire set of operating conditions as the optimization objective, the optimization objective function can be obtained from Equation (6) as

$$J = \min \sum_{k=0}^{N-1} E(k) \tag{26}$$

To ensure that all components of the tractor operate within a reasonable range, the following constraints are added:

$$\begin{cases} SOC_{\min} \le SOC(k) \le SOC_{\max} \\ P_{B\_\min} \le P_B(k) \le P_{B\_\max} \\ P_{e\_\min} \le P_e(k) \le P_{e\_\max} \\ n_{e\_\min} \le n_e(k) \le n_{e\_\max} \\ T_{e\_\min} \le T_e(k) \le T_{e\_\max} \end{cases} \tag{27}$$

where $SOC_{\min}$ and $SOC_{\max}$ are the minimum and maximum allowable values for the SOC of the power battery; $P_{B\_\min}$ and $P_{B\_\max}$ are the minimum and maximum power of the power battery during operation; $P_{e\_\min}$, $P_{e\_\max}$, $n_{e\_\min}$, $n_{e\_\max}$, $T_{e\_\min}$, and $T_{e\_\max}$ are the minimum and maximum power, minimum and maximum speed, and minimum and maximum torque of the engine during operation, respectively.

4.1.2. The Solution Process of Dynamic Programming Algorithm

The energy management strategy based on dynamic programming optimizes the operating state and output power of the engine throughout the entire set of working conditions using the dynamic programming algorithm, aiming to achieve the best fuel economy for the tractor. The solution process of dynamic programming is shown in Figure 5.

The specific steps are as follows:

1. The solution process is divided into N stages based on the operating conditions. The operational parameters of each component of the tractor at each stage are calculated using the tractor's mathematical model, and all control variables are solved. To further optimize the fuel consumption of the tractor, all control variables, including the calculated engine power P_e, corresponding engine speed ne, and engine torque T_e, are selected from the OOL fitted in Figure 2.
2. Select control variables that satisfy the constraints of the current stage and use dynamic programming to solve for the state variables and control variable parameter values that yield the minimum value of the objective function J for that stage.
3. Let $N = N - 1$, which enters the next stage of the solution operation. This process continues until $k = 0$, at which point the optimal control variables and corresponding SOC dataset are obtained, and the solution process is complete.

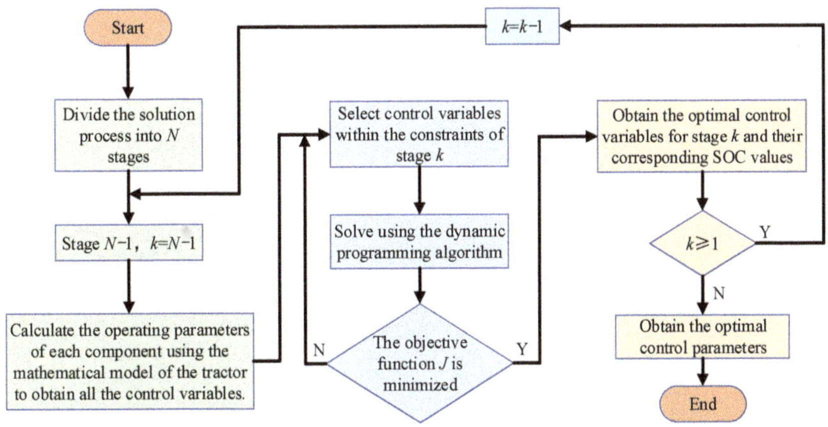

Figure 5. The solution process of dynamic programming.

4.2. Energy Management Strategy Based on Power Following

The power following-based energy management strategy is a rule-based control strategy. In a series hybrid tractor, this strategy uses the demand power of the drive motor and the remaining battery charge of the power battery to determine the start–stop and output power of the engine. The principle of this strategy is shown in Figure 6.

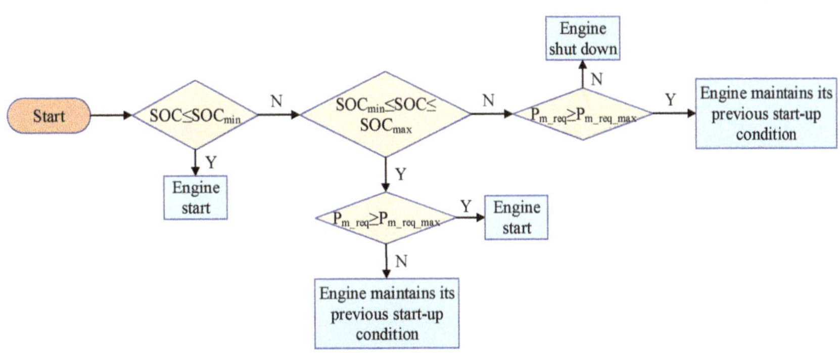

Figure 6. Power following schematic diagram.

Where P_{m_req} is the required power of the drive motor; $P_{m\_req\_max}$ is the maximum required power of the drive motor.

The specific steps are as follows:

1. When SOC $\leq$ SOC_{min}, the engine starts.
2. When SOC_{min} $\leq$ SOC $\leq$ SOC_{max}, if $P_{m\_req} \geq P_{m\_req\_max}$, the engine starts; otherwise, the engine maintains the started state from the previous moment.
3. When SOC $\geq$ SOC_{max}, if $P_{m\_req} \geq P_{m\_req\_max}$, the engine maintains the started state from the previous moment; otherwise, the engine shuts down.

To better compare the two control strategies, the engine control parameters output by the power following-based energy management strategy will also be selected from the OOL fitted in Figure 2.

5. Simulation and Results Analysis

5.1. Plowing Condition

As shown in Figure 7, the target vehicle speed tracking effect of the simulation model under the plowing condition is demonstrated. The results indicate that, under the plowing condition, the simulation model can effectively track the target vehicle speed with a maximum error of no more than 0.24 km/h, meeting the test requirements.

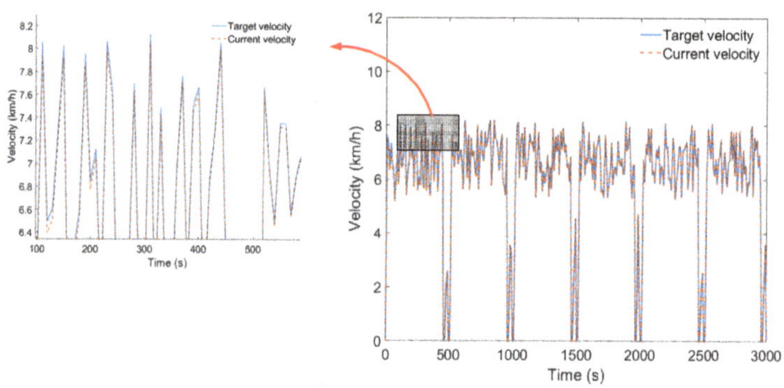

Figure 7. Vehicle speed tracking effect under plowing condition.

When the tractor performs plowing operations, the changes in drive motor power and battery power under two energy management strategies are shown in Figure 8. As can be observed from Figure 8a, under the plowing condition, the peak power of the drive motor is approximately 49.66 kW. As can be observed from Figure 8b, under the energy management strategy based on power following, the battery power will have negative values as the load of the plowing condition increases, indicating that the generator set will be activated to charge the battery during each plowing cycle. Under the energy management strategy based on dynamic programming, the battery power remains positive for approximately the first 2118 s. After approximately 2118 s, the battery power begins to decrease, at which point the generator set starts to operate and charge the battery. During the periods from approximately 2471 s to 2499 s and from 2971 s to 2999 s, the battery exhibits negative power, indicating that the entire tractor's load power is being supplied by the generator set.

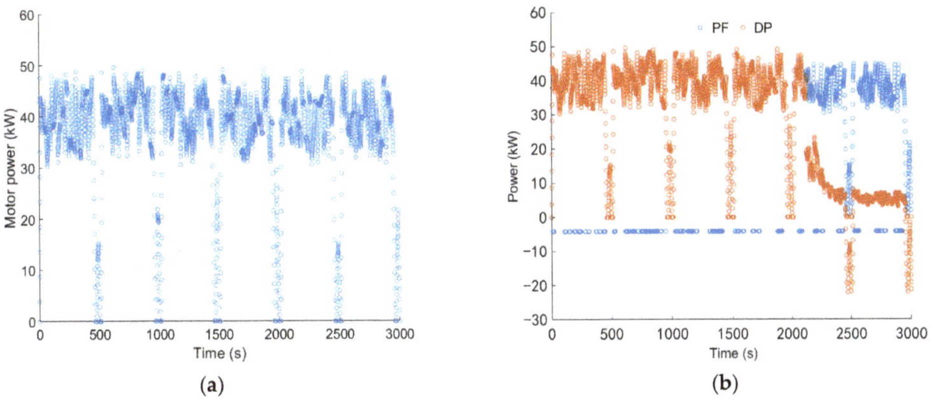

Figure 8. (**a**) Drive motor power; (**b**) battery power.

The operating state of the engine under the plowing condition is shown in Figure 9. As can be observed from Figure 9a, under the energy management strategy based on power following, the engine starts and stops multiple times throughout the entire plowing condition, with relatively short continuous operating times, and the peak power of the generator set is approximately 53.76 kW. On the other hand, under the energy management strategy based on dynamic programming, the engine starts to operate around 2118 s, and it does not frequently start and stop, with relatively concentrated operating times and a peak power of approximately 41.03 kW. As shown in Figure 9b, under both energy management strategies, the engine operates along the optimal operating curve, but, under the energy management strategy based on dynamic programming, the engine operates within a wider range.

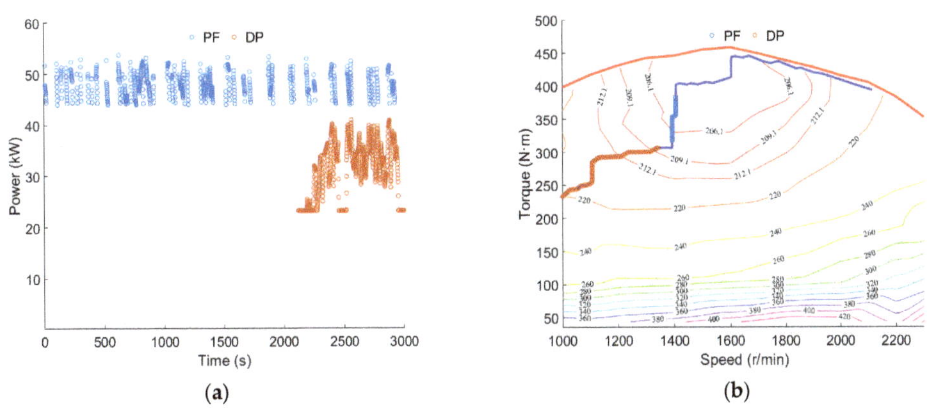

Figure 9. (**a**) Engine power; (**b**) engine operating point.

As can be observed from Figure 10a, the final SOC under the energy management strategy based on power following is approximately 50.54%, while the final SOC under the energy management strategy based on dynamic programming is approximately 46.31%. Compared to the energy management strategy based on power following, the battery SOC consumed approximately 8.37% more under the energy management strategy based on dynamic programming.

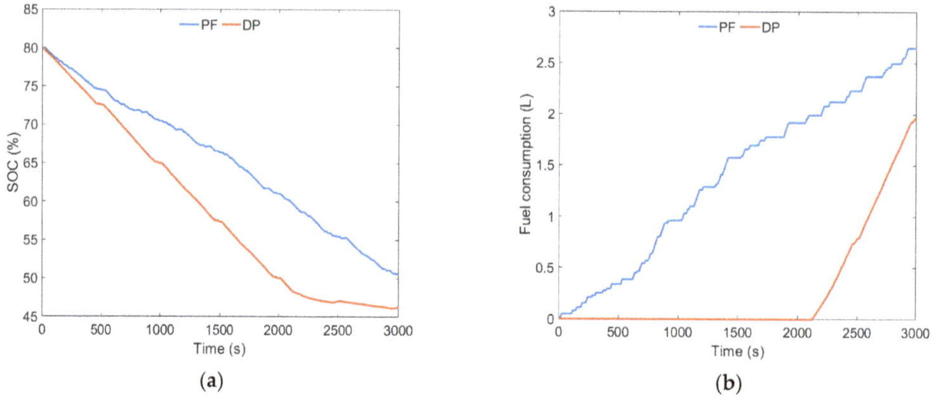

Figure 10. (**a**) SOC change curve; (**b**) fuel consumption change curve.

As shown in Figure 10b, the total fuel consumption under the energy management strategy based on power following is 2.65 L, while the total fuel consumption under the energy management strategy based on dynamic programming is 1.98 L. With the energy management strategy based on dynamic programming, the total fuel consumption is reduced by approximately 25.28%.

5.2. Rotary Tillage Condition

As shown in Figure 11, the target vehicle speed tracking effect of the simulation model under the rotary tillage condition is demonstrated. The results indicate that, under the rotary tillage condition, the simulation model can effectively track the target vehicle speed with a maximum error of no more than 0.22 km/h, meeting the test requirements.

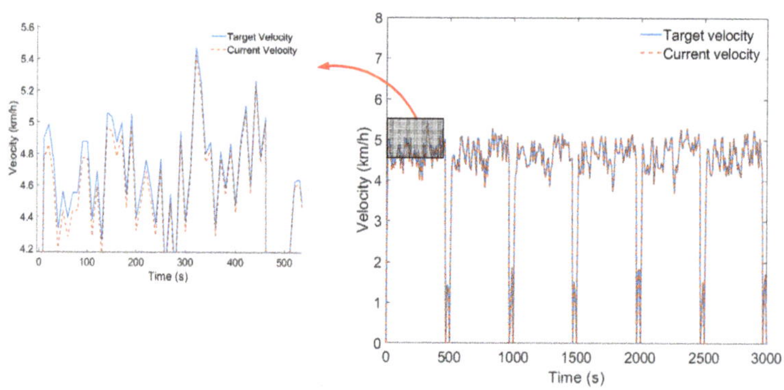

Figure 11. Vehicle speed tracking effect under rotary tillage condition.

When the tractor performs rotary tillage operations, the changes in drive motor power and battery power under two energy management strategies are shown in Figure 12. As can be observed from Figure 12a, under the rotary tillage condition, the peak power of the drive motor is approximately 46.10 kW. As can be observed from Figure 12b, under the energy management strategy based on power following, the trend regarding battery power changes is basically consistent with that of the plowing condition. In each rotary tillage cycle, the battery exhibits negative power, causing the generator set to start and charge the battery. Under the energy management strategy based on dynamic programming, the battery power remains positive for approximately the first 2145 s. After approximately 2145 s, the battery power begins to decrease, at which point the generator set starts to operate and charge the battery. During the periods from approximately 2464 s to 2506 s and from 2963 s to 3000 s, the battery exhibits negative power, indicating that the entire tractor's load power is being supplied by the generator set.

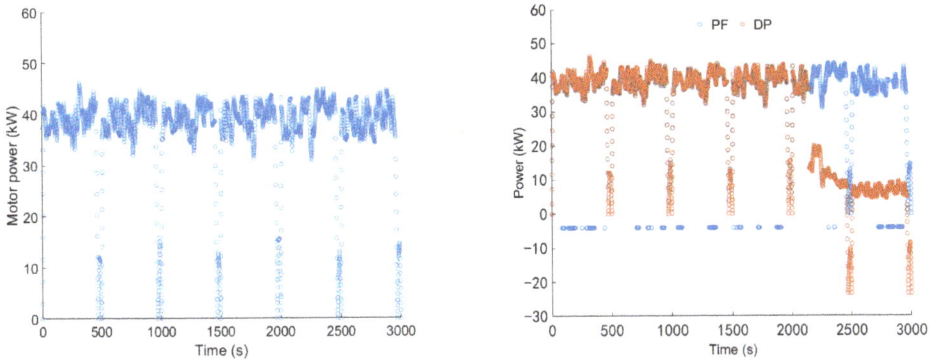

(a) (b)

Figure 12. (**a**) Drive motor power; (**b**) battery power.

The operating state of the engine under the rotary tillage condition is shown in Figure 13. As can be observed from Figure 13a, under the energy management strategy based on power following, the number of starts and stops of the engine during the entire rotary tillage condition has decreased compared to the plowing condition, but they remain relatively frequent, with short continuous operating times, and the peak power of the generator set is approximately 50.17 kW. On the other hand, under the energy management strategy based on dynamic programming, the engine starts to operate around 2145 s and continues to operate until the end of the rotary tillage operation, with relatively concentrated operating times and a peak power of approximately 35.39 kW. As shown in Figure 13b, under both energy management strategies, the engine operates along the optimal operating curve. However, under the energy management strategy based on dynamic programming, the engine operates within a wider range.

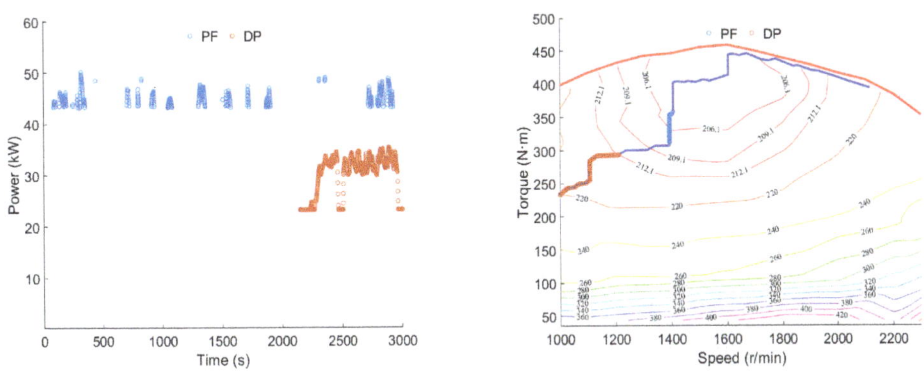

(a) (b)

Figure 13. (**a**) Engine power; (**b**) engine operating point.

As shown in Figure 14a, the final SOC under the power-following energy management strategy is approximately 48.76%, while the final SOC under the dynamic programming-based energy management strategy is approximately 45.23%. Compared to the power-following energy management strategy, the battery SOC consumed approximately 7.24% more under the dynamic programming-based energy management strategy.

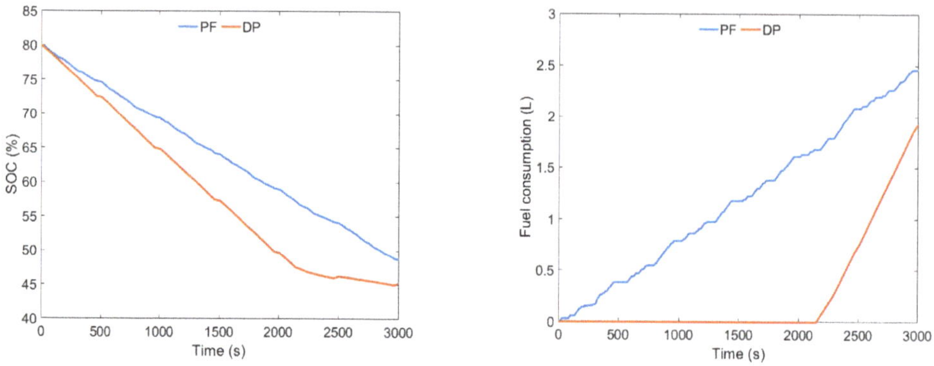

(a) (b)

Figure 14. (**a**) SOC change curve; (**b**) fuel consumption change curve.

As shown in Figure 14b, the total fuel consumption under the energy management strategy based on power following is 2.46 L, while the total fuel consumption under the energy management strategy based on dynamic programming is 1.93 L. With the energy management strategy based on dynamic programming, the total fuel consumption is reduced by approximately 21.54%.

5.3. Transportation Condition

The tractor transportation condition refers to the EUDC_Man driving cycle, and, based on the tractor powertrain system parameters, the maximum vehicle speed has been adjusted to 25.5 km/h. As shown in Figure 15, the target vehicle speed tracking effect of the simulation model under the transportation condition is demonstrated. The results indicate that, under the transportation condition, the simulation model can effectively track the target vehicle speed with a maximum error of no more than 0.27 km/h, meeting the test requirements.

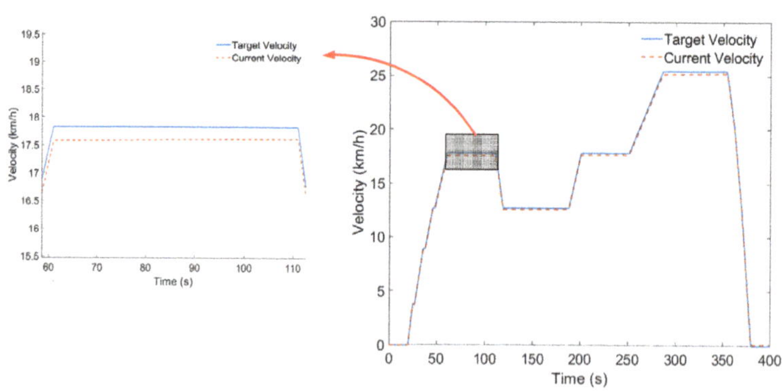

Figure 15. Vehicle speed tracking effect under transportation condition.

During the transportation operation of the tractor, the changes in the drive motor power and battery power under the two energy management strategies are shown in Figure 16. As can be observed from Figure 16a, under the transportation condition, the peak power of the drive motor is approximately 112.40 kW. As shown in Figure 16b, the trends in battery power changes are basically consistent under both the power-following energy management strategy and the dynamic programming-based energy management strategy, and no negative battery power occurs. Under the power-following energy management strategy, the generator set operates and the battery power decreases during approximately

60 s to 118 s and 200 s to 368 s. Under the dynamic programming-based energy management strategy, the generator set operates and the battery power decreases during approximately 53 s to 113 s and 197 s to 362 s.

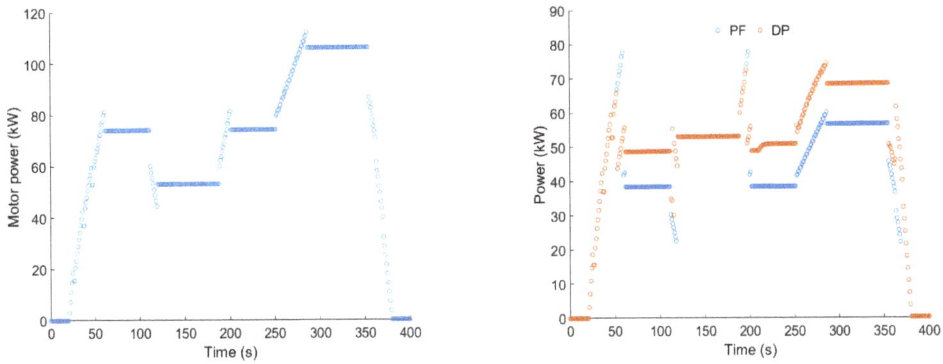

(a) (b)

Figure 16. (**a**) Drive motor power; (**b**) battery power.

The operating state of the engine under transportation conditions is shown in Figure 17. As can be observed from Figure 17a, under both the power-following energy management strategy and the dynamic programming-based energy management strategy, the number of starts and stops of the engine during the entire transportation condition is the same. Under the power-following energy management strategy, the peak power of the generator set is approximately 52.00 kW. Under the dynamic programming-based energy management strategy, the peak power of the generator set is approximately 37.74 kW. As shown in Figure 17b, under both energy management strategies, the engine operates along the optimal operating curve. However, under the dynamic programming-based energy management strategy, the engine operates within a wider range.

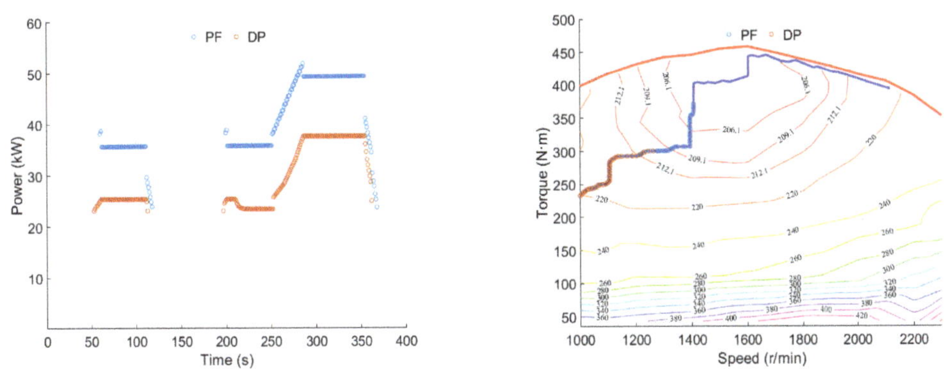

(a) (b)

Figure 17. (**a**) Engine power; (**b**) engine operating point.

As shown in Figure 18a, the final SOC under the power-following energy management strategy is approximately 73.01%, while the final SOC under the dynamic programming-based energy management strategy is approximately 72.45%. Compared to the power-following energy management strategy, the battery SOC consumed approximately 0.77% more under the dynamic programming-based energy management strategy.

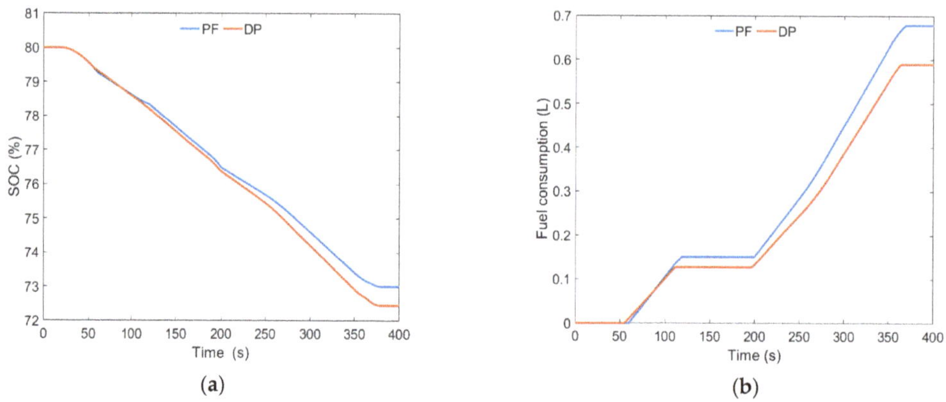

Figure 18. (**a**) SOC change curve; (**b**) fuel consumption change curve.

As shown in Figure 18b, the total fuel consumption under the power-following energy management strategy is 0.68 L, while the total fuel consumption under the dynamic programming-based energy management strategy is 0.59 L. The total fuel consumption decreased by approximately 13.24% under the dynamic programming-based energy management strategy.

6. Discussion

The results of this study emphasize the impact of the proposed energy management strategy for hybrid tractors on fuel economy under various typical operating conditions of tractors. Using the dynamic programming algorithm, the operating status of the engine in a series hybrid tractor was optimized under three typical operating conditions: plowing, rotary tilling, and transportation. Based on the simulation test results, this study provides a comprehensive reference for future related research. The main findings of the discussion are as follows.

The optimization effect of the energy management strategy based on the dynamic programming algorithm is closely related to the operating conditions of the tractor. By analyzing the simulation test results of three typical operating conditions, namely plowing, rotary tilling, and transportation, it is found that, the larger the load of the operating condition, the better the fuel-saving effect of the dynamic programming algorithm compared to the power-following energy management strategy.

According to the simulation test results, the power-following energy management strategy leads to frequent engine starts and stops in both plowing and rotary tilling test conditions. In actual tractor operation, frequent engine starts and stops can further increase fuel consumption. However, the dynamic programming-based energy management strategy does not exhibit frequent engine starts and stops. Future research on energy management strategies should also consider the issue of engine starts and stops.

To better compare the control effects of energy management strategies, it is necessary to consider not only fuel consumption but also the cost impact of battery power consumption. In the simulation tests conducted in this study, the dynamic programming-based energy management strategy optimizes the operating state of the engine based on the entire working conditions. Under the same working conditions, it is difficult to achieve the same final SOC value as the power-following energy management strategy, which has a certain impact on testing the optimization effect of the energy management strategy. Therefore, future research work should also consider issues related to electricity costs.

7. Conclusions

This study describes an energy management strategy for a series hybrid tractor, aiming to achieve optimal fuel consumption throughout the entire operating cycle of the tractor.

Firstly, the SOC of the power battery is considered as the state variable, and the engine power is the control variable. Then, the total fuel consumption of the engine throughout the entire set of operating conditions is taken as the objective function. Finally, a series hybrid tractor energy management strategy based on a dynamic programming algorithm is designed. The main conclusions are as follows.

Under the conditions of plowing, rotary tillage, and transportation operations, the total fuel consumption values for the power following-based energy management strategy are 2.65 L, 2.46 L, and 0.68 L, respectively. For the dynamic programming-based energy management strategy, the total fuel consumption values are 1.98 L, 1.93 L, and 0.59 L, respectively. Compared to the power-following energy management strategy, the dynamic programming-based energy management strategy results in an additional consumption of approximately 8.37%, 7.24%, and 0.77% in battery SOC for plowing, rotary tilling, and transportation operations, respectively. Simultaneously, the total fuel consumption of the tractor decreases by approximately 25.28%, 21.54%, and 13.24% for the respective operations.

In this paper, the total workload of the tractor during plowing operations is the highest, followed by rotary tillage operations, and transportation operations have the lowest workload. The trend of total workload change is consistent with the effect of reduced total fuel consumption observed in the simulation results. Specifically, the greater the total workload, the better the fuel-saving effect achieved by the dynamic programming-based energy management strategy compared to the power following-based strategy.

Compared to the power following-based energy management strategy, the dynamic programming-based strategy can better adjust the operating state of the engine, reasonably control the start–stop and output power of the engine, and keep the engine operating in a high-efficiency range.

Author Contributions: Conceptualization, X.Y. and J.Z.; methodology, X.Y. and J.Z.; software, Y.Z.; validation, Y.Z. and X.L.; formal analysis, Y.Z.; investigation, X.Y. and X.L.; resources, X.Y.; data curation, M.L. and Y.W.; writing—original draft preparation, J.Z. and Y.Z.; writing—review and editing, X.Y. and Y.Z.; visualization, Y.Z.; supervision, X.Y.; project administration, X.Y.; funding acquisition, J.Z. and X.Y. All authors have read and agreed to the published version of the manuscript.

Funding: This research has been awarded the "14th Five-Year" national key research and development Plan (2022YFD2001203, 2022YFD2001201B); Agricultural Key Core Technology GG project NK202216010103. This research was funded by the State Key Laboratory of Intelligent Agricultural Power Equipment Open Project (SKLIAPE2023006), Henan University of Science and Technology Innovation Team Support Program (24IRTSTHN029), and Henan Provincial Science and Technology Research Project (222102110233).

Data Availability Statement: The original contributions presented in the study are included in the article, further inquiries can be directed to the corresponding author.

Conflicts of Interest: The authors declare no conflicts of interest.

References

1. Mocera, F.; Somà, A.; Martelli, S.; Martini, V. Trends and Future Perspective of Electrification in Agricultural Tractor-Implement Applications. *Energies* **2023**, *16*, 6601. [CrossRef]
2. Wang, Z.; Zhou, J.; Yang, H.; Wang, X. Design and Test of Measurement and Control System for Rapid Prototype Platform Used in Electric Tractors. *Nongye Jixie Xuebao Trans. Chin. Soc. Agric. Mach.* **2022**, *53*, 412–420. [CrossRef]
3. Scolaro, E.; Beligoj, M.; Estevez, M.P.; Alberti, L.; Renzi, M.; Mattetti, M. Electrification of Agricultural Machinery: A Review. *IEEE Access* **2021**, *9*, 164520–164541. [CrossRef]
4. Mocera, F.; Martini, V.; Soma, A. Comparative Analysis of Hybrid Electric Architectures for Specialized Agricultural Tractors. *Energies* **2022**, *15*, 1944. [CrossRef]
5. Xie, B.; Wu, Z.; Mao, E. Development and Prospect of Key Technologies on Agricultural Tractor. *Nongye Jixie Xuebao Trans. Chin. Soc. Agric. Mach.* **2018**, *49*, 1–17. [CrossRef]
6. Chen, Y. Study on Design and Drive Control Methods of Powertrain for Electric Tractor. Ph.D. Thesis, China Agricultural University, Beijing, China, 2018.
7. Radrizzani, S.; Panzani, G.; Trezza, L.; Pizzocaro, S.; Savaresi, S.M. An Add-On Model Predictive Control Strategy for the Energy Management of Hybrid Electric Tractors. *IEEE Trans. Veh. Technol.* **2024**, *73*, 1918–1930. [CrossRef]

8. Medevepryt, U.K.; Makaras, R.; Lukoeviius, V.; Kerys, A. Evaluation of the Working Parameters of a Series-Hybrid Tractor under the Soil Work Conditions. *Teh. Vjesn.* **2022**, *29*, 45–50. [CrossRef]
9. Wang, B.; Qiao, M.; Chu, X.; Shang, S.; Wang, D. Design and Experiment on Extended-Range Electric Caterpillar Tractor. *Nongye Jixie Xuebao Trans. Chin. Soc. Agric. Mach.* **2023**, *54*, 431–439. [CrossRef]
10. Medževeprytė, U.K.; Makaras, R.; Lukoševičius, V.; Kilikevičius, S. Application and Efficiency of a Series-Hybrid Drive for Agricultural Use Based on a Modified Version of the World Harmonized Transient Cycle. *Energies* **2023**, *16*, 5379. [CrossRef]
11. Zhu, Z.; Zeng, L.; Chen, L.; Zou, R.; Cai, Y. Fuzzy Adaptive Energy Management Strategy for a Hybrid Agricultural Tractor Equipped with HMCVT. *Agriculture* **2022**, *12*, 1986. [CrossRef]
12. Li, T.; Cui, W.; Cui, N. Soft Actor-Critic Algorithm-Based Energy Management Strategy for Plug-In Hybrid Electric Vehicle. *World Electr. Veh. J.* **2022**, *13*, 193. [CrossRef]
13. He, H.; Meng, X. A Review on Energy Management Technology of Hybrid Electric Vehicles. *Beijing Ligong Daxue Xuebao Trans. Beijing Inst. Technol.* **2022**, *42*, 773–783. [CrossRef]
14. Yang, C.; Zha, M.; Wang, W.; Liu, K.; Xiang, C. Efficient Energy Management Strategy for Hybrid Electric Vehicles/Plug-in Hybrid Electric Vehicles: Review and Recent Advances under Intelligent Transportation System. *IET Intell. Transp. Syst.* **2020**, *14*, 702–711. [CrossRef]
15. Chen, Y.; Wei, C.; Li, X.; Li, Y.; Liu, C.; Lin, X. Research on Fuzzy Energy Management Strategy for Extended-Range Electric Vehicles with Driving Condition Identification. *Qiche Gongcheng Automot. Eng.* **2022**, *44*, 514–524+600. [CrossRef]
16. Yang, H.; Sun, Y.; Xia, C.; Zhang, H. Research on Energy Management Strategy of Fuel Cell Electric Tractor Based on Multi-Algorithm Fusion and Optimization. *Energies* **2022**, *15*, 6389. [CrossRef]
17. Zou, K.; Luo, W.; Lu, Z. Real-Time Energy Management Strategy of Hydrogen Fuel Cell Hybrid Electric Vehicles Based on Power Following Strategy–Fuzzy Logic Control Strategy Hybrid Control. *World Electr. Veh. J.* **2023**, *14*, 315. [CrossRef]
18. Zhu, Z.; Zeng, L.; Lin, Y.; Chen, L.; Zou, R.; Cai, Y. Adaptive Energy Management Strategy for Hybrid Tractors Based on Condition Prediction. *Hsi-Chiao Tung Ta Hsueh J. Xian Jiaotong Univ.* **2023**, *57*, 201–210. [CrossRef]
19. Cao, Y.; Yao, M.; Sun, X. An Overview of Modelling and Energy Management Strategies for Hybrid Electric Vehicles. *Appl. Sci.* **2023**, *13*, 5947. [CrossRef]
20. Xu, B.; Shi, J.; Li, S.; Li, H. A Study of Vehicle Driving Condition Recognition Using Supervised Learning Methods. *IEEE Trans. Transp. Electrif.* **2022**, *8*, 1665–1673. [CrossRef]
21. Wu, Y.; Tan, H.; Peng, J.; Zhang, H.; He, H. Deep Reinforcement Learning of Energy Management with Continuous Control Strategy and Traffic Information for a Series-Parallel Plug-in Hybrid Electric Bus. *Appl. Energy* **2019**, *247*, 454–466. [CrossRef]
22. Wang, Y.; Tan, H.; Wu, Y.; Peng, J. Hybrid Electric Vehicle Energy Management with Computer Vision and Deep Reinforcement Learning. *IEEE Trans. Ind. Inform.* **2021**, *17*, 3857–3868. [CrossRef]
23. Yu, P.; Li, M.; Wang, Y.; Chen, Z. Fuel Cell Hybrid Electric Vehicles: A Review of Topologies and Energy Management Strategies. *World Electr. Veh. J.* **2022**, *13*, 172. [CrossRef]
24. Zhu, Y.; Li, X.; Liu, Q.; Li, S.; Xu, Y. Review Article: A Comprehensive Review of Energy Management Strategies for Hybrid Electric Vehicles. *Mech. Sci.* **2022**, *13*, 147–188. [CrossRef]
25. Zhao, S.; Gao, Z.; Li, X.; Li, Y.; Xu, L. Research on Energy Management Strategy of Fuel Cell Tractor Hybrid Power System. *World Electr. Veh. J.* **2024**, *15*, 61. [CrossRef]
26. Dou, H.; Wei, H.; Ai, Q.; Zhang, Y. Optimal Energy Management Strategy for Dual-Power Coupling Tractor Based on the Adaptive Control Technology. *PLoS ONE* **2023**, *18*, e0292510. [CrossRef] [PubMed]
27. Curiel-Olivares, G.; Johnson, S.; Escobar, G.; Schacht-Rodriguez, R. Model Predictive Control-Based Energy Management System for a Hybrid Electric Agricultural Tractor. *IEEE Access* **2023**, *11*, 118801–118811. [CrossRef]
28. Du, C.; Huang, S.; Jiang, Y.; Wu, D.; Li, Y. Optimization of Energy Management Strategy for Fuel Cell Hybrid Electric Vehicles Based on Dynamic Programming. *Energies* **2022**, *15*, 4325. [CrossRef]
29. Pan, C.; Liang, Y.; Chen, L.; Chen, L. Optimal Control for Hybrid Energy Storage Electric Vehicle to Achieve Energy Saving Using Dynamic Programming Approach. *Energies* **2019**, *12*, 588. [CrossRef]
30. Jin, H.; Zhang, Z. Review of Research on HEV Energy Management Based on Adaptive Dynamic Programming. *Qiche Gongcheng Automot. Eng.* **2020**, *42*, 1490–1496. [CrossRef]
31. Wu, K. Simulation and Analysis of the Performance and Energy Management Strategies for Extended-Range Electric Tractors. Master's Thesis, Northwest A&F University, Xianyang, China, 2022.
32. Wang, Z.; Zhou, J.; Wang, X. Research on Energy Management Model for Extended-Range Electric Rotary-Tilling Tractor. *Nongye Jixie Xuebao Trans. Chin. Soc. Agric. Mach.* **2023**, *54*, 428–438. [CrossRef]
33. Li, T.; Xie, B.; Wang, D.; Zhang, S.; Wu, L. Real-Time Adaptive Energy Management Strategy for Dual-Motor-Driven Electric Tractors. *Nongye Jixie Xuebao Trans. Chin. Soc. Agric. Mach.* **2020**, *51*, 530–543. [CrossRef]
34. Zhou, R. Study on Energy Distribution Strategy of Tandem Hybrid Tractor under Different Working Conditions. Master's Thesis, Nanjing Agricultural University, Nanjing, China, 2020.
35. Dou, H.; Wei, H.; Ai, Q.; Zhang, Y. Control Strategy for Rotary Tillage Condition of Hybrid Electric Tractor with Coupled-Split Dynamic Configuration. *Nongye Jixie Xuebao Trans. Chin. Soc. Agric. Mach.* **2024**, *55*, 393–400+414. [CrossRef]

World Electric Vehicle Journal

MDPI

Article

Research on Power Optimization for Energy System of Hydrogen Fuel Cell Wheel-Driven Electric Tractor

Jingyun Zhang [1], Buyuan Wang [1], Junjiang Zhang [1,2], Liyou Xu [1,*] and Kai Zhang [3]

1 College of Vehicle and Traffic Engineering, Henan University of Science and Technology, Luoyang 471003, China; 9905070@haust.edu.cn (J.Z.); 210321030255@stu.haust.edu.cn (B.W.); 9906179@haust.edu.cn (J.Z.)

2 State Key Laboratory of Intelligent Agricultural Power Equipment, Luoyang 471039, China

3 YTO Group Corporation, Luoyang 471004, China; haitop0228@gmail.com

* Correspondence: xlyou@haust.edu.cn; Tel.: +86-136-638-73262

Abstract: Hydrogen fuel cell tractors are emerging as a new power source for tractors. Currently, there is no mature energy management control method available. Existing methods mostly rely on engineers' experience to determine the output power of the fuel cell and the power battery, resulting in relatively low energy utilization efficiency of the energy system. To address the aforementioned problems, a power optimization method for the energy system of hydrogen fuel cell wheel-driven electric tractor was proposed. A dynamic model of tractor ploughing conditions was established based on the system dynamics theory. From this, based on the equivalent hydrogen consumption theory, the charging and discharging of the power battery were equivalent to the fuel consumption of the hydrogen fuel cell, forming an equivalent hydrogen consumption model for the tractor. Using the state of charge (SOC) of the power battery as a constraint, and with the minimum equivalent hydrogen consumption as the objective function, an instantaneously optimized power allocation method based on load demand in the energy system is proposed by using a traversal algorithm. The optimization method was simulated and tested based on the MATLAB simulation platform, and the results showed under ploughing conditions, compared with the rule-based control strategy, the proposed energy system power optimization method optimized the power output of hydrogen fuel cells and power batteries, allowing the energy system to work in a high-efficiency range, reducing the equivalent hydrogen consumption of the tractor by 7.79%, and solving the energy system power distribution problem.

Keywords: hydrogen fuel cell; electric tractor; equivalent hydrogen consumption; energy system; power optimization

Citation: Zhang, J.; Wang, B.; Zhang, J.; Xu, L.; Zhang, K. Research on Power Optimization for Energy System of Hydrogen Fuel Cell Wheel-Driven Electric Tractor. *World Electr. Veh. J.* **2024**, *15*, 188. https://doi.org/10.3390/wevj15050188

Academic Editors: Fachao Jiang, Yongyu Li and Weiwei Kong

Received: 18 March 2024
Revised: 21 April 2024
Accepted: 25 April 2024
Published: 28 April 2024

1. Introduction

The International Energy Agency's "World Energy Outlook 2022" points out it is still very important to promote the energy revolution and build clean, low-carbon, safe, and efficient energy [1]. Hydrogen, although not an energy source itself, has gained extensive utilization as a readily available, environmentally friendly, and low-carbon secondary energy carrier. Hydrogen, being colorless and odorless, and exhibiting no toxic effects on humans or ecology [2], is emerging as a significant contributor to the global energy transition and development. It serves as one of the key carriers for sustainable energy advancement worldwide [3]. With the development of clean hydrogen production technologies such as water electrolysis, a hydrogen fuel cell (FC) utilizes carbon-free hydrogen as fuel during operation, with water being the main reaction product. This positions FC as a promising, important, and clean energy conversion technology for the future. Furthermore, since an FC operates based on electrochemical reactions converting chemical energy into electrical energy, its reaction process does not involve combustion.

Therefore, the energy conversion efficiency is not constrained by the limitations of the Carnot cycle [4,5], making it an ideal source of energy for power machinery.

There are few known studies on energy management of hydrogen FC tractors, mainly focusing on road vehicles [6,7]. Energy management approaches can be classified into two distinct categories: rule-based energy management and optimization-based energy management methodologies [8–10]. The rule-based energy management method can design power allocation rules based on expert experience and operational knowledge, so that the FC can operate in a high-efficiency range as much as possible. The main implementation methods include fuzzy logic, finite state machine, and power following [11]. To protect FCs and ensure their efficiency, Shen et al. [12] propose a fuzzy control energy management strategy that takes into account the characteristics of FCs. This strategy aims to achieve stable and efficient output from the FCs. However, it does not consider the state of charge (SOC) of the power battery. Considering the influence of the power battery SOC and the supercapacitor state voltage on its performance, Wen et al. [13] proposed an energy management strategy based on a "thermostat", which could maintain the power battery SOC and improve the whole machine economy. However, this method is not suitable for scenarios with more complex operating conditions. In response to this, Wang et al. [14] proposed a management strategy for the battery/supercapacitor/fuel cell system based on an effective state machine, which improved the power density of the power system and the net output power of the FC; it is capable of meeting the power requirements of most driving conditions. However, although rule-based energy management strategies have the advantages of low technical difficulty and small online computing volume, it is difficult to achieve control effects close to optimal. The optimization-based energy management methods optimize the target through optimization algorithms to make vehicle operation more energy-efficient. To realize the reasonable distribution of energy between different energy sources, Wang et al. [15] combined the dynamic programming global optimization method with fuzzy logic control to reduce the hydrogen consumption of the whole machine and improve the efficiency of the FC. However, this method has the issue of cumulative errors in battery SOC estimation. To solve this problem, Song et al. [16] proposed an FC vehicle energy management strategy based on dynamic programming, which improved the energy system efficiency and reduced the error accumulation in the control process. This dynamic programming algorithm has a large amount of calculations and takes a long time. Therefore, Zhou et al. [17] proposed a fast and unified method for solving dynamic programming problems, which effectively reduced the calculation time while reducing the energy consumption of the whole machine. Global optimization can greatly improve the energy-saving effect of the system, but it requires a large amount of calculation and has poor real-time performance. Instantaneous optimization can perform real-time optimization within the sampling period. Compared with global optimization, it has the advantages of smaller number of calculations and higher real-time performance. Therefore, it is an ideal online optimization method [18]. Zhou et al. [19] used a Markov pattern recognizer to divide driving modes into three types and used a multi-mode predictive controller to formulate a control strategy, which ensured the safe and stable operation of the FC while improving economy. This method did not consider the SOC of the power battery. Wang et al. [20] proposed an energy management strategy based on the Pontryagin's minimum principle, which can ensure both overall economy and stable SOC, but it cannot adapt to the dynamic changes in complex traffic environments. To solve this problem, Nie et al. [21] used the model predictive control method to carry out the energy scheduling of hybrid systems, which can effectively optimize driving performance and improve economy. In the field of non-road vehicles, there are few known studies on hydrogen FC power distribution. Regarding the energy management strategy of tractors, Xu et al. [22] developed an energy management strategy based on fuzzy control for FC power battery hybrid tractors; the simulation results showed this strategy reduced the depth of discharge of the dynamic battery and decreased the equivalent hydrogen consumption of the tractor, but this method also did not consider the impact of load on the strategy.

The joint operation of power batteries and hydrogen fuel cells to drive tractors, and determining the power distribution between the two, is crucial to achieving efficient operation of the energy system, representing an urgent problem that needs to be addressed. Traditional rule-based optimization strategies often directly determine a fixed proportion factor for the power allocation between the power battery and hydrogen fuel cell, which makes it difficult to achieve efficient output of the energy system. Furthermore, the load is crucial for the power allocation within the energy system of hydrogen fuel cell tractors. To tackle the challenge of instantaneous power optimization in hydrogen fuel cell electric tractors, this paper focuses on hydrogen fuel cell wheel-drive electric tractors. By integrating dynamics and the theory of equivalent hydrogen consumption, and introducing a charge retention strategy, a load-demand-based instantaneous optimization method for power allocation within the energy system is proposed, aiming to effectively distribute power between the power battery and hydrogen fuel cell. The main contributions of this paper are as follows: a real-time power allocation method based on instantaneous optimization is proposed to address the power allocation challenges faced by hydrogen fuel cell tractors under time-varying conditions.

The subsequent sections of this paper follow the following structure: in Section 2, the topology structure and main technical parameters of the hydrogen FC electric tractor are introduced. Section 3 models the main components of the tractor. In Section 4, an instantaneously optimized power allocation method based on load demand is proposed for the energy system. Section 5 conducts strategy verification on the MATLAB/Simulink simulation platform. Finally, Section 6 presents the research conclusions.

2. Tractor Topology Structure and Main Parameters

2.1. Hydrogen Fuel Cell Wheel-Driven Electric Tractor Topology Structure

Hydrogen FCs use carbon-free hydrogen as fuel and have advantages such as high energy density, long service life, clean and pollution-free process. Tractors that use hydrogen fuel cells/power cells effectively integrate the advantages of high-power density in power batteries with the high energy density in FCs, which is one of the directions for future research on energy-saving and environmentally friendly tractors.

The topology structure of the hydrogen FC wheel-driven electric tractor is shown in Figure 1. The tractor has two energy sources, hydrogen FCs and batteries. The output torque of the wheel-driven motor is directly transmitted to the drive wheel through the reducer.

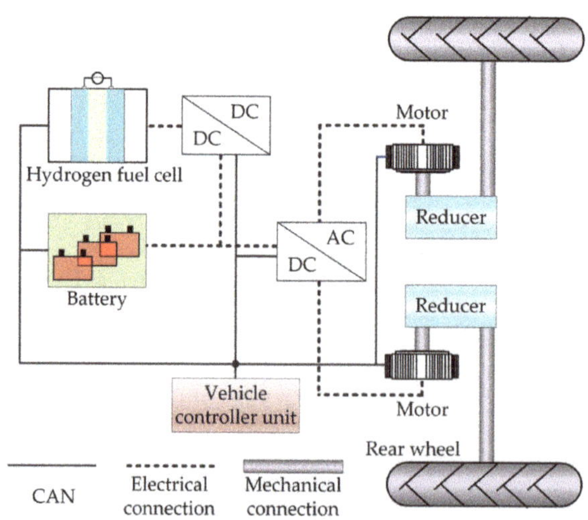

Figure 1. Topology of the hydrogen fuel cell wheel-driven electric tractor.

The vehicle control unit is connected to the hydrogen FC, battery, DC/DC module, DC/AC module, and two wheel-side driven electric motors through the Controller Area Network (CAN) bus. According to the power demand of the whole machine and the power battery SOC, the output power of the FC and the battery is dynamically allocated according to the power optimization method of the energy system, thereby enabling the tractor to achieve optimal power performance and economy.

2.2. Main Technical Parameters of Hydrogen Fuel Cell Wheel-Drive Electric Tractor

Expanding upon the preceding research findings of the research group, the main technical parameters of the hydrogen FC wheel-driven electric tractor used in this study are shown in Table 1.

Table 1. Topology of a diesel-electric parallel hybrid tractor.

Project	Parameters/Units	Value
Vehicle parameters	Tractor mass (kg)	2500
	Driving wheel rolling radius (m)	0.63
	Rolling resistance coefficient	0.07
	Reducer speed ratio	17
Motor	Rated speed (rpm)	1500
	Rated torque (N·m)	140
	Rated power (kW)	22
Power System	Energy capacity (kW·h)	41
	Battery voltage (V)	380
	Battery capacity (A·h)	108

3. Hydrogen Fuel Cell Tractor Model Construction

The key components of a hydrogen FC tractor include the hydrogen FC, power battery, motor, etc. The tractor model forms the foundation for the control of the entire machine. Therefore, the model established in this section includes the dynamics model, transmission system model, hydrogen FC model, power battery model, motor model, tyre model, and whole machine simulation mode.

3.1. Tractor Dynamics Model

By conducting the analysis of the tractor's driving force and driving resistance, the condition for the tractor to operate normally is the driving force equals the sum of all driving resistances. When the tractor is working in the field, the load of towing agricultural machinery is very large. Therefore, the driving resistance includes the resistance that must be overcome when towing agricultural machinery, that is, the towing resistance F_{TN}. Since the operating velocity of the tractor is low and close to uniform, the air resistance and acceleration resistance can be ignored [23]. Therefore, the driving equation for the ploughing condition of the tractor can be expressed as:

$$F_t = F_f + F_{TN} \tag{1}$$

where F_t is the total driving force, N. F_f is the rolling resistance, N. F_{TN} is the rated traction resistance, N.

During field operations, ploughing is a common high-load operation, so the driving force of the tractor should meet the needs of the tractor's ploughing operation. Considering the large load fluctuations caused by the working conditions of the tractor and the changes in the performance of agricultural machinery, a reserve of 10~20% should be left [24]. Therefore, the rated traction force when the tractor is ploughing is:

$$F_{TN} = (1.1 \sim 1.2)F_{Ta} \tag{2}$$

where F_{Ta} is the ploughing resistance, N.

The required torque and speed of the drive wheel during the hydrogen FC wheel-drive electric tractor ploughing operation are:

$$\begin{cases} T_q = \frac{F_L}{r} \\ n_q = \frac{v}{3.6 \times 0.377r} \end{cases} \tag{3}$$

where T_q is the torque required for the drive wheels, N·m. n_q is the required speed of the drive wheels, rpm. v is the tractor velocity, m/s. And, r is the driving wheel radius, m.

3.2. Transmission System Model

The transmission system of wheel-driven tractors mainly consists of wheel-side reducers, which belong to the gear transmission mechanism. Unlike urban vehicles that require frequent gear shifting, tractors experience relatively small variations in operating speeds during field work. Therefore, this paper simplifies the transmission system model to a fixed gear ratio gear model [25]:

$$\begin{cases} T_{reqm} = \frac{T_q}{i} \\ n_{reqm} = i n_q \end{cases} \tag{4}$$

where T_{reqm} is the required torque of the motor, N·m. n_{reqm} is the required speed of the motor, rpm. And, i is the speed ratio of the wheel-side reducer.

3.3. Motor Model

The motor uses a permanent magnet synchronous motor. Compared with other motors, the permanent magnet synchronous motor has advantages such as good reliability, high work efficiency, large torque, small size, and light weight. It can cope with the relatively complex operating environment when the tractor is working in outdoor fields. The efficiency model of the motor can be determined by establishing a quasi-static graph of output speed and torque. Subsequently, the efficiency of the motor can be obtained by referring to a table. The relationship between motor efficiency, speed, and torque is as follows:

$$\eta_m = f(n_m, T_m) \tag{5}$$

where η_m is the motor efficiency, %. n_m is the motor speed, rpm. And, T_m is the motor torque, N·m.

The required driving power (P_{reqm}) of the motor is:

$$P_{reqm} = \frac{n_{reqm} T_{reqm}}{9550\eta_m} \tag{6}$$

where P_{reqm} is expressed in kW.

When calculating the efficiency of the drive motor, since the trend of the motor's electric efficiency and power generation efficiency is similar, it is assumed in the model building that the power generation efficiency and electric efficiency of the permanent magnet synchronous motor used are the same. The efficiency numerical model of the permanent magnet synchronous motor used in this paper is shown in Figure 2.

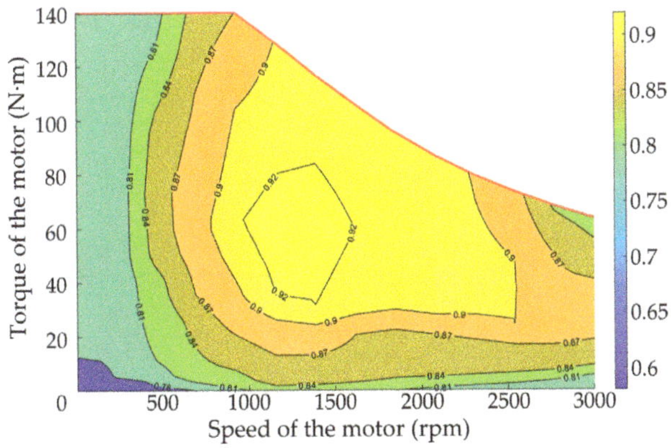

Figure 2. MAP diagram of a motor model.

3.4. Power Battery Model

The Rint model [26] is a common equivalent circuit model. The battery is modeled as an ideal voltage source connected in series with an internal resistance. The structure is simple, and the parameters are easy to identify, making it suitable for whole vehicle energy management algorithms. In the Rint model, the effects of temperature on the battery pack are ignored. The open circuit voltage and charge-discharge internal resistance of the battery are affected by the state of charge (SOC) and the charge-discharge current of the battery.

In accordance with the Rint model [27,28] of the battery, the power of the battery, denoted as P_{bat}, can be expressed by the following equation:

$$P_{bat} = V_{oc}(\mathrm{SOC})I_{bat}(t) - I_{bat}(t)^2 R_{\mathrm{int}}(\mathrm{SOC}) \tag{7}$$

where P_{bat} is the battery power, kW. V_{oc} is the battery open circuit voltage, V. R_{int} is the battery internal resistance, Ω. And, I_{bat} is the battery current, A.

The SOC of the battery can be represented by the ampere-hour method [29] as follows:

$$I_{bat}(t) = \frac{V_{oc}(\mathrm{SOC}) - \sqrt{V_{oc}(\mathrm{SOC})^2 - 4R_{\mathrm{int}}(\mathrm{SOC})P_{bat}(t)}}{2R_{\mathrm{int}}(\mathrm{SOC})} \tag{8}$$

$$\mathrm{SOC}(t) = \mathrm{SOC}(0) - \frac{\int_0^t I_{bat}(t)dt}{Q_{bat}} \tag{9}$$

where Q_{bat} is the battery rated capacity, A·h.

The instantaneous discharge efficiency/charge efficiency, denoted as η_{dis} and η_{chg}, of the battery [30] can be represented as follows:

$$\begin{cases} \eta_{dis} = \frac{V_{oc} - I_{bat}(t)R_{\mathrm{int}}}{V_{oc}} \times 100\%, P_{bat} \geq 0 \\ \eta_{chg} = \frac{V_{oc}}{V_{oc} - I_{bat}(t)R_{\mathrm{int}}} \times 100\%, P_{bat} < 0 \end{cases} \tag{10}$$

where η_{dis} is the instantaneous discharge efficiency of the battery, %. And, η_{chg} is the instantaneous charge efficiency of the battery, %.

Substituting (8) into (10) yields the following result:

$$\begin{cases} \eta_{dis} = \frac{1+\sqrt{1-\frac{4R_{\mathrm{int}}P_{bat}(t)}{V_{oc}^2}}}{2} \times 100\%, P_{bat} \geq 0 \\ \eta_{chg} = \frac{2}{1+\sqrt{1-\frac{4R_{\mathrm{int}}P_{bat}(t)}{V_{oc}^2}}} \times 100\%, P_{bat} < 0 \end{cases} \tag{11}$$

3.5. Hydrogen Fuel Cell Model

Hydrogen FCs convert the chemical energy in hydrogen into electricity and are a clean and efficient power generation device. This paper uses the proton exchange membrane fuel cell (PEMFC), which is widely used in engineering, and has the advantages of fast cold start and high energy conversion rate and can be used as a mobile power supply. Hydrogen FC models can be divided into mechanism models and numerical models. Although the mechanism model has good dynamic response, the model is more complex and requires a large amount of calculation. The operating characteristics of the hydrogen FC system used in this paper are shown in Figure 3, which shows the optimal operating area of the hydrogen FC system is within 10% to 80% of its power range.

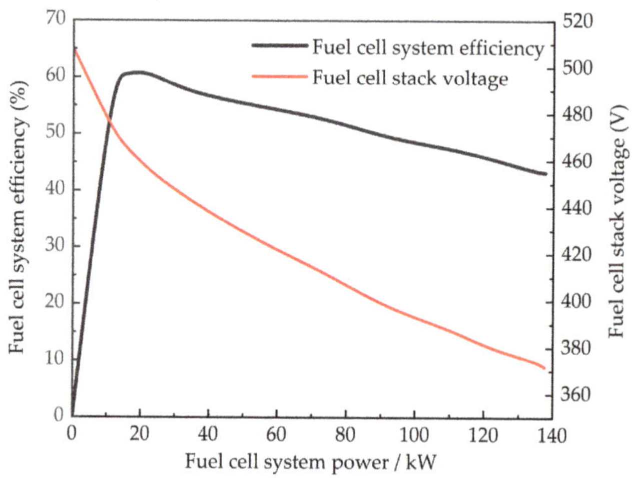

Figure 3. Operational characteristics of the hydrogen FC system.

Through polynomial fitting, the voltage model of the FC is obtained as follows:

$$V_{fc} = \xi_1 P_{fc}^3 + \xi_2 P_{fc}^2 + \xi_3 P_{fc} + \xi_4 \quad (12)$$

where V_{fc} is the FC output voltage, V. P_{fc} is the FC system power, kW. ξ_j is the fitting parameters, j = 1, 2, 3, 4.

In practical use, a hydrogen FC requires auxiliary equipment to support its operation. The auxiliary equipment mainly includes air circulation pumps, cooling water circulation pumps, exhaust fans, hydrogen supply pumps, and electronic control equipment. The efficiency decreases at high power because a larger voltage drop occurs in the hydrogen FC stack at high power. On the other hand, the lower efficiency at low power is due to the increased percentage in energy consumed by auxiliary equipment at low power [31]. The hydrogen FC efficiency used in this paper is fitted by polynomial and yielded as follows:

$$\eta_{fc} = \left(\gamma_1 P_{fc}^8 + \gamma_2 P_{fc}^7 + \gamma_3 P_{fc}^6 + \gamma_4 P_{fc}^5 + \gamma_5 P_{fc}^4 + \gamma_6 P_{fc}^3 + \gamma_7 P_{fc}^2 + \gamma_8 P_{fc} + \gamma_9\right) \times 100\% \quad (13)$$

where η_{fc} is the FC system efficiency, %. And, γ_j is the fitting parameters, j = 1–9.

3.6. Tyre Model

Commonly used tyre models include theoretical models such as the Gim model and Finala model, as well as empirical or semi-empirical models such as the power exponential formula, semi-empirical model, UniTire model, and magic tyre model [32]. The magic formula model not only has a simple structure but also has high simulation accuracy. Its general expression is:

$$\begin{cases} Y(x) = y(x) + S_V \\ y(x) = D\sin\{C\arctan[Bx(1-E) + E\arctan Bx]\} \\ x = X + S_H \end{cases} \tag{14}$$

where X is the tyre slip rate. $Y(x)$ is the longitudinal force of the tyre(F_x), N. B is the stiffness coefficient. C is the curve shape coefficient. D is the peak adhesion coefficient. E is the curve curvature coefficient. S_V is the vertical drift of the curve. S_H is the horizontal drift of the curve.

3.7. Whole-Machine Simulation Model

Combining the characteristics of the fuel cell wheeled drive tractor system, a full vehicle simulation model was established based on MATLAB, as shown in Figure 4. According to the working conditions of the tractor, the controller acquires signals and provides the plow resistance and plow speed (F_t, v) of the tractor as inputs to the controller. Inside the controller, calculations and processing are performed according to Equations (1), (3) and (4), etc., and corresponding power demand (P_{reqbat}) from the power battery, and motor required speed (n_{reqm}) are output according to the predetermined control strategies (including comparative strategies and the strategy proposed in this paper). P_{reqfc} is provided as an input to the fuel cell model, and the fuel cell model works according to the instructions to output corresponding power (P_{fc}); P_{reqbat} is provided as an input to the power battery model, and the power battery model works according to the instructions to output corresponding power (P_{bat}). P_{bat} and n_{reqm} are provided as inputs to the dynamical system, which outputs power (F_t, F_x). F_t is provided as input to the tractor dynamics model, and F_x is also provided as input to the tyre model.

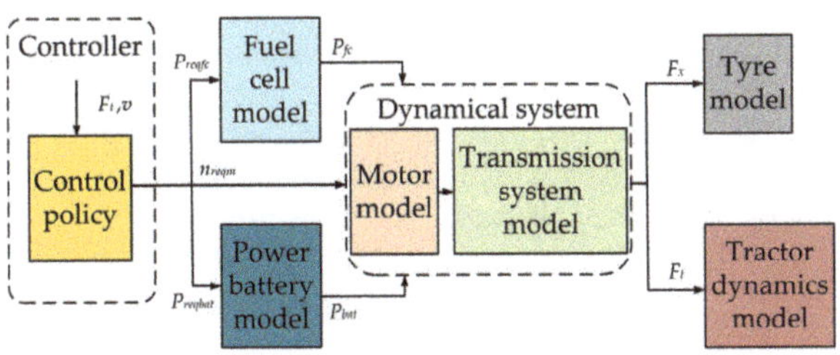

Figure 4. Schematic diagram of the whole-machine simulation mode.

4. Hydrogen Fuel Cell Tractor Energy System Power Optimization Method

The hydrogen FC tractor incorporates two different energy sources and establishes an equivalent hydrogen consumption model based on the calorific value method. Combining the equivalent hydrogen consumption model, an instantaneously optimized power allocation method based on load demand is proposed. The aim is to minimize equivalent hydrogen consumption by optimizing the rational distribution of power output between the hydrogen FC and the power battery.

4.1. Equivalent Hydrogen Consumption Model

The hydrogen consumption of the FC system is:

$$C_{fc} = \frac{P_{fc}}{E_{H_2,low}\eta_{fc}\left(P_{fc}\right)} \tag{15}$$

where C_{fc} is the hydrogen consumption of the FC system, kg/s. $E_{H_2,low}$ is the low calorific value of hydrogen, 1.2×10^8 J/kg.

The equivalent hydrogen consumption during battery charging/discharging is represented as follow:

$$C_{bat} = \begin{cases} \frac{P_{bat}}{E_{H_2,low}\eta_{fc}(P_{fc})} \cdot \frac{1}{\overline{\eta_{dis}}\overline{\eta_{chg}}}, P_{bat} \geq 0 \\ \frac{P_{bat}}{E_{H_2,low}\eta_{fc}(P_{fc})} \cdot \overline{\eta_{chg}}\overline{\eta_{dis}}, P_{bat} < 0 \end{cases} \tag{16}$$

where C_{bat} is the equivalent hydrogen consumption during battery charging/discharging, kg/s. $\overline{\eta_{dis}}, \overline{\eta_{chg}}$ are the average discharge and charging efficiency of power battery, %.

After introducing the battery charge holding strategy [33], the equivalent hydrogen consumption of the whole machine can be expressed by the following function:

$$C_{sys} = C_{fc} + \alpha(\mathrm{SOC})C_{bat} \tag{17}$$

where α is the linear compensation coefficient used to keep the battery SOC within a certain range. Its value is represented as follows:

$$\alpha(\mathrm{SOC}) = 1 - \beta\left(\frac{2\mathrm{SOC} - (SOC_H + SOC_L)}{(SOC_H - SOC_L)}\right) \tag{18}$$

where SOC_H is the given maximum limit of SOC. SOC_L is the given minimum limit of SOC. β is the adjustment coefficient.

By calibrating β, the battery's SOC can be effectively kept within a certain range. After calibration, β equals 0.5.

4.2. Instantaneously Optimized Power Allocation Method Based on Load Demand in the Energy System

Tractor ploughing demonstrates significant time-varying nonlinear load changes, which profoundly influence the energy management strategy, so it is crucial to study the power distribution strategy in energy systems in response to load demand. This study starts with load input as the basis, optimizing the power distribution between the hydrogen FC and the power battery in the energy system of the hydrogen FC tractor. The optimization goal of the instantaneous optimized energy system power allocation method based on load demand is to improve the FC efficiency on the basis of meeting the power required by the tractor load. After taking the equivalent hydrogen consumption of the system as the optimization target and introducing the charge retention strategy of the power battery, the optimization objective function $f(P_{fc}, P_{bat})$ can be obtained by combining (15)–(18) as follows:

$$f\left(P_{fc}, P_{bat}\right) = \min(C_{sys}) = \min\left[\begin{cases} \frac{P_{fc}}{E_{H_2,low}n_{fc}(P_{fc})} + 1 - \alpha\frac{P_{bat}}{E_{H_2,low}n_{fc}(P_{fc})}\frac{1}{\overline{\eta_{dis}}\overline{\eta_{chg}}}, P_{bat} \geq 0 \\ \frac{P_{fc}}{E_{H_2,low}n_{fc}(P_{fc})} + 1 - \alpha\frac{P_{bat}}{E_{H_2,low}n_{fc}(P_{fc})}\overline{\eta_{chg}}\overline{\eta_{dis}}, P_{bat} < 0 \end{cases}\right] \tag{19}$$

Considering the working capabilities of hydrogen FCs and power batteries are limited by their output characteristics; the following constraints need to be met:

$$\begin{cases} P_{reqm} = P_{reqfc} + P_{reqbat} \\ SOC_{\min} \leq \mathrm{SOC} \leq SOC_{\max} \\ P_{dis\max} \leq P_{bat} \leq P_{chg\max} \\ 10\text{ kW} \leq P_{fc} \leq P_{fc\cdot\max} \end{cases} \tag{20}$$

The power allocation method in the energy system adopts a cyclic iterative approach, starting from load demand and constraining with the power battery SOC. It traverses and searches for the output power of the FC and power battery, to reduce the equivalent hydrogen usage of the whole machine to the minimum possible. The scheme process is shown in Figure 5. When the power battery SOC is less than 0.3: when $P_{reqm} \leq p_0$ kW, the hydrogen FC works at the highest efficiency point, meeting the tractor's power demand and charging the battery at the same time. When $P_{reqm} > p_0$ kW, the power needed for the tractor is supplied solely by the hydrogen FC. When the power battery SOC is between 0.3 and 0.8 when $P_{reqm} \leq p_0$ kW, the hydrogen FC works at the highest efficiency point, meeting the tractor's power demand and charging the battery at the same time. When $P_{reqm} > p_0$ kW, the hydrogen FC and the battery simultaneously provide the required power for the tractor, and the power ratio is determined by the optimization result. When the power battery SOC is higher than 0.8: when $P_{reqm} \leq p_0$ kW, The hydrogen FC outputs the minimum output power, and the battery output power meets the difference between the hydrogen FC and the tractor's power demand. When $P_{reqm} > p_0$ kW, the hydrogen FC and the battery simultaneously provide the required power for the tractor, and the power ratio is determined by the optimization result; where p_0 is the power corresponding to the highest efficiency point of the hydrogen FC.

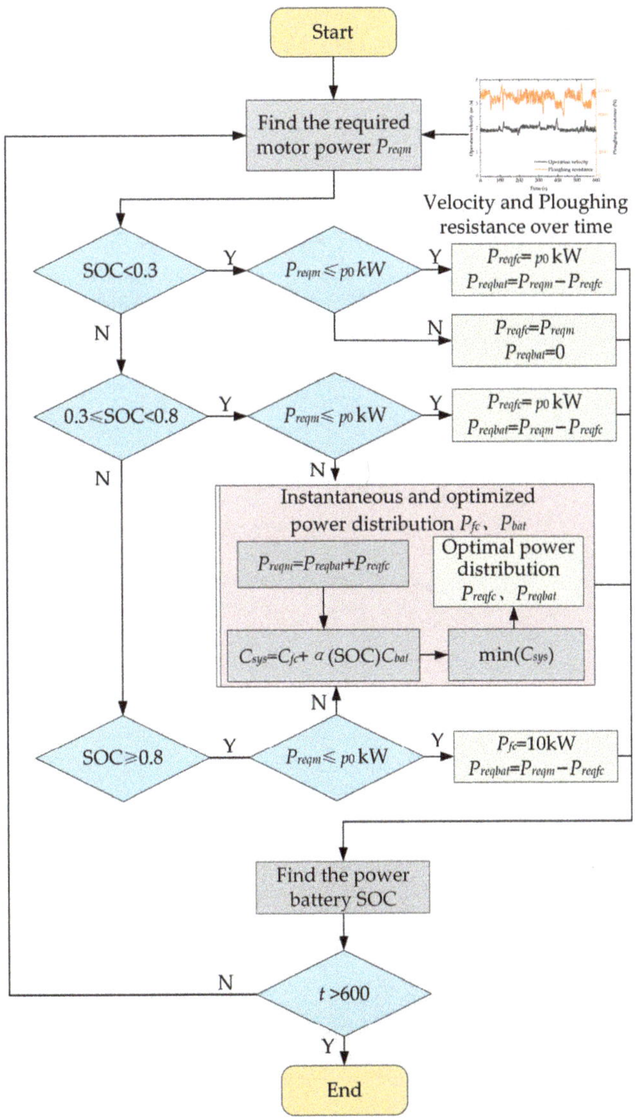

Figure 5. Scheme process of energy system power optimization method.

5. Simulation Analysis

To demonstrate the effectiveness of the optimization approach outlined in this paper, we choose the rule-based control strategy for comparison. In the rule-based control strategy, when the power battery SOC is less than 0.3, the hydrogen FC provides the required power of the tractor. When the power battery SOC is greater than 0.8, the hydrogen FC stops working, and the power battery provides the required power of the tractor. When the power battery SOC is between 0.3 and 0.8, if the power required by the tractor is higher than the rated power of the power battery, the hydrogen FC starts to work. If the power required by the tractor is lower than the rated power of the power battery, the hydrogen FC maintains its previous state until the power battery SOC is no longer between 0.3 and 0.8, and then changes its state.

We built the control strategy and whole machine model based on the MATLAB/Simulink simulation platform, and conducted experimental verification. The changes in velocity and resistance of the tractor during ploughing operations in the field are shown in Figure 6.

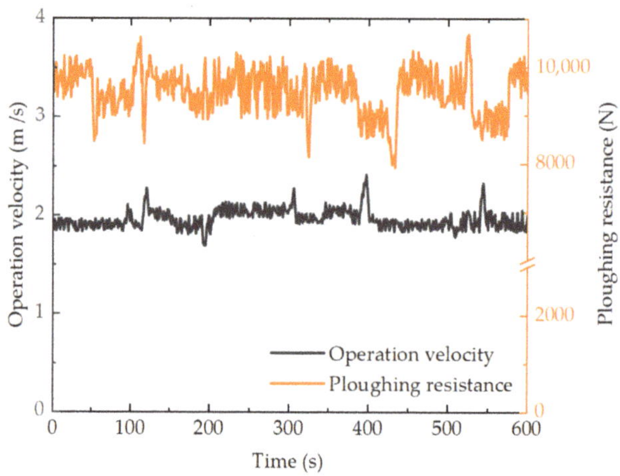

Figure 6. Ploughing velocity and resistance over time.

Figure 7 shows the distribution of the working efficiency of the hydrogen FC system under field ploughing conditions.

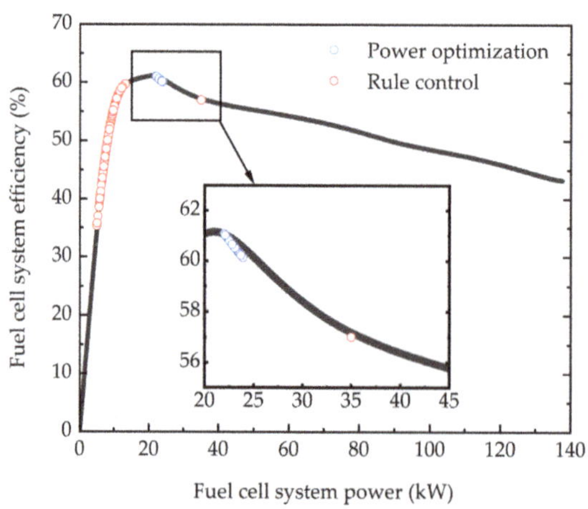

Figure 7. Hydrogen fuel cell operating point.

From Figure 7, it can be seen the instantaneously optimized power allocation method based on load demand is more efficient in power allocation for the energy system than the rule-based control strategy. Under the rule-based control strategy, the efficiency points of the FC are distributed within a range of 35–60% efficiency, with an average output efficiency of 53.93% and an overall variance of 18.63%. However, using the power optimization method allows for a tighter distribution of the hydrogen fuel cell's efficiency points closer to the best efficiency point on the efficiency curve, resulting in an average output efficiency of 60.50% and an overall variance of just 0.14%. From the data presented, it can be concluded the method proposed in this paper effectively increases the output efficiency of the fuel cell system while reducing the fluctuations in system output, thereby contributing to the improvement of the overall performance and stability of the fuel cell system.

Figure 8 shows the instantaneous efficiency of the hydrogen FC system under field ploughing conditions.

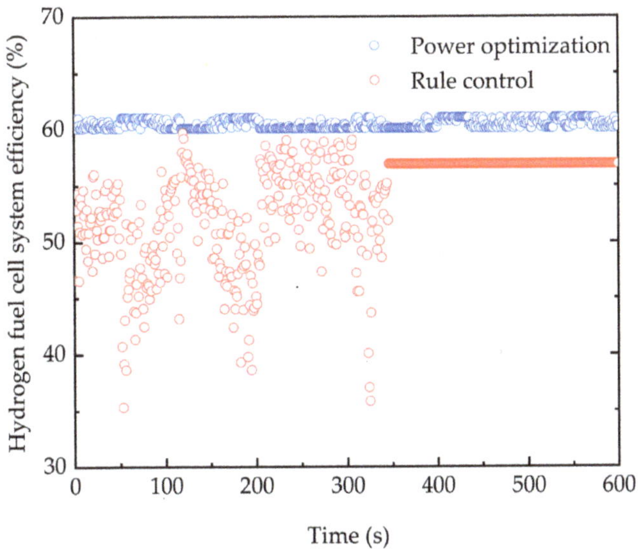

Figure 8. Hydrogen fuel cell system instantaneous efficiency over time.

From Figure 8, it can be seen the hydrogen FCs under the control of the instantaneously optimized power allocation method based on load demand can always operate in the high-efficiency region during operation. However, the rule-based control strategy cannot guarantee the hydrogen FCs work in the high-efficiency region when the SOC is in the normal range. When the SOC is too low, the hydrogen FCs also cannot operate in the highest efficiency region to provide the required power for the entire machine and ensure charging efficiency.

Figure 9 shows the power distribution between the hydrogen FC and the power battery under the control of two control strategies.

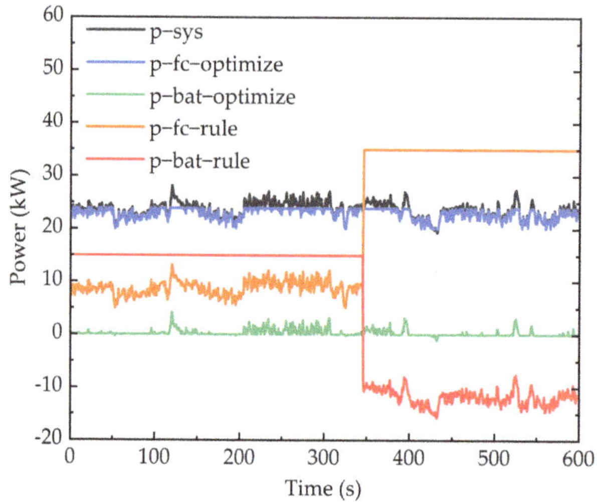

Figure 9. The output power of the fuel cell and power battery over time.

From Figure 9, it can be seen during the ploughing operation, the rule-based control strategy ensures continuous power output from the power battery before 346 s. After 346 s, when the power battery SOC becomes too low, the required power for the entire machine is provided by hydrogen FCs, which also charge the power battery. Under the control of the instantaneously optimized power allocation method based on load demand, the power output curve of the hydrogen FCs remains relatively stable, with the operating power consistently staying around 22 kW without significant sudden changes.

Figure 10 shows the relationship between power battery SOC with time under field ploughing conditions.

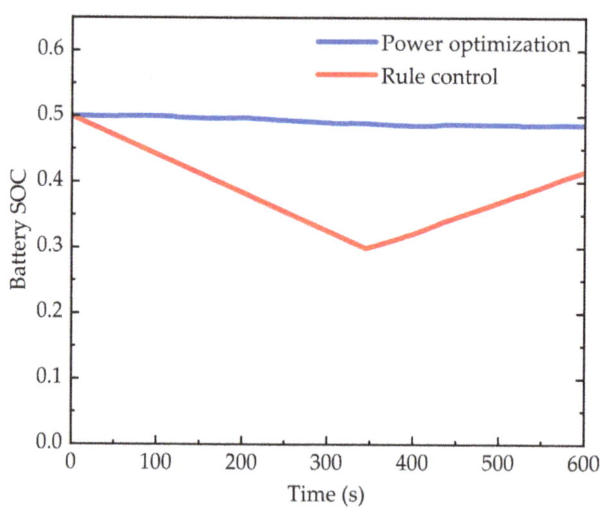

Figure 10. The power battery SOC over time.

From Figure 10, it can be observed under the control of the rule-based control strategy, the variation of the SOC of the power battery is significant, decreasing to around 0.3 at 346 s. Then, the required power for the entire machine is supplied by hydrogen FCs, and they also charge the power battery. However, due to the introduction of the power battery charge maintenance strategy, the variation of the power battery SOC, when using the instantaneously optimized power allocation method based on load demand, is relatively stable, and it remains around 0.48 even after the simulation ends.

Figure 11 shows the variation of equivalent hydrogen consumption over time under field ploughing conditions.

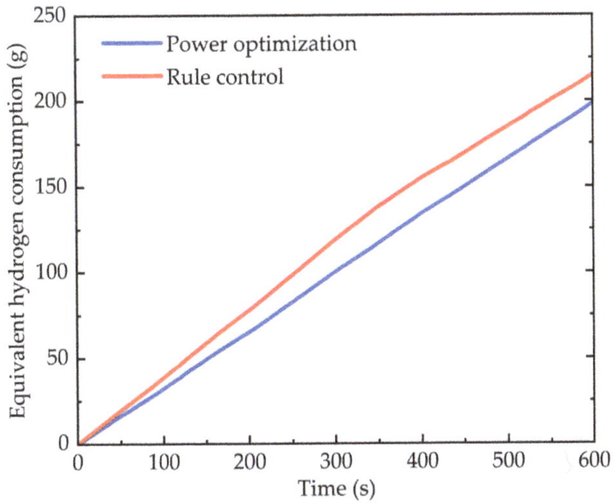

Figure 11. Variation of equivalent hydrogen consumption over time.

Combining Figure 7, Figure 8, and Figure 11, it can be clearly seen the rule-based control strategy results in low hydrogen FC efficiency and a faster growth rate of equivalent hydrogen consumption. At 346 s, when the hydrogen FC provides the required power for the whole machine and charges the power battery, the hydrogen FC efficiency is improved, and the slope of the equivalent hydrogen consumption curve decreases. However, the instantaneously optimized power allocation method based on load demand controls the FC operation point to focus on the high-efficiency range of the FC, resulting in a lower growth rate of equivalent hydrogen consumption compared with the rule-based control strategy, and thus a lower overall equivalent hydrogen consumption of the machine.

As shown in Figure 12, under the rule-based control strategy, the actual hydrogen consumption at the end of the operating condition is 1.75×10^{-1} kg, with an equivalent hydrogen consumption of 2.15×10^{-1} kg. However, when the tractor adopts the instantaneous optimized power allocation method based on load deceleration, the actual hydrogen consumption at the end of the operating condition is 1.95×10^{-1} kg, and the equivalent hydrogen consumption decreases to 1.98×10^{-1} kg, which is 7.79% lower compared to the rule-based control strategy. The increase in the actual hydrogen consumption and the decrease in the equivalent hydrogen consumption suggest that after optimization, the system more efficiently and stably outputs power from the fuel cell, while less frequently drawing power from the battery. This benefits in maintaining the battery's state of operation within an ideal range, thus helping to extend the service life of the power battery.

As shown in Figure 12, the rule-based control strategy results in an equivalent hydrogen consumption of 2.15×10^{-1} kg at the end of the operating condition. However, when the tractor uses the instantaneously optimized power allocation method based on load demand, the equivalent hydrogen consumption at the end of the operating condition is reduced to 1.98×10^{-1} kg, which is 7.79% lower than that of the rule-based control strategy.

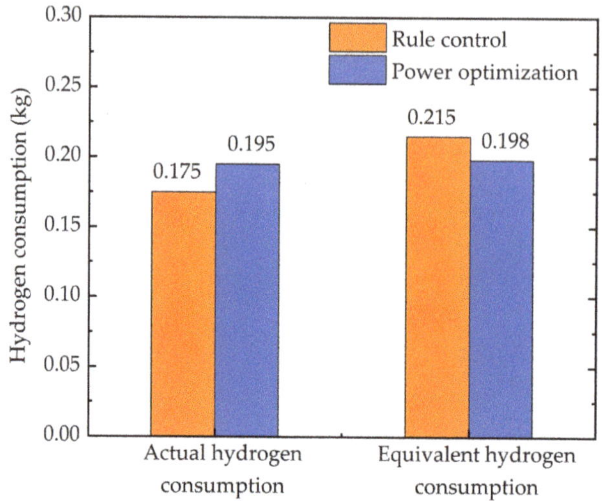

Figure 12. Hydrogen consumption under two control strategies.

In conclusion, starting from the load demand, the simulation results analysis of tractor plowing operation can verify that compared to rule-based control strategies, the instantaneously optimized power allocation method based on load demand ensures the FC operates in the high-efficiency range and controls the power battery SOC within a reasonable range. This reduces the equivalent hydrogen consumption of the tractor and improves the overall efficiency of the machine.

6. Conclusions

To address the power allocation challenge of the energy system of a hydrogen FC electric tractor under time-varying operating conditions, this paper takes a hydrogen FC wheel-side drive electric tractor as the research object and proposes an instantaneously optimized power allocation method based on load demand, which effectively improves the energy utilization efficiency.

By simulating and comparing with a rule-based control strategy [22], this paper ultimately proves the strategy proposed herein: effectively improves the output efficiency of the fuel cell system, reduces the fluctuation of the system's output, enhances the overall performance and stability of the fuel cell system. This simultaneously causes the energy system to rely more on the efficient and stable output of the fuel cell while utilizing the power battery less, which is beneficial for maintaining the power battery's SOC within an ideal range and for extending the lifespan of the power battery.

There are many factors that affect the performance of hydrogen FCs, which are an important source of energy for the entire machine. The future research goals will focus on the following issues: the threshold power value at which the FC begins to intervene and the minimum operating power set to prevent frequent start-stop of the FC; these values can be further optimized. Furthermore, the strategies in this paper did not take into account the impact of motor efficiency on the overall energy utilization of the system. In the future, further optimization can be conducted on transmission system parameters such as gear ratios of the reducers to improve motor efficiency, thereby enhancing the overall energy utilization efficiency of the system.

Author Contributions: Conceptualization, J.Z. (Jingyun Zhang) and B.W.; methodology, L.X.; software, B.W; validation, J.Z. (Jingyun Zhang), B.W. and L.X.; formal analysis, B.W.; investigation, J.Z. (Junjiang Zhang); resources, L.X.; data curation, K.Z.; writing—original draft preparation, J.Z. (Jingyun Zhang); writing—review and editing, J.Z. (Junjiang Zhang); visualization, K.Z.; supervision,

J.Z. (Jingyun Zhang); project administration, L.X.; funding acquisition, K.Z. All authors have read and agreed to the published version of the manuscript.

Funding: This research was funded by the National Key R&D Program of China [2022YFD2001203, 2022YFD2001201B], Henan University of Science and Technology Innovation Team Support Program [24IRTSTHN029], Zhongyuan Science and Technology Innovation Leading Talent Support Program Project [244200510043].

Data Availability Statement: The original contributions presented in the study are included in the article, further inquiries can be directed to the corresponding author.

Conflicts of Interest: Kai Zhang is an employee of YTO Group Corporation, Luoyang 471004, China. The paper reflects the views of the scientists, and not the company.

Nomenclature

B	Stiffness coefficient.
C	Curve shape coefficient.
CAN	Controller Area Network. It is a serial communication protocol used for communication between electronic control units in the automotive and industrial fields.
C_{fc}	Hydrogen consumption of the fuel cell system (kg/s).
C_{bat}	Equivalent hydrogen consumption of the power battery (kg/s).
D	Peak adhesion coefficient.
DC/AC	Direct Current-Alternating Current converter. It is a device that transforms a direct current power source (typically the input power) into another voltage, current, or power level of alternating current.
DC/DC	Direct Current-Direct Current converter. It is a device that transforms a direct current power source (typically the input power) into another voltage, current or power level of direct current.
E	Curve curvature coefficient.
$E_{H_2,low}$	Low calorific value of hydrogen (1.2×10^8 J/kg).
FC	Fuel cell.
F_t	Total driving force (N).
F_f	Rolling resistance (N).
F_{TN}	Rated traction resistance (N).
F_{Ta}	Ploughing resistance (N).
F_x	Longitudinal force of the tyre (N).
i	Speed ratio of the wheel-side reducer.
I_{bat}	Battery current (A).
MAP	A graph that describes the performance characteristics of an electric motor. It is usually used to express the relationship between the speed, torque and efficiency of the motor under certain conditions.
n_q	Required speed of the drive wheels (rpm).
n_{reqm}	Required speed of the motor (rpm).
PEMFC	Proton Exchange Membrane Fuel Cell. It is a type of fuel cell that uses a proton exchange membrane as the electrolyte.
P_{bat}	Battery power (kw).
P_{fc}	Fuel cell system power (kw).
P_{reqbat}	Power battery required power (kw).
P_{reqfc}	Fuel cell required power (kw).
P_{reqm}	Required driving power of the motor (kw).
Q_{bat}	Battery rated capacity (A·h).
r	Driving wheel radius (m).
R_{int}	Battery internal resistance (Ω).
S_V	Vertical drift of the curve.
S_H	Horizontal drift of the curve.
SOC	State of charge.
SOC_H	Given maximum limit of SOC.

SOC_L	Given minimum limit of SOC.
T_m	Motor torque (N·m).
T_q	Torque required for the drive wheels (N·m).
T_{reqm}	Required torque of the motor (N·m).
v	Tractor velocity (m/s).
V_{oc}	Battery open circuit voltage (V).
V_{fc}	Fuel cell output voltage (V).
X	Tyre slip rate.
$Y(x)$	Longitudinal force of the tyre (N).
α	Linear compensation coefficient.
β	Adjustment coefficient.
γ_j	Fitting parameters, j = 1–9.
η_{dis}	Instantaneous discharge efficiency of power battery (%).
η_{chg}	Instantaneous charging efficiency of power battery (%).
$\overline{\eta_{dis}}$	Average discharge efficiency of power battery (%).
$\overline{\eta_{chg}}$	Average charging efficiency of power battery (%).
η_{fc}	Fuel cell system efficiency (%).
η_m	Motor efficiency (%).
ξ_j	Fitting parameters, j = 1, 2, 3, 4.

References

1. Group, C.W. World Energy Outlook 2022. *Chem. Wkly.* **2022**, *68*, 165–170.
2. Technical Committee 4.4 Tunnels. *Impact of New Propulsion Technologies on Road Tunnel Operations and Safety—A PIARC Technical Report*; The World Road Association: Paris, France, 2023; p. 18.
3. National Development and Reform Commission. The National Development and Reform Commission issued the "Medium and Long-term Plan for the Development of Hydrogen Energy Industry (2021–2035)". *Rare Earth Inf.* **2022**, *4*, 26–32.
4. Sulaiman, N.; Hannan, M.; Mohamed, A.; Majlan, E.; Daud, W.W. A review on energy management system for fuel cell hybrid electric vehicle: Issues and challenges. *Renew. Sustain. Energy Rev.* **2015**, *52*, 802–814. [CrossRef]
5. Hassan, F. Novel fuel cell/battery/supercapacitor hybrid power source for fuel cell hybrid electric vehicles. *Energy* **2018**, *143*, 467–477.
6. Zhao, S.; Wang, B.; Xie, Y.H.; Han, M.; Jia, J.B. Linear Temperature Sweep Experimental Study on Proton Exchange Membrane Fuel Cell Without External Humidification. *Proc. CSEE* **2014**, *34*, 4528–4533.
7. Li, Q.; Chen, W.R.; Liu, S.K.; Cheng, Z.L.; Liu, X.Q. Application of Multivariable H∞ Suboptimal Control for Proton Exchange Membrane Fuel Cell Pressure Control System. *Proc. CSEE* **2010**, *30*, 123–128.
8. Zhang, H.; Li, Q.; Wang, H.; Li, Q.; Qin, G.; Wu, Q. A review of energy management optimization based on the equivalent consumption minimization strategy for fuel cell hybrid power systems. *Fuel Cells* **2022**, *22*, 116–130. [CrossRef]
9. Peng, J.K.; He, H.W.; Xiong, R. Rule based energy management strategy for a series–parallel plug-in hybrid electric bus optimized by dynamic programming. *Appl. Energy* **2016**, *185*, 1633–1643. [CrossRef]
10. Lü, X.; Wu, Y.; Lian, J.; Zhang, Y.; Chen, C.; Wang, P.; Meng, L. Energy management of hybrid electric vehicles: A review of energy optimization of fuel cell hybrid power system based on genetic algorithm. *Energy Convers. Manag.* **2020**, *205*, 112474. [CrossRef]
11. Fernandez, A.M.; Kandidayeni, M.; Boulon, L.; Chaoui, H. An Adaptive State Machine Based Energy Management Strategy for a Multi-Stack Fuel Cell Hybrid Electric Vehicle. *IEEE Trans. Veh. Technol.* **2019**, *69*, 220–234. [CrossRef]
12. Shen, Y.; Cui, P.; Wang, X.; Han, X.; Wang, Y.-X. Variable structure battery-based fuel cell hybrid power system and its incremental fuzzy logic energy management strategy. *Int. J. Hydrogen Energy* **2020**, *45*, 12130–12142. [CrossRef]
13. Wen, P.M.; Song, K.; Zhang, T. Energy management strategy of thermostat for fuel cell vehicles based on Pontriagin minimum principle. *Mechatronics* **2017**, *23*, 7–12. [CrossRef]
14. Wang, Y.J.; Sun, Z.D.; Chen, Z.H. Energy management strategy for battery/supercapacitor/fuel cell hybrid source vehicles based on finite state machine. *Appl. Energy* **2019**, *254*, 113707. [CrossRef]
15. Wang, Z.F.; Xu, S.; Luo, W. Research on energy management strategy of fuel cell vehicle based on dynamic programming. *Acta Energiae Solaris Sin.* **2023**, *44*, 550–556.
16. Song, K.; Zhang, T.; Niu, W.X.; Zhang, T. Error accumulation problem and solution of dynamic programming algorithm for energy management of fuel cell electric vehicles. *Automot. Eng.* **2017**, *39*, 249–255.
17. Zhou, W.; Yang, L.; Cai, Y.; Ying, T. Dynamic programming for new energy vehicles based on their work modes Part II: Fuel cell electric vehicles. *J. Power Sources* **2018**, *407*, 92–104. [CrossRef]
18. Zhou, Y.; Li, H.; Ravey, A.; Péra, M.-C. An integrated predictive energy management for light-duty range-extended plug-in fuel cell electric vehicle. *J. Power Sources* **2020**, *451*, 227780. [CrossRef]
19. Zhou, Y.; Ravey, A.; Péra, M.-C. Multi-mode predictive energy management for fuel cell hybrid electric vehicles using Markov driving pattern recognizer. *Appl. Energy* **2020**, *258*, 114057. [CrossRef]

20. Wang, Z.; Xie, Y.; Zang, P.F.; Wang, Y. Energy management strategy of fuel cell bus based on Pontryagin's minimum principle. *J. Jilin Univ. (Eng. Technol. Ed.)* **2020**, *50*, 36–43.
21. Nie, Z.; Jia, Y.; Wang, W.; Chen, Z.; Outbib, R. Co-optimization of speed planning and energy management for intelligent fuel cell hybrid vehicle considering complex traffic conditions. *Energy* **2022**, *247*, 123476. [CrossRef]
22. Xu, L.Y.; Liu, E.Z.; Liu, M.N.; Zhao, X.P.; Wang, T. Energy Management Strategy for Hybrid Fuel Cell/Battery Electric Tractor. *J. Henan Univ. Sci. Technol. (Nat. Sci.)* **2019**, *40*, 80–86. [CrossRef]
23. Gao, X.L.; Shi, S.B. Tractor vehicle dynamics. In *Tractor Automobile Engineering Volume 2: Vehicle Chassis and Theory*; China Agriculture Press: Beijing, China, 2009; pp. 178–206.
24. Liu, M.N. Study on Design Theory and Control Strategy of Electric Tractor. Ph.D. Thesis, Vehicle Engineering, Xi'an University of Technology, Xi'an, China, 2020.
25. Wang, Z.Z.; Zhou, J.; Wang, X. Research on Energy Management Model of Extended Range Electric Tractor Rotary Tiller Unit. *Trans. Chin. Soc. Agric. Mach.* **2023**, *54*, 428–438. [CrossRef]
26. Xu, L.F.; Lu, L.G.; Li, J.Q.; Ouyang, M.G. Modeling and Simulation of a Hybrid Fuel Cell System and Energy Management Strategy. *J. Mech. Eng.* **2009**, *45*, 141–147. [CrossRef]
27. Yang, Y.; Hu, X.; Pei, H.; Peng, Z. Comparison of power-split and parallel hybrid powertrain architectures with a single electric machine: Dynamic programming approach. *Appl. Energy* **2016**, *168*, 683–690. [CrossRef]
28. van Harselaar, W.; Hofman, T.; Brouwer, M. Automated Dynamic Modeling of Arbitrary Hybrid and Electric Drivetrain Topologies. *IEEE Trans. Veh. Technol.* **2018**, *67*, 6921–6934. [CrossRef]
29. Zhang, J.J. Research on Control Strategy of Four-Wheel Drive Pure Electric Vehicles Taking into Account Regenerative Braking and Ride Comfort. Ph.D. Thesis, Vehicle Engineering, Chongqing University, Chongqing, China, 2021.
30. Wang, T.; Li, Q.; Han, Y.; Hong, Z.; Liu, T.; Chen, W. Fuel Cell Hybrid Power Generation System Equivalent Hydrogen Consumption Instantaneous Optimization Energy Management Method. *Proc. CSEE* **2018**, *38*, 4173–4182+4323.
31. Ehsani, M. Modern Electric, Fuel cell. In *Hybrid Electric, and Fuel Cell Vehicles Fundamentals, Theory and Design*, 2nd ed.; China Machine Press: Beijing, China, 2010; pp. 363–385.
32. Pacejka, H.B.; Bakker, E. The magic formula tyre model. *Veh. Syst. Dyn.* **1992**, *21*, 1–18. [CrossRef]
33. Gan, X.Q.; Zhang, H.B.; Zhang, N. An Equivalent Fuel Consumption Minimization Strategy for Fuel Cell Vehicles Based on Variable Equivalent Coefficient. *J. Tongji Univ. (Nat. Sci.)* **2021**, *49*, 224–230. [CrossRef]

World Electric Vehicle Journal

Article

Free Riding of Vehicle Companies under Dual-Credit Policy: An Agent-Based System Dynamics Model

Zhong Zhou and Yuqi Shen *

School of Economics & Management, Shanghai Institute of Technology, Shanghai 201418, China; zzhou@sit.edu.cn
* Correspondence: 216121143@mail.sit.edu.cn

Abstract: The dual-credit policy promotes green transition in automobile companies. This paper investigates the dual-credit policy framework in the Chinese automotive industry, with a focus on the phenomenon of free riding. This occurs when traditional vehicle manufacturers within an alliance benefit from the excess credits generated by a transitioning vehicle company without fully committing to their own green transitioning. The focus of this study lies on an alliance constituted by a transitioning vehicle company in partnership with two traditional vehicle manufacturers, all interconnected via equity ties. Utilizing an agent-based system dynamics model, this study explores the strategic behaviors emerging from such credit collaborations and their consequent effects on operational efficiency and financial performance. The findings reveal that 1. free riding negatively impacts the transitioning company's revenue but benefits the alliance by easing transition pressures and boosting collective performance; 2. stricter policies increase intra-alliance credit transfers and performance, while lower credit prices reduce transfer value and harm the transitioning company's earnings. This study implies that transitioning vehicle companies with equity-linked partners can benefit from a nuanced understanding of how policy mechanisms interact with alliance dynamics under free riding. By adjusting credit transfer strategies in line with market conditions and policy trends, they can better navigate the dual-credit policy landscape, balancing individual profitability with the needs of the broader alliance and long-term sustainability goals.

Keywords: dual-credit policy; system dynamics; horizontal alliance; free riding

Citation: Zhou, Z.; Shen, Y. Free Riding of Vehicle Companies under Dual-Credit Policy: An Agent-Based System Dynamics Model. *World Electr. Veh. J.* **2024**, *15*, 227. https://doi.org/10.3390/wevj15060227

Academic Editor: Michael Fowler

Received: 12 April 2024
Revised: 17 May 2024
Accepted: 22 May 2024
Published: 23 May 2024

1. Introduction

"Parallel Management Regulation for Corporate Average Fuel Consumption and New Energy Vehicle Credits for Passenger Cars", conventionally referred to as the dual-credit policy, was jointly issued and enforced by the government of China in 2017 [1]. This policy is specifically designed for passenger vehicle companies, introducing a groundbreaking regulatory framework that measures automotive energy efficiency through a credit system. Furthermore, it imposes credit limits and implements stringent punitive actions to enforce compliance with energy efficiency standards [2]. The dual-credit policy boosts technological innovation and transition within vehicle companies. It holds significant importance for the development of the global new energy vehicle industry [3].

The dual-credit policy permits credit transfers among affiliated enterprises [4], thereby indirectly promoting the formation of horizontal alliance among equity-related enterprises [5]. Within alliances, the transfer of credits between companies serves as a primary mode of collaboration. However, credit cooperation could potentially give rise to free riding behavior, wherein one party gains resources without bearing the associated costs [6]. For example, cooperation among the Shanghai Automotive Group, the Volkswagen Group of Germany, and General Motors of the United States has formed the horizontal alliance [7]. Within this alliance, Chinese vehicle companies display a pronounced tendency towards transitioning to new energy vehicles. The production and sales share of new energy vehicles within joint venture (JV) portfolios remains significantly diminutive, resulting in their substantial need to acquire large amounts of credits externally. Chinese vehicle companies confront the dilemma of whether to transfer surplus credits at no cost to JV partners. Research on the dual-credit policy has primarily concentrated on the strategic production decisions of vehicle manufacturers, underscoring the significance of collaborative efforts that harness complementary resources to amplify overall performance and expedite the transition to cleaner energy solutions [8,9]. Nonetheless, a gap exists in understanding the impacts that arise when the collaborative dynamics within horizontal alliances composed of vehicle companies become intricate due to free riding behaviors.

This study adopts an agent-based perspective to delineate the internal structure and interactive traits of equity-related horizontal alliances. It utilizes system dynamics to create a model for technology management, production operations, and credit settlement in passenger vehicle companies under the influence of the dual-credit policy. Through this process, it devises several comparative scenarios that effectively simulate how traditional companies might exhibit free riding behavior within a horizontal alliance context, as well as examining the strategic options available to those companies undergoing transition. The purpose of this study was to answer the following research questions:

(1) Faced with the free riding behavior of partners, would the supportive actions of transitioning vehicle companies harm their own and alliance performance?
(2) How does corporate transition and policy change affect performance under free riding behavior?

The innovative contributions and significance of this research lie in the following: ① It focus on free riding behavior and member strategies within equity-related horizontal alliances, which will aid in enriching the theoretical framework of strategic alliances. ② This study employs modeling tools to simulate annual rollovers, credit transfers among affiliates, and secondary market transactions under the framework of the dual-credit policy. These comprehensive simulations and the strategic decisions offer valuable groundwork for comprehending complex industrial policies analogous to the dual-credit policy. ③ This study provides guidance for enterprises to optimize their production strategies under the scrutiny of policy assessment, offering valuable feedback and suggestions for policymakers on the effective implementation of policies. The research positively contributes to the ongoing promotion of sustainable development within the new energy vehicle industry.

The remainder of this paper is structured as follows. Section 2 provides a summary of the literature relevant to this study. In Section 3, based on the analysis of policy mechanisms, we employed the system dynamics methodology to construct a model, defining the parameters of the model accordingly. Section 4 conducts a simulation analysis to examine the free riding behavior of conventional vehicle manufacturers within a horizontal alliance and evaluate the strategic options of transitioning vehicle companies in this context. Section 5 summarizes the findings and implications, along with suggesting avenues for future research.

2. Literature Review

2.1. The Dual-Credit Policy

Since 2017, academic research has extensively engaged with the intricacies of China's dual-credit policy, a regulatory framework designed to incentivize the production and sale of new energy vehicles (NEVs) while promoting overall energy efficiency in the automotive industry [10]. On a macroeconomic level, studies have delved into the far-reaching implications of this policy, including its facilitation of industry-wide growth and market expansion [11,12], its role in mitigating greenhouse gas emissions and contributing to climate change mitigation targets [13], and its stimulation of technological progress and innovation in both NEV and conventional vehicle technologies [14,15]. Collectively, this body of research underscores the dual-credit policy's multifaceted nature, its ability to reshape industry dynamics, and its contribution to the broader objectives of environmental sustainability.

From a more granular, micro-economic perspective, the literature has explored the policy's influence on specific aspects of automobile manufacturing and market dynamics. These inquiries have ranged from the direct impact on vehicle production volumes and the resultant changes in energy efficiency standards of gasoline vehicles [16] to strategic timing considerations for manufacturers' investments in electric vehicle (EV) production capacity and portfolio transitions [17]. The policy's effect on financial decisions regarding green technology investments and research and development (R&D) [18], as well as the strategic interplay between competition and cooperation among manufacturers in their production and credit trading activities [1,8], have also been meticulously analyzed. Moreover, the implications for supply chain management, particularly the coordination and alignment of efforts between upstream suppliers and downstream manufacturers to meet the dual-credit criteria [9], have been a significant focus of investigation.

Most studies on the dual credit policy typically take into account the purchase of NEV credits as an integral aspect of the policy [19]. The rollover, transfer, and trade of credits are pivotal features of the dual-credit policy, which are integrated into the decision-making framework along with technology management and production operations. Despite this, this specific aspect has not received sufficient attention, particularly from the viewpoint of alliances forged by equity-related enterprises.

2.2. Free Riding Behavior

Existing research commonly asserts that the formation of strategic alliances serves as a pivotal approach for companies to establish competitive advantages and elevate performance [5]. Horizontal alliances comprising competitors within the same industry segment typically necessitate external facilitation. However, they exhibit intricate game-theoretic dynamics among members due to technological spill-over effects.

There is a substantial amount of research dedicated to vertical strategic alliances, predominantly focusing on how supply chain contract choices affect costs and profits [20]. Regarding horizontal alliances, scholars have investigated the collaborative mechanisms for sharing resources and the associated arrangements for distributing benefits among participating parties [21], the pricing decisions of company alliances in the supply chain [22], and the influence of partner resemblance on horizontal alliances [23]. Under the dual-credit policy, researchers have looked at coordination contracts from the vantage point of component suppliers and vehicle companies [24]. Resources obtained through complementary partnerships enable enterprises to enhance alliance performance [25]. However, there is a dearth of inquiry into production collaboration within enterprise alliances when free riding behavior is present.

The phenomenon of free riding in common-interest entities garners attention from experts. Free riding refers to a situation where one party benefits from the sales efforts or resources of another party without incurring the corresponding cost [6]. Differentiated pricing in dual distribution channels results in varying degrees of consumer free riding behavior, affecting sales efforts and service levels across channels [26,27]. Researchers like Liu et al. [28] and Guo et al. [29] have approached the topic from a consumer angle, addressing pricing and service level issues for businesses and supply chains across different channel combinations. Inter-firm free riding also occurs, such as instances where both companies and consumers exploit services provided by traditional retailers [30]. Scholars have used evolutionary game theory and similar methods to analyze supplementary benefits attached to free riding behavior in supply chain structures [6] and identify key determinants that weaken free riding behavior in supply chains [31] to mitigate the detrimental effects of free riding on corporate interests. While most existing literature addresses free riding from consumer and supply chain viewpoints, supply chain members need to carefully design incentive mechanisms and coordination strategies to balance individual interests with collective emission reduction goals, while mitigating the detrimental effects of free riding [32]. However, in the field of the new energy vehicle industry, existing research fails to take into account the impact of free riding behavior on corporations' manufacturing choices within the framework of a horizontal alliance.

In conclusion, current research on the dual-credit policy mostly emphasizes production decisions made by companies within the supply chain. in the context of the dual-credit policy, cooperation mechanisms and performance improvements via resource complementary partnerships are emphasized, with a gap in understanding how these dynamics affect production collaboration amidst free riding. This behavior, characterized by one party benefiting without equal investment, garners interest in management circles, especially concerning its implications across dual distribution channels and in balancing individual and collective goals within supply chains. Thus, there is a clear need to bridge this gap by examining free riding's influence on manufacturing strategies within horizontal alliances more closely.

3. Agent-Based System Dynamics Modeling Approach

3.1. Model Framework

The dual-credit policy plays a pivotal role in forging a community of interests among certain Chinese vehicle companies, with its policy framework consisting of three core mechanisms [4]: (1) the calculation and offset rules for corporate average fuel consumption credits (CAFC) and new energy vehicle credits (NEV); (2) the carryover and inter-affiliate transfer of CAFC; and (3) the carryover and trading of NEV credits. If a company incurs negative credits that cannot be cleared in a given year, it faces penalties such as the suspension of new product registration and production. After clearing the negative credits, excess NEV can be sold in the credit market. According to the dual-credit policy, enterprises are considered related if they hold direct or indirect equity totaling 25% or more in another domestic passenger vehicle company, or if they are both held by a third party with direct or indirect equity totaling 25% or more.

Considering the reality of China's automotive industry, as depicted in Figure 1, this study focuses on a horizontal alliance consisting of three equity-related enterprises. All three companies are domestic passenger vehicle companies engaged in both internal combustion engine vehicles and new energy vehicles, with equity types classified as joint venture, joint venture, and Chinese enterprise. This assumption aligns with "China Passenger Vehicle Enterprise Average Fuel Consumption and New Energy Vehicle Credits Calculation Sheet", as well as the equity relationships among major Chinese passenger vehicle producers.

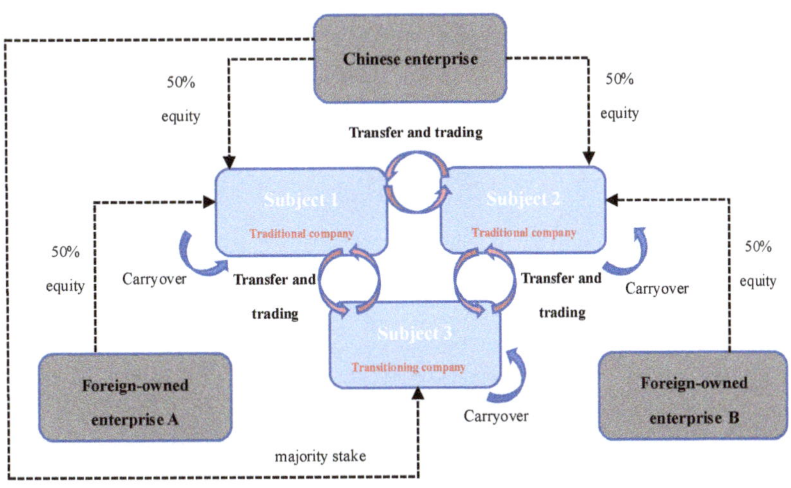

Figure 1. Structure of equity-linked horizontal alliance system.

Adhering to the agent-based modeling philosophy, each entity is an independent decision-maker despite complex equity linkages. Each agent in the model can possess distinct decision-making rules and behavioral logic, reflecting the diversity and complexity of real-world market participants [33]. System dynamics is a cross-disciplinary research method applied to comprehend and analyze complex systems with dynamic characteristics [34]. It involves designing stable information feedback structures to describe and analyze the behavior of time-varying systems, thereby exploring the effectiveness of strategies and their optimization. The combination of agents and modeling is intended to solve complex systems [35]. This approach allows for the modeling of intricate systems, the prediction of future trends, and the evaluation of policy interventions under different scenarios, contributing to a deeper understanding and more informed decision-making in managing complex real-world problems. As a pioneering integrated industrial policy that directly influences the management and operation of both emerging and established technologies, the dual-credit policy is designed around fuel consumption targets linked to research and development endeavors, driving range indicators, and production volume, which collectively constitute several pivotal credit calculation parameters. Based on the dual-credit policy mechanisms and drawing upon existing research, a system dynamics modeling method is adopted. The affected single decision-making entity is decomposed into four primary sub-models: the production sub-model, the R&D sub-model, the credit calculation sub-model, and the credit interaction sub-model, with the causal loop diagram illustrated in Figure 2, where arrows denote the interdependencies among key variables.

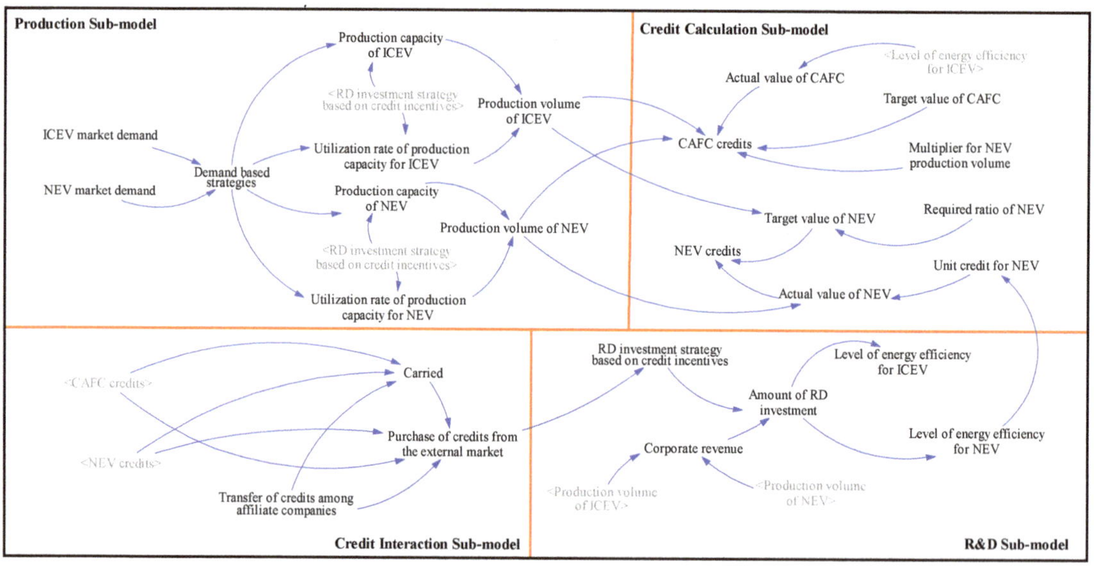

Figure 2. Causal loop diagram.

From the causal loop diagram, it is evident that in the production and R&D input sub-models, R&D strategies influence vehicle performance, which, combined with production decisions, determines the CAFC and NEV credit values in the credit calculation sub-model. The surplus or deficit of credits dictates the amount needed to offset and zero out annually. When internal credit transfers and rollovers fail to meet the required credit offset, companies must purchase credits from external markets. This feedback loop influences the company's production and R&D strategies, prompting adjustments in vehicle production allocation and investment in R&D.

3.2. Stock and Flow Diagrams of Sub-Models

3.2.1. Production Sub-Model

The production sub-model reflects the automotive production processes and decision-making within a vehicle company. Fundamentally, vehicle companies engage in profit-seeking behavior through external market activities and refine their strategies internally. considering the external characteristics of the dual credit policy, there exists a more intimate relationship of response between enterprises and the market as two main stakeholders. This is largely due to the distinct preferences of consumers towards traditional ICEVs and NEVs in terms of attributes and purchase decisions [36]. Consequently, under the impetus of respective market demands, automobile manufacturers primarily adjust the production volume of both vehicle types by manipulating production capacity and capacity utilization rates [11]. As shown in Figure 3, the company produces traditional fuel vehicles and new energy vehicles according to market demands and adjusts production based on purchasing credits from the external credit market. This study assumes that over time, market demand changes, and companies respond to these changes, indicating their operational strategies in both internal combustion engine vehicles and new energy vehicle sectors.

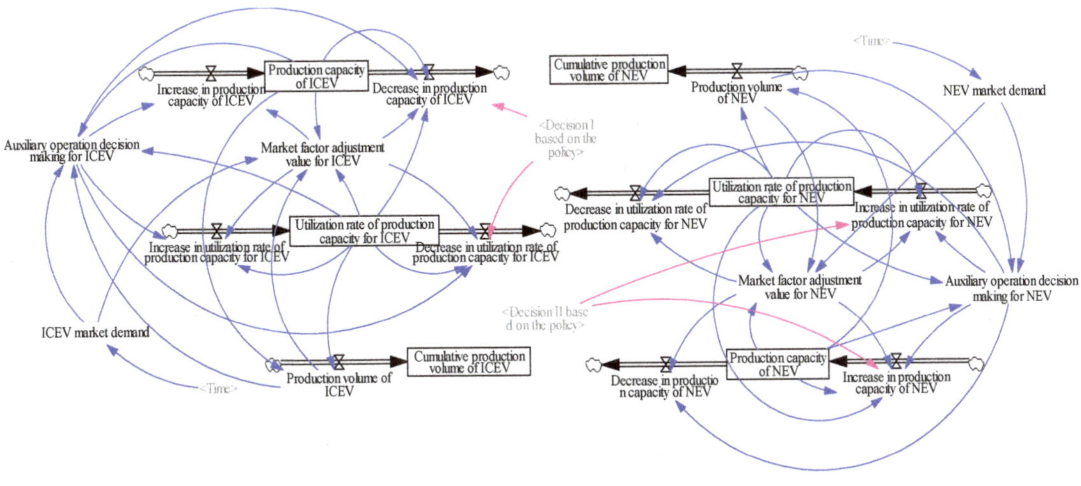

Figure 3. Production sub-model stock flow diagram.

The operating mechanisms for internal combustion engine vehicles and new energy vehicles are detailed in Table 1. When market demand exceeds current production capacity, if the demand increase is less than the potential utilization rate increase, the vehicle company first boosts the utilization rate; if higher, it simultaneously raises capacity while increasing the utilization rate. Conversely, when demand falls below current production, a threshold for capacity utilization is established. If actual utilization is lower than the threshold, it suggests severe overcapacity, warranting a rational decision to reduce capacity. If the adjusted utilization rate after considering market decline is above the threshold, the vehicle company merely adjusts the utilization rate; otherwise, it reduces both utilization and capacity.

Table 1. Analysis of the decision-making mechanism of passenger car producers in response to market demand.

Market Situation			Decision-Making Mechanisms	Market Factor Adjustments
Market demand ≥ Production		Increase in demand < 1—Utilization rate	Increased capacity utilization	(Demand—production)/capacity
		Increase in demand ≥ 1—Utilization rate	Increased capacity utilization increase production capacity	Increased capacity utilization to 100% Capacity increase: (demand—capacity)/capacity
market <Yield	Utilization < threshold	/	Reduce production capacity	Capacity—demand/utilization
	Utilization > threshold	Decrease in demand < Utilization rate—threshold	Reduced capacity utilization	(Production—demand)/capacity

Under the influence of the dual-credit policy, companies give priority to producing more new energy vehicles and, based on credit status, may opt for a moderate overproduction of new energy vehicles before considering reducing internal combustion engine vehicle production. The extent of new energy vehicle production increase and internal combustion engine vehicle production decrease is determined by the strictness of the dual-credit policy execution and the company's credit situation. The auxiliary variable "Policy-based Decision" will be further elaborated in the R&D sub-model.

3.2.2. R&D Sub-Model

R&D on engine technology is a primary means to improve fuel economy [22]. Improved fuel economy translates to a lower actual average fuel consumption, and similar logic applies to R&D investments in battery ranges for new energy vehicles [37]. The R&D sub-model is shown in Figure 4. In response to the dual-credit policy, vehicle companies can also consider boosting automotive R&D investment. In the R&D sub-model, the company invests in R&D based on a percentage of revenue. However, the uncertainty in R&D investment and the non-linear nature of learning curves can complicate the analysis of their impact on enterprise strategies. Therefore, this model assumes that passenger vehicle companies adopt a fixed R&D investment strategy. Drawing from annual reports of automotive groups and historical data from "China Passenger Vehicle Enterprise Average Fuel Consumption and New Energy Vehicle Credits Calculation Sheet", the relationship between R&D investment and vehicle performance is linearly modeled as follows:

$$\text{Actual value of CAFC} = -0.0005 \times \text{Cumulative R\&D investment for ICEV} + 7 \tag{1}$$

$$\text{Unit credit for NEV} = (2 \times \text{Cumulative R\&D investment for NEV} + 3600)/1900 \tag{2}$$

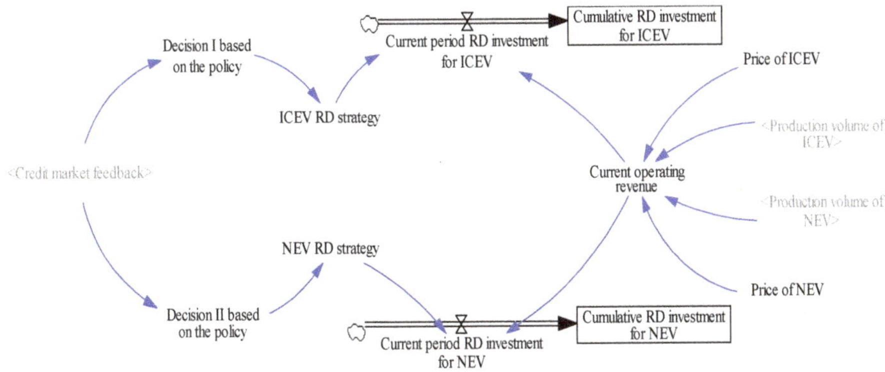

Figure 4. R&D sub-model stock flow diagrams.

Moreover, if the two types of negative credits cannot be offset, the vehicle company must purchase NEV credits from the external market. This transaction provides feedback to the company, shaping a policy-based decision that not only adjusts the production configuration in the production sub-model but also increases the revenue percentage allocated to R&D expenses for the current period.

3.2.3. Credit Calculation Sub-Model

In the credit calculation sub-model, the credits for CAFC and NEV are computed based on the company's produced vehicle quantities and performances, following policy-regulated formulas [38]. Since the introduction of the dual-credit policy in 2017, it has achieved positive outcomes in supporting new energy vehicle development. To optimize policy effectiveness and promote high-quality development of energy-saving and new energy vehicle industries in line with carbon neutrality goals, relevant parameters for credit calculations have become stricter over the years [39]. For instance, the multiplication factor for NEV production declined from 5 times in 2017 to 1.6 times in 2023, and the required proportion of NEV credits rose from 10% in 2019 to 18% in 2023. The two formulas for calculating CAFC and NEV credits as per the dual-credit policy are listed in Table 2. Using vehicle production quantities, average fuel consumption rates, and policy-controlled parameters, the model calculates the company's annual CAFC and NEV credits.

Table 2. Average fuel consumption and new energy vehicle points calculation rules.

Formula Number	Key Function Expressions and Interpretations	Variable Description
(3)	Actual value of CAFC $A_{CAFC} = \frac{CAFC \times P_{ICEV}}{P_{ICEV} + W \times P_{NEV}}$	A_{CAFC}—Actual value of CAFC CAFC—Corporate average fuel consumption of ICEV P_{ICEV}—Annual production of ICEV P_{NEV}—Annual production of NEV W—NEV production multiplier (policy control)
(4)	Corporate average fuel consumption credits $Credit - ICEV = (T_{CAFC} - A_{CAFC}) \times P_{ICEV}$	Credit-ICEV—Corporate average fuel consumption credits TCAFC—Target value of CAFC (policy control)
(5)	New energy vehicle credits $Credit - NEV = P_{NEV} \times U - P_{ICEV} \times R$	Credit-NEV—New energy vehicle credits R—Required ratio of NEV (policy control) U—Unit credit for NEV (policy control)

3.2.4. Credit Interaction Sub-Model

In the credit interaction sub-model, the vehicle company's current CAFC positive credits can carryover or transfer, while CAFC negative credits can be offset by carryover credits, transferred credits, or NEV positive credits. Current NEV credits can carryover, or trade within related companies, and NEV negative credits can be offset by carryover credits or traded credits. As shown in Figure 5, this study stipulates that the order of credit settlement for companies is carryover, transfer between related companies, and then trading in the external credit market. Vehicle companies first use carryover credits to offset negative credits. If there are still CAFC (NEV) negative credits after settling carryovers, the priority for transferring (trading) related companies' CAFC (NEV) positive credits is from Chinese-owned to joint venture companies. If there are still negative credits after settling with related companies, credits are purchased from the credit market. This assumption conforms to the parallel management rules of the dual-credit policy and reflects the actual conditions of passenger vehicle companies.

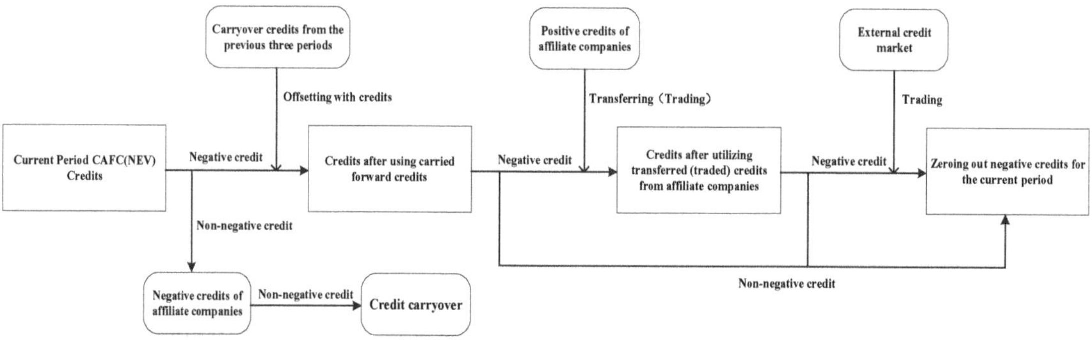

Figure 5. Credit accounting path for a single decision subject.

3.3. Model Parameter Settings

Data obtained from sources such as group annual reports and credit accounting disclosure tables reveal distinct characteristics in production configurations and credit situations between these two types of companies under the same group. As a representative case, this research selects the Shanghai Automotive Group (SAIC). Specifically, data for conventional companies A and B come from SAIC Volkswagen and SAIC-GM, while data for transitioning company C come from SAIC Motor Passenger Car Company. Table 3 presents the initial and fixed values of the variables for these three companies, providing reasonable estimates for their parameter values.

Table 3. Model initial values and fixed value settings.

Variable	Type	Value			Unit	Descriptive
		A	B	C		
Fuel vehicle production capacity	Initial value	2000	2000	1000	thousand	Estimated based on existing capacity data disclosed in the 2022 Annual Report of SAIC
Fuel vehicle capacity utilization	Initial value	80%	80%	80%	/	Derived from the proportion of ICEV production capacity utilization reported in the 2022 Annual Report of SAIC
Total R&D investment in fuel vehicles	Initial value	800	800	400	billion	Estimated based on the cumulative R&D investment figures from the annual reports of SAIC between 2013 and 2022
New energy vehicle production capacity	Initial value	100	100	200	thousand	Estimated using NEV production data compiled by the China Association of Automobile companies for SAIC
New energy vehicle capacity utilization rate	Initial value	10%	10%	30%	/	Based on the ratio of production volume in 2022 to total capacity
Total R&D investment in new energy vehicles	Initial value	10	10	100	billion	Estimated from the cumulative R&D investment figures in the Annual Reports of SAIC between 2013 and 2022
Average selling price of fuel vehicles	Constant	150	150	100	thousand	Estimated using the sales volume and revenue of ICEV as reported in the 2022 Annual Report of SAIC
Average selling price of new energy vehicles	Constant	200	200	150	thousand	Estimated using the sales volume and revenue of NEV as reported in the 2022 Annual Report of SAIC

Regarding the equity-related horizontal alliance under the dual-credit policy, this study explores the credit transfer strategies of transitioning vehicle companies in the context of free riding behavior. This study designs four different scenarios through parameter adjustments, as shown in Tables 4 and 5. The specific ideas and differences behind these scenarios are explained below:

- Scenario 1: Baseline scenario. Transitioning vehicle companies settle their own credits independently, and there is no credit transfer between equity-related vehicle companies.
- Scenario 2: Free riding scenario. The transitioning vehicle company considers choosing to partially or fully transfer its surplus credits to traditional vehicle companies. This eases the credit pressure on the traditional companies, weakening their motivation to cut down on fuel vehicle production, thereby exhibiting free riding behavior.
- Scenario 3: Impact of corporate transition. Building upon Scenarios 1 and 2, this scenario examines the credit transfer strategies of transitioning vehicle companies at varying degrees of transition. Three different situations are set up to form Scenario 3, with parameter adjustments detailed in Table 5.
- Scenario 4: Influence of policy regulation. Also building on Scenarios 1 and 2, this scenario investigates the impacts of varying stringency levels of the dual-credit policy. The policy adjustment parameter chosen is the requirement for the proportion of NEV credits. Details are presented in Table 5, where "time" represents the simulation time in the model. Considering the actual price of NEV credits in recent years, a base unit credit price of 2000 yuan is assigned in the baseline scenario. When the policy tightens, the availability of tradable credits in the market becomes scarce, leading to an increase in the unit credit price. Conversely, when the policy relaxes, the unit credit price decreases.

Table 4. Scenarios 1 and 2 vehicle decision adjustment parameters.

	Scenario 1—Separate Settlements	Scenario 2—Half Transfers	Scenario 2—Total Transfer
Transfer coefficient	0	0.25	0.5
Adjustment of ICEV production for ICEV	0%	+10%	+20%

Table 5. Transitioning vehicle company C's production strategy and policy credit adjustment parameters.

Transitional car decision-making	Slow	Base	Fast
Adjustment of NEV production	−20%	0%	+20%
Policy adjustment	**Base**	**Austerity**	**Easing**
NEV points ratio requirements	IF THEN ELSE (Time $\leq$ 18, IF THEN ELSE (Time $\leq$ 2, 0, 0.1 + 0.02 $\times$ (Time − 3)), 0.4)	IF THEN ELSE (Time $\leq$ 18, IF THEN ELSE (Time $\leq$ 2, 0, 0.1 + 0.03 $\times$ (Time − 3)), 0.55)	IF THEN ELSE (Time $\leq$ 18, IF THEN ELSE (Time $\leq$ 2, 0, 0.1 + 0.01 $\times$ (Time − 3)), 0.25)
Credit price	0.2	0.3	0.1

4. Simulation Analysis

4.1. Sensitivity Analysis

This paper conducts relevant tests on the model. Based on a thorough review of existing research results and actual conditions, it meticulously revises and enhances the causal diagrams and flowcharts within the model, ensuring that extraneous research variables are not included. With an annual time step, a total of 60 simulation periods are executed, starting from an initial time of 0. The system dynamics model constructed in this research is simulated using Vensim PLE 7.3.5.

Sensitivity analysis examines whether conclusions significantly change in a manner relevant to the objective when assumptions vary within a reasonable range of uncertainty. To conduct a sensitivity analysis on the revenues of automobile manufacturers, three indicators—NEV market demand, ICEV market demand, and the credit transfer coefficient—were each increased by 5%. The simulation forecasts obtained using Vensim PLE 7.3.5 were compiled into a dataset, represented as Table 6. Observations from the table indicate that despite varying influencing factors, the trends in revenue changes align consistently. The simulation outcomes closely match the known patterns observed in the real world, thereby validating the system dynamics model established in this paper.

Table 6. Sensitivity analysis.

Time	Revenue	NEV Market Demand	ICEV Market Demand	Transfer Coefficient
	–	+5%	+5%	+5%
1	463.947	463.947	463.947	463.947
2	474.948	478.945	494.698	474.948
3	480.334	485.132	499.737	480.334
4	490	495.25	509.25	490
5	500	506	519	500
6	510	516.75	528.75	510
7	520	527.5	538.5	520
...	...	...	...	...
53	1012.67	1058.49	1019.52	1012.27
54	1024.4	1071.07	1031.01	1023.99
55	1036.18	1083.7	1042.55	1035.76
56	1048	1096.37	1054.13	1047.56
57	1059.86	1109.09	1065.76	1059.41
58	1071.77	1121.85	1077.43	1071.31
59	1083.72	1134.67	1089.15	1083.25
60	1095.72	1147.53	1100.91	1095.24

4.2. Free Riding Behavior and Credit Transfer Decisions

Under Scenario 1 and Scenario 2, the annual operating revenue of transitioning vehicle companies C is shown in Figure 6. As a company involved in both traditional fuel cars and new energy vehicles, this revenue includes the performance of its own two types of vehicle businesses, along with income from selling surplus NEV credits or expenses from purchasing NEV credits. Additionally, the Chinese capital group holds a 50% stake in each of the two traditional companies, thereby gaining a corresponding share of their operational performance. In Scenario 1, where all three companies settle credits independently, the annual revenue of the transitioning company rapidly rises during the first 20 periods and slows down after the transition, as it accumulates more credits by improving vehicle performance and increases corporate performance through credit trading. In Scenario 2, the transitioning company C transfers credits to its equity-affiliated horizontal alliance partners, traditional companies A and B, offsetting their negative credits free of charge.

As seen in Figure 6, when transitioning company C transfers surplus credits to its equity-related companies in Scenario 2, its operating revenue is lower compared with settling credits individually. The underlying reason is that traditional companies A and B benefit from the credit transfer, reducing their pressure on credits and consequently slowing down the pace of cutting production of fuel cars. With reduced demand for purchasing credits from the external market, there is also a decreased incentive effect on adjusting the production strategies of traditional companies A and B, leading to a decrease in the annual average proportion of new energy vehicle production. This causes a higher amount of negative credits annually compared with Scenario 1, necessitating the transitioning company C to transfer more credits over a longer period, which reduces the number of NEV credits available for sale in the credit trading market. If all credits can be transferred, the annual average revenue of the transitioning company C decreases by 3.63% compared with Scenario 1. Conversely, as shown in Figure 7, the overall operating performance of the group indicates that in Scenario 1, traditional companies A and B purchase credits externally. This causes them to reduce production and hence lower their operating income, reducing the total profits attributable to the Chinese capital group. Under Scenario 2, with smaller production cuts by traditional companies, the difference in sales revenues from fuel cars outweighs the income gap from the transitioning company C selling credits. When transitioning company C transfers all its surplus credits, the group attains the highest profit.

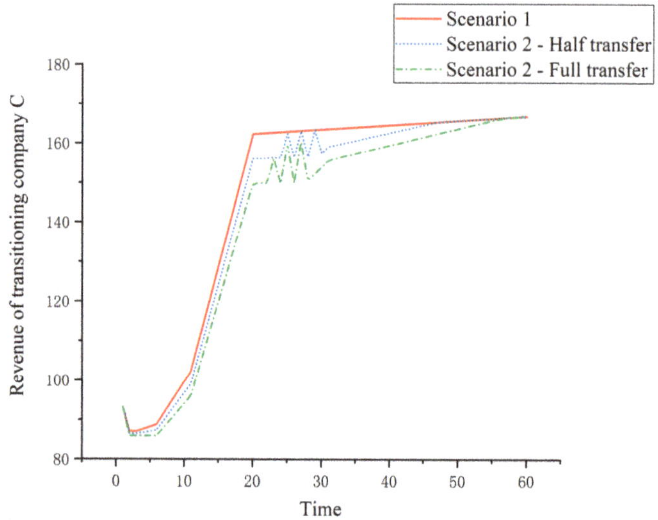

Figure 6. Revenue of transitioning company C in Scenarios 1–2.

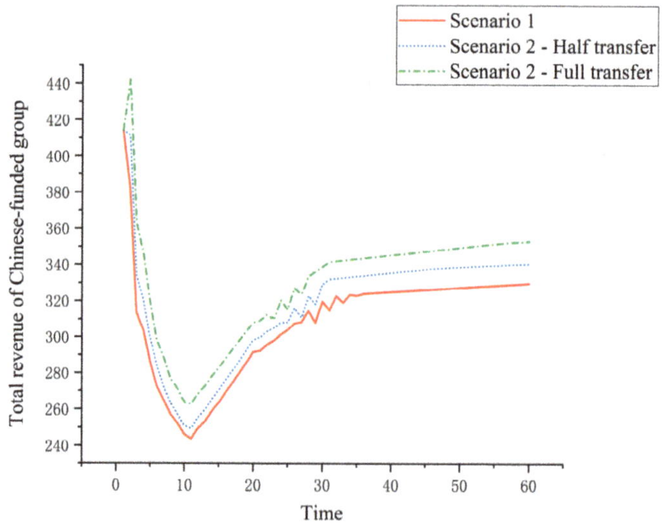

Figure 7. Total revenue of Chinese group in Scenarios 1–2.

4.3. Impacts of Company Transition

Scenario 3 compares the annual revenue results of transitioning company C under different transition speeds. The annual revenue of the transitioning company is illustrated in Figure 8. When the transitioning company C undergoes rapid transition and the credits are settled independently among the three companies, company C has the highest revenue. On the other hand, when the transitioning company C transitions slowly and transfers all current-period credits, its revenue is at its lowest. Regardless of the transition speed, the revenue of transitioning company C still inversely correlates with the transfer coefficient. Furthermore, when transitioning quickly, not transferring any credits leads to a 3.71% higher average annual revenue for company C compared with fully transferring credits, and this percentage increases to 3.45% when transitioning slowly. Therefore, the higher the proportion of surplus credits transferred by transitioning company C, the lower its revenue, and the faster the transition speed, the greater the cumulative revenue disparity. Additionally, the speed of the transition of company C does not affect the timing at which the average annual revenue plateaus across scenarios.

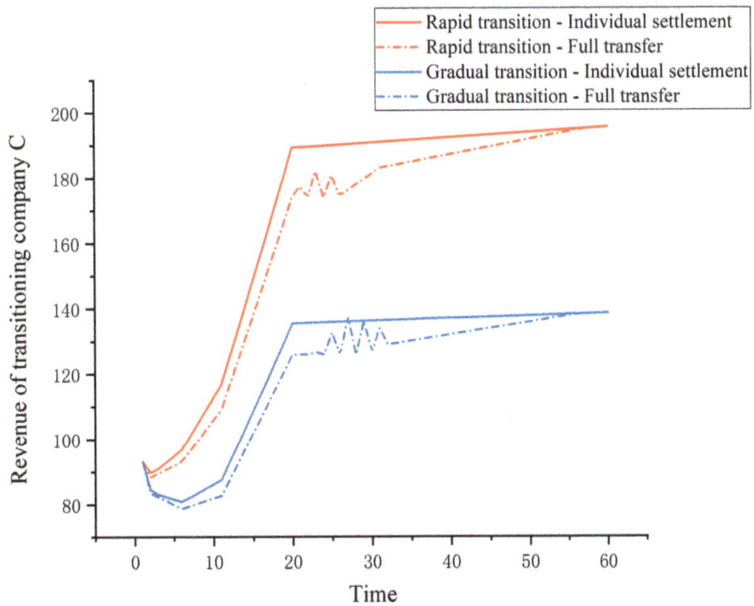

Figure 8. Revenue of transitioning company C in Scenario 3.

For the Chinese capital group, the total operating revenue resulting from the transition of company C is depicted in Figure 9. Similar to Scenarios 1 and 2, under the rapid or slow transition of company C, the higher the credit transfer ratio, the higher the group's total operating income. This is because the transition of company C does not reverse the degree of free riding by traditional companies A and B, who continue to adjust production according to the credit transfer ratio. Moreover, the compression of fuel car production by traditional companies A and B affects the group's total revenue in the early stages, with the allowance of credit transfer enhancing early-stage performance. An increase in the transition speed of company C brings about more revenue in later stages; thus, when company C transforms rapidly and transfers all surplus credits, the group achieves the highest total revenue. Independently settling credits does not affect traditional companies' purchases of credits due to the transition speed of the transitioning company. However, when all credits are transferred by the transitioning company, the faster the transition speed, and the fewer credits the group needs to buy from the market.

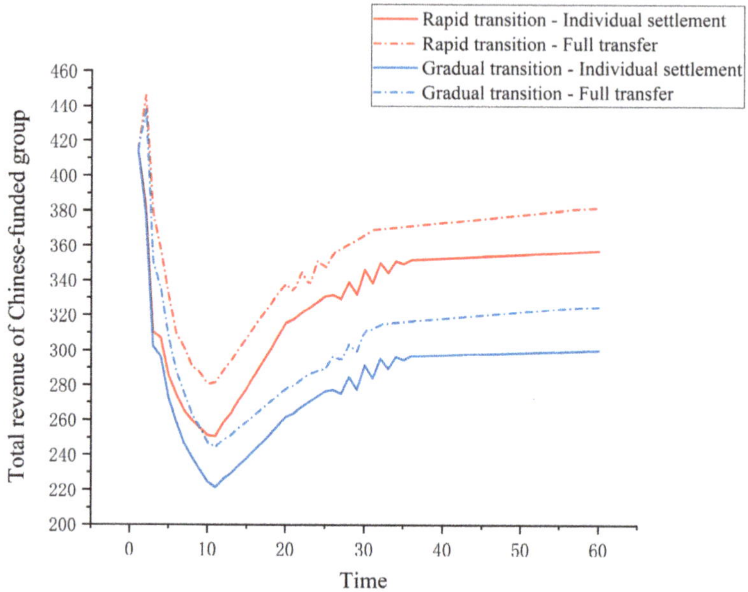

Figure 9. Total revenue of Chinese group in Scenario 3.

4.4. Impact of Policy Regulation

Scenario 4 compares the annual revenue of transitioning company C under varying degrees of policy tightness, as shown in Figure 10. When policies become stringent, the NEV credit ratio requirement increases, and the three companies settle credits independently, and the revenue of transitioning company C is at its highest. When policy requirements tighten, companies receive fewer credits, and the supply of credits in the external market also shrinks, causing credit prices to rise. This increased credit price benefits transitioning company C, which does not need to assist affiliated companies, resulting in a 6.22% higher average annual revenue when settling credits independently compared with fully transferring, while under loose policy conditions, the difference is only 1.63%. As seen in Figure 10, under different levels of policy tightness, the revenue of transitioning company C continues to form an inverse relationship with the transfer coefficient, with credit prices amplifying differences in corporate revenues. Furthermore, if transitioning company C chooses to transfer all surplus credits to traditional companies A and B, its revenue stays relatively constant in the first 20 periods. Due to policy adjustments, the significant negative credit balance from the large-scale production of fuel cars by traditional companies A and B leaves transitioning company C with no surplus credits to sell and thus no related revenue.

Figure 11 shows the total operating revenue of the Chinese capital group. Based on the parameters of Scenarios 1 and 2, under both relaxed and tightened policy regulations, if a horizontal alliance strategy involves the full transfer of surplus credits, the group realizes higher revenue performance. Under tightened policy conditions, compared with relaxed or standard policy scenarios, joint ventures generate more negative credits, leading to a lower level of total revenue in the early to mid-simulation periods due to a combination of increased credit prices caused by a reduction in market supply and the compressed production of fuel cars following the purchase of credits. Once the three companies complete their transitions and contribute credits to the market, the total revenue of the Chinese capital group rebounds and exceeds that under other policy regulation scenarios.

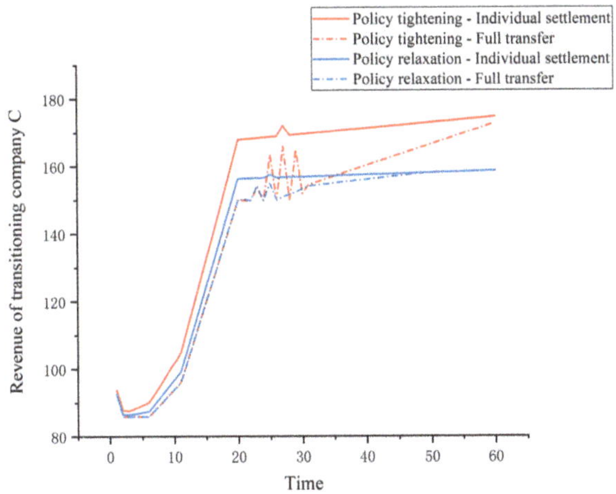

Figure 10. Revenue of transitioning company C in Scenario 4.

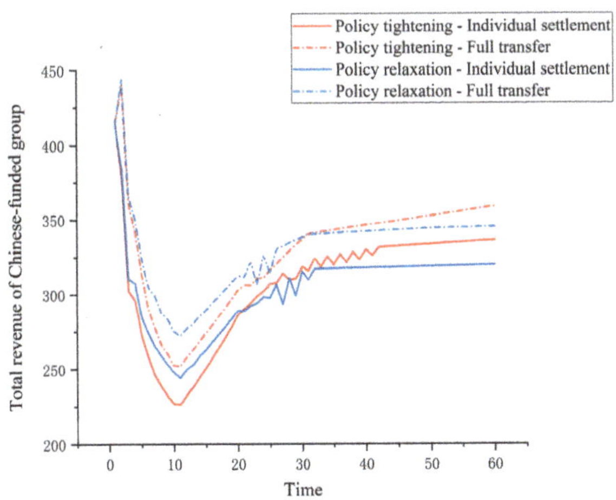

Figure 11. Total revenue of Chinese group in Scenario 4.

5. Conclusions and Implications

5.1. Conclusions

This study focuses on a specific form of affiliation that arises in response to the dual-credit policy—a horizontal equity-based alliance. In this alliance, a representative Chinese-funded transitioning vehicle company collaborates with two joint venture traditional company collaborates. This study employs agent-based modeling concepts and develops a system dynamics model to explore the effects of policy mechanisms and credit free riding behavior on company operations.

Through multi-scenario simulation analysis, the following is demonstrated: (1) When the transitioning vehicle company allows credit transfers, an increase in the transfer ratio negatively impacts the transitioning company's revenue performance, but it weakens the willingness for drastic production cuts among traditional auto companies, thus indirectly contributing to improved overall alliance performance. (2) Under free riding, accelerating the transition towards new energy vehicles amplifies the effect on the transitioning company's performance. Stringent policy requirements enhance the role of increasing the credit transfer ratio in boosting overall alliance performance. A decline in credit prices mitigates the extent to which a high credit transfer ratio depresses the transitioning company's revenue.

According to the findings, traditional vehicle companies, considering factors such as market demand for fossil-fuel vehicles and their business strategies, face difficulties in transitioning to green technology. And stringent policy requirements could lead to profit losses due to insufficient credit generation. However, the policy allowing credit transfers provides traditional vehicle companies within the alliance an opportunity to "free-ride" on the credits of transitioning companies. It alleviates their immediate transition pressures under the policy, potentially providing a positive effect on the alliance. It is also noted that such free riding behavior might dampen the transitioning company's enthusiasm for green innovation and transition.

5.2. Implications

For practitioners, our research underscores the importance of strategic decision-making in credit allocation within alliances. Transitioning vehicle companies must delicately balance individual profitability against the collective performance of the alliance. Our findings suggest that, practically, these companies could adopt flexible credit transfer strategies that adjust based on market conditions and policy stringency. By doing so, they can maximize the value of their credits while supporting the gradual transition of traditional partners towards producing electric vehicles.

In the domain of electric vehicle industry evolution, this study enriches the understanding of strategic management and organizational behavior in the context of environmental policy-driven collaborations. It introduces a new dimension to the analysis of coordination strategies within horizontal alliances, especially under regulatory frameworks that incentivize green transition. The integration of agent-based modeling and system dynamics offers a powerful toolset for analyzing complex systems influenced by policy mechanisms, setting a foundation for future theoretical advancements.

The present study has certain limitations that call for further research. It assumes that the transitioning vehicle company transfers credits gratuitously and with a fixed transfer ratio. It does not consider contractual arrangements for credit transfers or the dynamic adjustment of the transfer ratio. Moreover, the technical performance of vehicles plays a critical role in the effectiveness of the dual-credit policy. Future studies could delve into innovation decisions made by enterprises in horizontal strategic alliances with free riding behavior under the dual-credit policy.

Author Contributions: Methodology, Z.Z.; resources, Z.Z. and Y.S.; data curation and analysis, Z.Z. and Y.S.; writing—original draft preparation, Z.Z. and Y.S.; writing—review and editing, Z.Z. and Y.S.; supervision, Z.Z. All authors have read and agreed to the published version of the manuscript.

Funding: This research was funded by the National Social Science Fund of China, grant number 19CGL056.

Data Availability Statement: Data available on request due to restrictions eg privacy or ethical. The data presented in this study are available on request from the corresponding author.

Conflicts of Interest: The authors declare no conflicts of interest.

References

1. Ma, M.; Meng, W.; Huang, B.; Li, Y. The influence of dual credit policy on new energy vehicle technology innovation under demand forecast information asymmetry. *Energy* **2023**, *271*, 127106. [CrossRef]
2. Li, B.; Chen, Y.; Cao, S. Carrot and stick: Does dual-credit policy promote green innovation in auto firms? *J. Clean. Prod.* **2023**, *403*, 136863. [CrossRef]
3. Li, Y.; Zhang, Q.; Li, H.; Tang, Y.; Liu, B. The impact of dual-credit scheme on the development of the new energy vehicle industry. *Energy Procedia* **2019**, *158*, 4311–4317. [CrossRef]
4. Li, J.; Ku, Y.; Yu, Y.; Liu, C.; Zhou, Y. Optimizing production of new energy vehicles with across-chain cooperation under China's dual credit policy. *Energy* **2020**, *194*, 116832. [CrossRef]
5. Panico, C. Strategic interaction in alliances. *Strateg. Manag. J.* **2017**, *38*, 1646–1667. [CrossRef]

6. Yan, N.; Zhang, Y.; Xu, X.; Gao, Y. Online finance with dual channels and bidirectional free-riding effect. *Int. J. Prod. Econ.* **2021**, *231*, 107834. [CrossRef]
7. Li, R.; Yan, J.-J.; Wang, X.-Y. Horizontal cooperation strategies for competing manufacturers in a capital constrained supply chain. *Transp. Res. Part E Logist. Transp. Rev.* **2024**, *181*, 103369. [CrossRef]
8. Wang, Y.; Zhang, X.; Cheng, T.C.E.; Wu, T.H. Choice of the co-opetition model for a new energy vehicle supply chain under government subsidies. *Transp. Res. Part E. Logist. Transp. Rev.* **2023**, *179*, 103326. [CrossRef]
9. Cong, L.; Jie, L.; Lu, S.; Yan, Z. Contract Design to Incentive Supplier Innovation under Dual-Credit Policy. *Chin. J. Manag.* **2022**, *19*, 928.
10. Yang, D.-X.; Yang, L.; Chen, X.-L.; Wang, C.; Nie, P.-Y. Research on credit pricing mechanism in dual-credit policy: Is the government in charge or is the market in charge? *Environ. Dev. Sustain.* **2023**, *25*, 1561–1581. [CrossRef]
11. Li, Y.; Zhang, Q.; Liu, B.; McLellan, B.; Gao, Y.; Tang, Y. Substitution effect of New-Energy Vehicle Credit Program and Corporate Average Fuel Consumption Regulation for Green-car Subsidy. *Energy* **2018**, *152*, 223–236. [CrossRef]
12. Ou, S.; Lin, Z.; Qi, L.; Li, J.; He, X.; Przesmitzki, S. The dual-credit policy: Quantifying the policy impact on plug-in electric vehicle sales and industry profits in China. *Energy Policy* **2018**, *121*, 597–610. [CrossRef]
13. He, X.; Ou, S.; Gan, Y.; Lu, Z.; Przesmitzki, S.V.; Bouchard, J.L.; Sui, L.; Amer, A.A.; Lin, Z.; Yu, R.; et al. Greenhouse gas consequences of the China "dual-credit" policy. *Nat. Commun.* **2020**, *11*, 5212. [CrossRef] [PubMed]
14. Chen, K.; Zhao, F.; Hao, H.; Liu, Z.; Liu, X. Hierarchical Optimization Decision-Making Method to Comply with China's Fuel Consumption and New Energy Vehicle Credit Regulations. *Sustainability* **2021**, *13*, 7842. [CrossRef]
15. Zhang, H.; Zhao, F.; Hao, H.; Liu, Z. Effect of Chinese Corporate Average Fuel Consumption and New Energy Vehicle Dual-Credit Regulation on Passenger Cars Average Fuel Consumption Analysis. *Int. J. Environ. Res. Public Health* **2021**, *18*, 7218. [CrossRef] [PubMed]
16. Lou, G.; Ma, H.; Fan, T.; Chan, H.K. Impact of the dual-credit policy on improvements in fuel economy and the production of internal combustion engine vehicles. *Resour. Conserv. Recycl.* **2020**, *156*, 104712. [CrossRef]
17. He, H.; Zhang, C.; Li, S.; Sun, Y.; Zhang, J.; Sun, Q. Dual-credit price variation and optimal electrification timing of traditional automakers: A dynamic programming approach. *J. Clean. Prod.* **2022**, *353*, 131593. [CrossRef]
18. Zhou, D.; Yu, Y.; Wang, Q.; Zha, D. Effects of a generalized dual-credit system on green technology investments and pricing decisions in a supply chain. *J. Environ. Manag.* **2019**, *247*, 269–280. [CrossRef] [PubMed]
19. Xiao, L.; Chen, Z.-S.; Hou, R.; Mardani, A.; Skibniewski, M.J. Greenness-based subsidy and dual credit policy to promote new energy vehicles considering consumers' low-carbon awareness. *Comput. Ind. Eng.* **2023**, *185*, 109620. [CrossRef]
20. Huang, X.; He, J.; Mao, L. Carbon Reduction Incentives under Multi-Market Interactions: Supply Chain Vertical Cooperation Perspective. *Mathematics* **2024**, *12*, 599. [CrossRef]
21. Abou Mjahed, M.; Ben Abdelaziz, F.; Tarhini, H. A multiobjective coalition formation in facility and fleet sharing for resilient horizontal supply chain collaboration. *Ann. Oper. Res.* **2023**, 1–26. [CrossRef]
22. Lin, P.; Shi, M.; Kai, L.; Ye, C. Manufacturer Optimal Alliance Pricing Decision in Many to One Supply Chain with Cross Elasticity of Demand Taken into Account. *Manag. Rev.* **2021**, 33, 313–324.
23. Gao, H.; Yang, J.; Yin, H.; Ma, Z. The impact of partner similarity on alliance management capability, stability and performance. *Int. J. Phys. Distrib. Logist. Manag.* **2017**, *47*, 906–926. [CrossRef]
24. Ma, H.; Lou, G.; Fan, T.; Chan, H.K.; Chung, S.H. Conventional automotive supply chains under China's dual-credit policy: Fuel economy, production and coordination. *Energy Policy* **2021**, *151*, 112166. [CrossRef]
25. Liu, X.; Wang, W.; Su, Y. Leveraging complementary resources through relational capital to improve alliance performance under an uncertain environment: A moderated mediation analysis. *Sustainability* **2022**, *15*, 310. [CrossRef]
26. Zhou, Y.-W.; Guo, J.; Zhou, W. Pricing/service strategies for a dual-channel supply chain with free riding and service-cost sharing. *Int. J. Prod. Econ.* **2018**, *196*, 198–210. [CrossRef]
27. Pu, X.; Gong, L.; Han, X. Consumer free riding: Coordinating sales effort in a dual-channel supply chain. *Electron. Commer. Res. Appl.* **2017**, *22*, 1–12. [CrossRef]
28. Liu, Y.; Lin, C.-X.; Zhao, G. A pricing strategy of dual-channel supply chain considering online reviews and in-sale service. *J. Bus. Ind. Mark.* **2024**. *ahead-of-print*. [CrossRef]
29. Guo, J.; Zhou, Y.; Li, B. The Optimal Pricing and Service Strategies of A Dual-Channel Retailer under Free Riding. *J. Ind. Manag. Optim.* **2022**, *18*, 2049. [CrossRef]
30. Cao, Y.; Yi, C.; Wan, G.-Y. Inventory Competition and Promotion Strategy in a Dual-channel Supply Chain with Free Riding Behavior. *Chin. J. Manag. Sci.* **2019**, *27*, 106–115.
31. Liu, Z.; Qian, Q.; Hu, B.; Shang, W.-L.; Li, L.; Zhao, Y.; Zhao, Z.; Han, C. Government regulation to promote coordinated emission reduction among enterprises in the green supply chain based on evolutionary game analysis. *Resour. Conserv. Recycl.* **2022**, *182*, 106290. [CrossRef]
32. Yan, L.; Hong, P.; Wu, Z. Dynamic pricing and emission reduction efforts in a dual-channel green supply chain under bidirectional free riding. *J. Clean. Prod.* **2024**, *438*, 140713. [CrossRef]
33. Westphal, R.; Sornette, D. How Market Intervention can Prevent Bubbles and Crashes: An Agent Based Modelling Approach. *Comput. Econ.* **2023**, 1–42. [CrossRef]

34. Deng, R.; Shen, N.; Zhao, Y. Diffusion model to analyse the performance of electric vehicle policies: An evolutionary game simulation. *Transp. Res. Part D Transp. Environ.* **2024**, *127*, 104037. [CrossRef]
35. Wu, J.; Ohya, T.; Sekiguchi, T. Applications of agent-based modeling and simulation in organization management: A quarter-century review through bibliometric mapping (1998–2022). *Comput. Math. Organ. Theory* **2024**, *30*, 1–31. [CrossRef]
36. Carley, S.; Siddiki, S.; Nicholson-Crotty, S. Evolution of plug-in electric vehicle demand: Assessing consumer perceptions and intent to purchase over time. *Transp. Res. Part D Transp. Environ.* **2019**, *70*, 94–111. [CrossRef]
37. Ma, L.; Zhong, W.; Mei, S. Study on subsidy strategy of new energy automobile industry chain based on endurance demand. *Syst. Eng. Theory Pract.* **2018**, *38*, 1759–1767.
38. Zhao, D.; Wang, J.; Li, Y.-K.; Tang, J.-H.; Zhang, S.-W. How to promote the transition of fuel vehicle enterprises under dual credit policy? An improved tripartite evolutionary game analysis with time delay. *Energy* **2024**, *293*, 130696. [CrossRef]
39. Wu, D.; Zhang, Z.; Tu, Q. Research on decision optimization of new energy vehicle supply chain considering demand disruptions under dual credit policy. *J. Ind. Manag. Optim.* **2024**. [CrossRef]

MDPI AG
Grosspeteranlage 5
4052 Basel
Switzerland
Tel.: +41 61 683 77 34

World Electric Vehicle Journal Editorial Office
E-mail: wevj@mdpi.com
www.mdpi.com/journal/wevj

www.ingramcontent.com/pod-product-compliance
Lightning Source LLC
Chambersburg PA
CBHW040826250926
51850CB00007BA/1273
* 9 7 8 3 7 2 5 8 1 6 9 7 2 *